INFORMATION MODELLING AND KNOWLEDGE BASES XVI

Frontiers in Artificial Intelligence and Applications

FAIA covers all aspects of theoretical and applied artificial intelligence research in the form of monographs, doctoral dissertations, textbooks, handbooks and proceedings volumes. The FAIA series contains several sub-series, including "Information Modelling and Knowledge Bases" and "Knowledge-Based Intelligent Engineering Systems". It also includes the biannual ECAI, the European Conference on Artificial Intelligence, proceedings volumes, and other ECCAI – the European Coordinating Committee on Artificial Intelligence – sponsored publications. An editorial panel of internationally well-known scholars is appointed to provide a high quality selection.

Volume 121

Recently published in this series

Vol. 120. T.F. Gordon (Ed.), Legal Knowledge and Information Systems – JURIX 2004: The Seventeenth Annual Conference
Vol. 119. S. Nascimento, Fuzzy Clustering via Proportional Membership Model
Vol. 118. J. Barzdins and A. Caplinskas (Eds.), Databases and Information Systems – Selected Papers from the Sixth International Baltic Conference DB&IS'2004
Vol. 117. L. Castillo et al. (Eds.), Planning, Scheduling and Constraint Satisfaction: From Theory to Practice
Vol. 116. O. Corcho, A Layered Declarative Approach to Ontology Translation with Knowledge Preservation
Vol. 115. G.E. Phillips-Wren and L.C. Jain (Eds.), Intelligent Decision Support Systems in Agent-Mediated Environments
Vol. 114. A.C. Varzi and L. Vieu (Eds.), Formal Ontology in Information Systems – Proceedings of the Third International Conference (FOIS-2004)
Vol. 113. J. Vitrià et al. (Eds.), Recent Advances in Artificial Intelligence Research and Development
Vol. 112. W. Zhang and V. Sorge (Eds.), Distributed Constraint Problem Solving and Reasoning in Multi-Agent Systems
Vol. 111. H. Fujita and V. Gruhn (Eds.), New Trends in Software Methodologies, Tools and Techniques – Proceedings of the Third SoMeT_W04
Vol. 110. R. López de Mántaras and L. Saitta (Eds.), ECAI 2004 – Proceedings of the 16th European Conference on Artificial Intelligence
Vol. 109. E. Onaindia and S. Staab (Eds.), STAIRS 2004 – Proceedings of the Second Starting AI Researchers' Symposium
Vol. 108. V. Stefanuk and K. Kaijiri (Eds.), Knowledge-Based Software Engineering – Proceedings of the Sixth Joint Conference on Knowledge-Based Software Engineering
Vol. 107. F. Darses et al. (Eds.), Cooperative Systems Design – Scenario-Based Design of Collaborative Systems
Vol. 106. D. Bourcier (Ed.), Legal Knowledge and Information Systems – JURIX 2003: The Sixteenth Annual Conference
Vol. 105. Y. Kiyoki et al. (Eds.), Information Modelling and Knowledge Bases XV
Vol. 104. A. Abraham et al. (Eds.), Design and Application of Hybrid Intelligent Systems

ISSN 0922-6389

Information Modelling and Knowledge Bases XVI

Edited by

Yasushi Kiyoki

Keio University, Japan

Benkt Wangler

University of Skövde, Sweden

Hannu Jaakkola

Tampere University of Technology, Finland

and

Hannu Kangassalo

University of Tampere, Finland

IOS Press

Amsterdam • Berlin • Oxford • Tokyo • Washington, DC

ISBN 1 58603 497 9
Library of Congress Control Number: 2005920425

Publisher
IOS Press
Nieuwe Hemweg 6B
1013 BG Amsterdam
The Netherlands
fax: +31 20 687 0019
e-mail: order@iospress.nl

Distributor in the UK and Ireland
IOS Press/Lavis Marketing
73 Lime Walk
Headington
Oxford OX3 7AD
England
fax: +44 1865 750079

Distributor in the USA and Canada
IOS Press, Inc.
4502 Rachael Manor Drive
Fairfax, VA 22032
USA
fax: +1 703 323 3668
e-mail: iosbooks@iospress.com

PRINTED IN THE NETHERLANDS

Preface

Information modelling and knowledge bases are becoming very important topics in academic communities related to information systems today. The amount and complexity of information itself, the number of abstraction levels of information, and the size of databases and knowledge bases are continuously growing.

The aim of this series of Information Modelling and Knowledge Bases is to provide research communities in information modelling and knowledge bases with scientific results and experiences achieved using innovative methods and approaches in computer science and other disciplines, which have a common interest in understanding and solving problems on information modelling and knowledge bases, as well as applying the results of research to practice. The research topics in this series are mainly concentrated on a variety of themes in the important domains:

- theoretical and philosophical basis of concept modelling and conceptual modelling,
- conceptual modelling, information modelling and specification,
- conceptual models in intelligent activity,
- collections of data, knowledge, and descriptions of concepts,
- human-computer interaction and modelling,
- database and knowledge base systems,
- software engineering and modelling and
- applications for information modelling and knowledge bases.

The important aim is to recognize and study new areas of information modelling and knowledge bases to which more attention should be given. Therefore, cognitive science, knowledge management, linguistics, philosophy, logic, and management science are relevant areas, too.

To achieve these aims in this series, an international program committee has selected 15 full papers, 12 short papers and 2 position papers out of 44 contributions, in a rigorous reviewing process. The selected papers cover many areas of information modelling, concept theories, database semantics, knowledge bases and systems, WWW information systems, context-based models and systems, requirement modelling in software engineering, multimedia databases, document data managements and so on.

The Program Committee consisted of 30 well-known researchers from the areas of information modelling, concept theories, conceptual modelling, databases, knowledge bases, information systems, as well as linguistics, philosophy, logic, and related fields important for information modelling. We are very grateful for their great work in reviewing the papers.

We hope that the series of Information Modelling and Knowledge Base will be productive and valuable in the advancement of research and practice of those academic areas.

The Editors

Yasushi Kiyoki
Benkt Wangler
Hannu Jaakkola
Hannu Kangassalo

Program Committee

Organizing Committee

Permanent Steering Committee

Additional Reviewers

Kazuhiro Asami, Tokyo Institute of Technology, Japan
Per Backlund, University of Skövde, Sweden
Sven Casteleyn, Vrije Universiteit Brussel, Belgium
Thomas Feyer, Brandenburg University of Technology at Cottbus, Germany
Paula Gouveia, Lisbon Institute of Technology (IST), Portugal
Ingi Jonasson, University of Skövde, Sweden
Steffen Jurk, Brandenburg University of Technology at Cottbus, Germany
Makoto Kondo, Shizuoka University, Japan
Stephan Lechner, Johannes Kepler University, Austria
Michele Melchiori, University of Brescia, Italy
Erkki Mäkinen, University of Tampere, Finland
Jyrki Nummenmaa, University of Tampere, Finland
Günter Preuner, Johannes Kepler University, Austria
Roope Raisamo, University of Tampere, Finland
Jaime Ramos, Lisbon Institute of Technology (IST), Portugal
Joao Rasga, Lisbon Institute of Technology (IST), Portugal
Yutaka Sakane, Shizuoka University, Japan
Jun Sakaki, Iwate Prefectural University, Japan
Mattias Strand, University of Skövde, Sweden
Tetsuya Suzuki, Tokyo Institute of Technology, Japan
Eva Söderström, University of Skövde, Sweden
Mitsuhisa Taguchi, Tokyo Institute of Technology, Japan
Shiro Takata, ATR, Japan
Yoshimichi Watanabe, Yamanashi University, Japan

Contents

Information Modelling and Knowledge Bases XVI
Y. Kiyoki et al. (Eds.)
IOS Press, 2005

Evolution Patterns to Elicit Compliance Requirements

Anne ETIEN, Camille SALINESI
Centre de Recherche en Informatique
Université Paris 1 – Panthéon Sorbonne
90, rue de Tolbiac
75013 Paris – France
{aetien, camille}@univ-paris1.fr

Abstract. Eliciting change requirements in the context of system evolution seems to us different from eliciting requirements for a system developed from scratch. Indeed, the issue is not to specify new functions but to understand differences with the current situation. For this purpose, we proposed a gap driven approach that (i) allows to express change requirements through gap operators and (ii) avoids re-specifying what remains stable in the system.

Furthermore, new change requirements can be elicited through compliance i.e. by analysing the consequences of the business evolutions on the running process and system instances. We have already shown that depending the modification policy applied, the compliance change requirement is different.

This paper identifies a collection of evolution patterns drawn from our experience to structure and systematise the elicitation of compliance requirement. The structure of three particular patterns is presented in details and illustrated.

1 Introduction

Our experience in several industrial projects, in a context of Information System (IS) evolution showed that two important issues should be dealt with, in this kind of project: (i) eliciting change requirements in the context of IS evolution is different from eliciting requirements for a system developed from scratch and (ii) specific guidance of the requirements elicitation process in the context of software evolution is needed.

To deal with the first issue, we adopted a gap driven approach allowing to synthetically express the differences between the future and the current situation. The advantage of specifying change requirements with gaps is that it allows to avoid re-specifying what remains stable in the system. For example, it is possible to specify that a number of required IS changes will appear through addition, removal, merging or splitting of a number of elements from the As-Is models. We have developed a requirements language based on a typology of gap operators to systematise the formalisation of such change requirements.

To address the second issue, we proposed (a) an approach to uncover system change requirements from the study of the business processes evolution and (b) an approach to elicit new change requirements through compliance, i.e. by analysing the consequences of the business evolutions on the running process and system instances. In this last approach, the question of what happens with the current instances of the corresponding processes, classes, etc. is raised each time a change is required by the business. Several policies that answer this question have been proposed in the literature [20], [15], [2]. Their principle is to specify what should technically be done at the instance level to reach a situation that complies with the To-Be models. Each time such a policy is adopted, adaptations have to be carried out on the business and on the system. In our approach, such adaptations correspond to new requirements that can also be specified under the form of change requirements [24]. So far, our approach includes a typology of modification policies and proposes to express the associated change requirements with gaps called *compliance gaps*. We have developed a number of case studies where compliance gaps were issued from modification policies [24]. However, the way to specify compliance gaps, once a modification policy is adopted, is still not systematic. This paper proposes a number of re-occurring evolution patterns that we identified in our case studies while solving this issue. Each pattern is specified under the form of a heuristic rule that aims at helping stakeholders and engineers selecting a modification policy.

The remainder of the paper is structured as follows: section 2 introduces our typology of modification policies. The structure of our evolution patterns is presented in section 3; then three of these patterns are defined in details in section 4. Section 5 evokes lessons learned and future works in our research programme.

2 Overview of the approach

Our approach to elicit compliance requirements is based on two concepts: the gaps and the modification policies. These two perspectives are described in the following sub-sections.

2.1 A gap oriented approach

Change creates a movement from an existing situation captured in *As-Is models* to a new one captured in *To-Be models* [13]. *As-Is* models describe the current situation whereas *To-Be* models describe the future one. We believe that software evolution should be *gap* driven: gaps express what has to be changed/adapted to the new situation. Our approach distinguishes two types of gaps: business gaps and compliance gaps. Both of them express change requirements, but their origin is different.

- *Business gaps* express the requirements for change implied by the organisational business expectations and the needs for the future.
- *Compliance gaps* define, at the requirement level, the impact expected from modification policies at the instance level. Therefore, these gaps tell how to ensure continuity when putting into practice the business gaps.

2.2 Modification policies

Modification policies offer technical solutions to support continuity during the As-Is / To-Be transition. In the literature [20], [15], [2], [25], [6], [19], [14], the proposed modification policies can be organised in five groups: Flush, Abort, Migrate, Adapt and Build. These two last policies do not correspond to system evolutions that can be analysed through business evolution propagation. Therefore, they are not taken into account in the following. It is for example possible to *migrate* an instance of the As-Is model to an instance of a To-Be model i.e. to consider that the As-Is model instance is an instance of the To-Be model too (e.g.: although the contracting process has changed, the contract structure remains the same). Another possibility is to use the *flush* policy i.e. to let the As-Is instance terminates its lifecycle according to the As-Is model (e.g. terminate contracting process as they were defined when they started). The current instance can also be *aborted*, removed, or abandoned in the case of a process, etc.

Based on this classification, a typology of modification policies has been defined in [24] and is shown in Figure 1. It organises nine different policies into three groups.

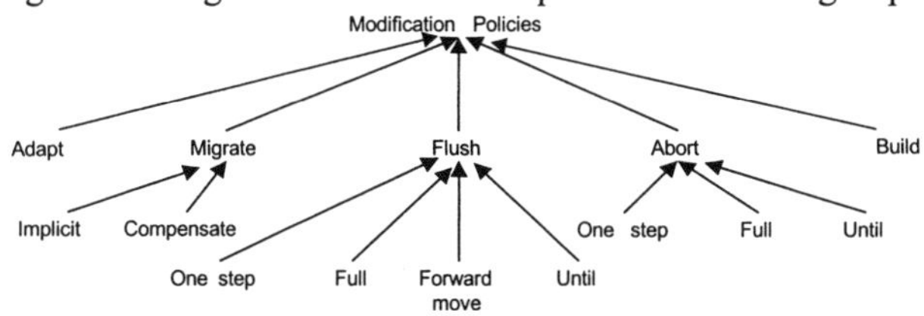

Figure 1: Modification policies typology

The nine policies can be summarized as follows (more detailed definitions and examples of application can be found in [24]):

- *Implicit migration* occurs when all conditions to proceed as defined in the To-Be are fulfilled. The As-Is instances can terminate their lifecycle according to the To-Be model.
- *Migration by compensation* consists in transforming As-Is instances so that they match the To-Be model and then making the instances migrate.
- *One-step flush* allows to deal with As-Is instances that are in an unstable state, i.e. for which no migration can immediately be achieved. This policy specifies that the As-Is instances terminate the in progress process before a possible migration.
- The *Until flush* policy is similar to the previous one. However, in the case of the until flush, a specific state that has to be reached using the As-Is model is chosen in advance
- In the *Full flush*, the As-Is instances are never migrated towards the To-Be model and finish their lifecycle according to the As-Is model.
- *Forward move flush*: this policy occurs when As-Is instances are in a situation that is not enough informative to decide on how to proceed to comply with the To-Be models. Looking at what happens in the next step before migrating, completely flushing or aborting is then necessary.
- The *Full Abort* policy provides a way to remove/cancel/abort/rollback current instances of an As-Is model, e.g. when their value is considered insufficient to justify the cost of a migration.
- *Until Abort* has for purpose to find back in the history of As-Is instances a state that can be exploited in the To-Be and to abort the part of the instances life cycle until there.
- *One-step abort* is a policy quite similar to the previous one except that only the last state is cancelled.

The choice of modification policies creates new requirements that can be considered as changes, and therefore be modelled at their turn using gaps. Thus, from the definitions of the policies, one can notice that the requirements engendered by the enactment of the policies are different according to each of them. For instance in all of the four flush policies, the lifecycle of existing instances proceeds as defined in As-Is models. This implies that (i) elements of the As-Is models must be introduced into the To-Be models, so that they define how to flush; if these elements were removed or replaced, compliance gaps must be specified to re-introduce them, and (ii) compliance gaps should also specify how to check that there is no new instance of the re-introduced elements. Similarly, in the case of compensation migration, compliance gaps have to be defined to specify the compensation.

Each policy thus leads to a specific reasoning on what should be done to ensure a consistent transition from As-Is instances to To-Be instances. It is also possible that for a same modification policy different situations or requirements imply different evolution patterns. Thus several patterns can be associated to each policy depending of the situation and the requirements.

3 Overview of the evolution patterns

Patterns have been widely used in different domains such as object-oriented modelling and design [5], software architecture design [4], programming [3], or even method engineering [9]. The aim of patterns is to facilitate the reuse of knowledge about adequate solutions to recurring issues or families of issues. For example, [1] gives the following definition: *"a pattern describes a problem which occurs over and over again in our environment and then describes the core of the solution to that problem, in such a way that you can use this solution a million times over, without ever doing it the same way twice"*. Other definitions [7], [11], [10] share two aspects with Alexander's: (i) a pattern relates a way of working with a recurring problem, and (ii) each problem has unique characteristics that distinguish it from other problems. Besides, a lesson learned from these definitions is that it is important to specify the situation in which patterns can be used, and how they should be adapted in specific situations.

The patterns that we define provide methodological solutions to common requirements engineering issues in the IS evolution domain. Based on the aforementioned works, we chose to develop patterns that include: (i) a specification of the *problems* for which the patterns are adequate, (ii) the *solutions* describing the reusable knowledge provided by the patterns, and (iii) the *contexts of use* specifying a set of project situations in which the patterns apply. These thus constructed patterns help in eliciting compliance gaps.

3.1 Structure of the evolution patterns

We adapted the approach presented in [8], which proposes a pattern structure composed of two parts: (i) a formal signature and (ii) an informal signature. In our particular case, the former identifies the problem that requires specifying compliance gaps. The latter defines the heuristics that solves the problem at hand and the requirements for adapting this solution in various contexts. This pattern structure is modelled in Figure 2 with the UML notation.

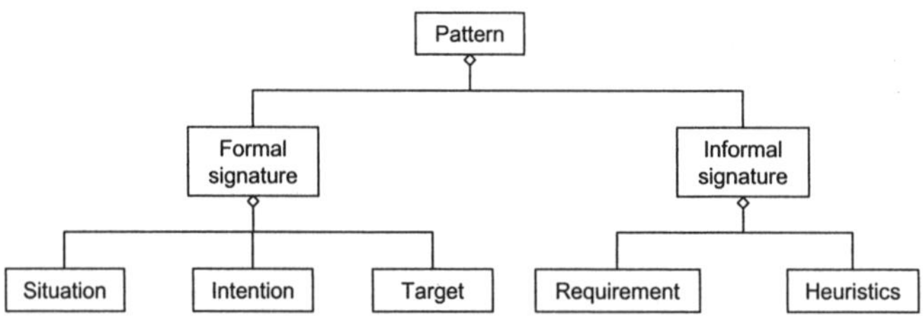

Figure 2: Evolution pattern meta-model

The *formal signature* of a pattern specifies why the decision could be taken to use it. As in [12], we define the context of such a decision as composed of three elements: (i) the situation in which the pattern can be applied, (ii) the intention that the pattern user has when he/she takes the decision to use it, and (iii) the target that this intention aims at producing. Therefore, the formal signature of patterns is formed by a triplet <situation, intention, target>. More precisely, in our approach:

- *Situations* are defined under the form of first order logic predicates about the As-Is models, the To-Be models, and about the business gaps between those.

- *Intentions* provide under the form of structured high-level goal statements a synthetic overview of the processes that are achieved when the process heuristics are applied.
- *Targets* present the kind of compliance gaps that can be elicited and enacted by application of the patterns.

While the formal signature adopts a black-box view on the pattern (there is no precise information on how the pattern works), *informal signatures* go into detail. Informal signatures have two parts: (i) the requirements and (ii) the heuristics.

- Pattern *heuristics* provide a detailed description of how to proceed to achieve the pattern intention when the pattern situation is at hand. Such heuristics are presented under the form of a sequence of steps.
- These steps can be adapted depending on how the pattern *requirements* are fulfilled.

3.2 Collection of evolution patterns

From our experience in various industrial projects [21], [22], [18] and by empirical derivation from the patterns found in industry, we identified five criteria that can have an impact on an evolution pattern. As shown in figure 3, one of these criteria is naturally the migration policy that the pattern helps to realise. The pattern is applied on an *element of interest*. It can sometimes depend on the *business change requirement*. The pattern can correspond to a *span of migration*. Finally, a pattern can impact only the current state or also the history; this criterion is called the *backward distance*.

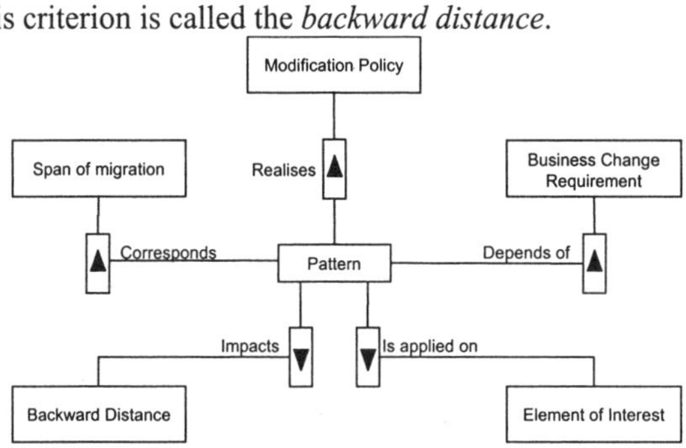

Figure 3: The five criteria and their impact on the pattern

These criteria have the following meaning:

- *Element of interest*: different kinds of objects can be migrated in an IS. Different kinds of compliance requirements are generated when different kinds of elements are analysed. For example, migrating dynamic items requires a different approach than migrating static items. Different evolution patterns are thus proposed depending on the object of interest, 'process' or 'object'.
- *Type of business change requirement*: compliance requirements are identified by analysing how process or object instances comply with business change requirements. Specific kinds of instances management are possible for a given kind of business change requirement. Therefore, the type of business change requirement at hand is a criterion that characterises the evolution patterns. For example, the evolution patterns are different depending on whether they result from the modification of a property, from the replacement of an element by another one, or from its transformation through split, merge or other kind of change operation. However, it can sometimes be difficult to specify the business change requirement because it results from the application of a combination of gap operators. This occurs in particular for the flush and abort policies.

- *Backward distance*: when a change is required on an IS element, the question of what happens with the past history of that element can be raised. Sometimes it doesn't matter because the new version of the element is not dependent at all on the element's history. However, it can also happen that the new definition of a changed element refers to its backward history (e.g. in a state transition diagram, in temporal logic process constraints, etc). In this case, the evolution pattern should propose to migrate the element together with its historic. The backward distance of a migration can be large (in which case the compliance requirements affect the complete historic of the element), or reduced (in which case, the compliance requirements only affect the last part of the element's historic). This criterion takes sense only in a case of migration. However, it can be required to keep, in the new system, information concerning instances that fully flush in the legacy system. It can so be speak about "large" full flush.
- *Span of migration*: in the case of migration, sometimes, changes have to be achieved instantaneously; sometimes it is required that the old situation coexists with the new one. Different kinds of evolution patterns are proposed depending on the span of the migration. One shot evolution patterns require specific compliance requirements to control an instantaneous transition between the old situation and the new one. When the legacy instances of As-Is keeps co-existing with the instances of To-Be, then a synchronisation can be required between them. We identified evolution patterns that propose: (i) a one-way synchronization from the legacy to the new system, and (ii) a two-way synchronisation between those. In the first case the modifications during the span of migration can only be made in the As-Is system but are visible in the To-Be one whereas in the second case, modifications can indifferently occur in the old or the new system.

  In the two other types of modification policies (full and abort), a synchronisation between the legacy and the new system has no sense insofar as the instances remain in the As-Is system without taking benefit of the novelties brought by the To-Be model.

From the definition of these criteria, it can be noticed that some criteria have no sense with some modification policies. We propose a pattern each time a compliance requirement had to be elicited because of an evolution issue. So far, we have been able to identify 42 patterns in Table 1 that we hope to be general enough to be applicable in various projects (this shall be evaluated soon as shown in conclusion). Each retained combination defines a pattern and thus also a compliance gap. A detailed analysis of the link between the criteria and the pattern parts is foreseen in our research agenda.

In the evolution patterns that we propose, As-Is models and To-Be models were defined with intention/strategy maps [16], [17]. *Intention/strategy maps* abstract the IS functionalities and identify what users want to do with the IS (the intentions) and how they want to proceed (the strategies). The change requirements resulting from the patterns are expressed using a specific set of gap operators as shown in [18], [24].

The 42 evolution patterns can easily be named after their location in the table that organises them. For example, the seventh pattern proposes to guide an "implicit migration of an object with its historic". The twenty-seventh pattern indicates how to achieve the "implicit migration of an object when it is merged without its historic", etc.

Three of these 42 evolution patterns are presented in more details in the next section: "implicit migration of an object without its historic", "migration by compensation of an object when it is replaced without its historic", and "fully flush a process with persistence of some information".

Table 1: Collection of evolution patterns for compliance requirement elicitation.

Modification Policy	Element of interest	Business change requirement	Backward distance	Span of migration
Implicit migration	Process		Large	one shot
				one way
				two ways
			Reduced	one shot
				one way
				two ways
	Object		Large	
			Reduced	
Migration by compensation	Process	Modification of Property	Large	one shot
				one way
			Reduced	one shot
				one way
		Splitting	Large	one shot
				one way
			Reduced	one shot
				one way
		Merging	Large	one shot
				one way
			Reduced	one shot
				one way
		Replacement	Large	one shot
				one way
			Reduced	one shot
				one way
	Object	Modification of Property	Large	
			Reduced	
		Splitting	Large	
			Reduced	
		Merging	Large	
			Reduced	
		Replacement	Large	
			Reduced	
One-Step Flush	Process	Splitting		
		Removal		
		Change Origin		
Full Flush	Process		Large	
			Reduced	
Until Flush	Process			
Forward move flush	Process			
One Step abort	Process			
Until abort	Process			
Full Abort	Process			

4 Three patterns to elicit compliance gaps

For sake of space, all the aforementioned patterns are not presented in details in the following. Three very different patterns were chosen in order to highlight the diversity of the patterns. Each of these three patterns is described using the meta-model presented in Figure 2 and illustrated using an example of room booking developed in [24] and described below.

A system handles room booking for several hotels in a centralised way. The main service provided by the system is the support of the sales processes. A project is undertaken to change the hotel booking business process in order to improve competitiveness.

In the current situation, products are independently designed in a flat list which is augmented each time a new product is conceived. Once products are in the list, they are used for contracts. Any product of this list can be removed to achieve its lifecycle. Booking contracts are either created directly in a hotel or in a travel agency. The contract management process ends up either by cancellation of the contract, or by consumption of the associated product by the consumer.

A number of evolutions were required. Three major evolutions can be highlighted: (i) From now on, the system should be customer centric; (ii) It should be possible to propose to customers complex products such as packages including tourist activities; (iii) The sales channels have to be diversified.

These evolutions were traduced by different business gaps presented in [24]. The three sub-sections present the three patterns aforementioned in the case where the room booking system is modelled with intention /strategy maps [16], [17].

4.1 Implicit migration of an object without its historic

Requirement of the engineer

This pattern deals with the migration of elements that results from the enactment of an intention I on which no change is required by the business. The instances of intention I in the As-Is can be considered as instances of I in the To-Be model. This means in particular that the data, that are used to achieve I, have the same format in the As-I and in the To-Be models.

Situation

$$[(\text{As-Is.I} = \text{To-Be.I}) \vee (\text{To-Be.I} = \text{Rename(As-Is.I)})]$$
$$\wedge [(\text{As-Is.I.precondition} = \text{To-Be.I.precondition}) \wedge (\text{As-Is.I.postcondition} = \text{To-Be.I.postcondition})]$$

The I intention has not been modified in the To-Be model or it has just been renamed. The pre-conditions for achieving intention I and the post-conditions are the same in the As-Is and in the To-Be.

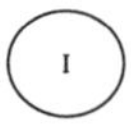

Intention I in the As-Is model = Intention I in the To-Be model

Figure 4: situation of the "Implicit migration of an object without its historic" pattern

Intention

Implicit migration of an object without its historic

Heuristics

The As-Is and the To-Be models both contain intention I. To make the instances of this intention migrate, the implicit migration policy is used.

The application of this pattern has the following consequences:

- The data corresponding to the result of the achievement of intention I migrate from the old system to the new one. The instances of the As-Is having the same data format than to those of the To-Be instances, the data can be automatically transferred from the old system to the new one for example by creating a trigger that invoques a procedure.
- Each instance resulting from this migration is considered as instance of the To-Be model and can thus continue their lifecycle according to this new model.

Target

$[S_{cg} = \text{AddStrategy(I, I)}] \wedge [S_{cg}.\text{name} =$ "implicit migration of data corresponding to As-Is instances"$]$

The compliance gap resulting from the achievement from this pattern consists in the addition of the strategy "implicit migration of data corresponding to As-Is instances" from I to I. It thus expresses a new way to achieve I once I has already been achieved once. The first time, I is reached according the As-Is model. The second time the instances of I belong to the To-Be model.

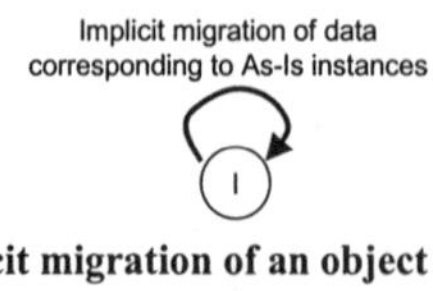

Figure 5: target of the "Implicit migration of an object without its historic" pattern

Example

In the case of the hotel room booking example described sooner, the current and the future system contain the creation of the booking. The required data and their format are the same in the two systems. So once a booking has been created in the As-Is system, it can be migrated in the To-Be system without any transformation and then be managed according this new system. The compliance gap elicited with this pattern allows to specify that the As-Is data can be automatically transferred.

4.2 Migration by compensation of an object when it is replaced without its historic

Requirement of the engineer

This pattern deals with the migration of elements that result from the enactment of an intention I which has been replaced between the As-Is and the To-Be model. In order to consider the instances of intention I in the As-Is as instances of I in the To-Be model, some compensation are needed corresponding for example to a data format change, to an addition of new data, etc.

Situation

To-Be.J = ReplaceIntention(As-Is.I)

The I intention of the As-Is model has been replaced by the J intention in the To-Be model.

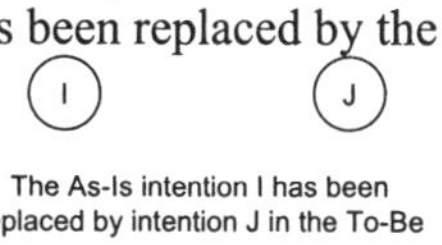

Figure 6: situation of the "Migration by compensation of an object when it is replaced without its historic" pattern

Intention

Migration by compensation of an object when it is replaced without its historic

Heuristics

The As-Is intention I has been replaced by the J intention in the To-Be. The migration of the instances of I implies some compensations.

The application of this pattern has for consequence:

- The data contained in the As-Is instances have to be modified in order for these transformed instances of intention I to be considered as instances of the To-Be intention J. For this purpose, data of the As-Is instances have to be reorganised to correspond to the To-Be format. Some additional data can perhaps be indispensable in the new system. A procedure to collect them should thus be defined.
- Each instance resulting from this migration is considered as instance of the To-Be model and can thus continue their lifecycle according to this new model.

Target

[S_{cg} = AddStrategy(Start, J)] $\wedge$ [S_{cg}.name = « migration by compensation resulting from a replacement »]

In this case, as instances of I in the As-Is cannot be considered as instances of J in the To-Be, the compliance gaps elicited by this pattern allow to specify a new manner to reach J from Start. Start is a special intention that exists in every map and begins the process. The strategy S_{cg} traducing the migration by compensation can be considered as a new way for the migrating instances to begin their lifecycle according to the To-Be model.

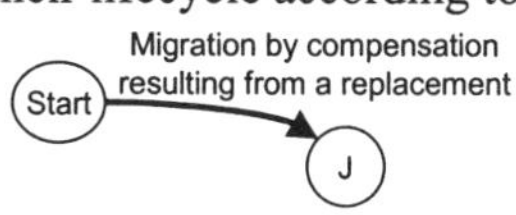

Figure 7: target of the "Migration by compensation of an object when it is replaced without its historic" pattern

Example

The structures of products in the list of the As-Is hotel room booking system and in the catalogue of the To-Be system are radically different. However, salesmen expect that customers and third parties keep referring to the old products list for a while before the transition to the new catalogue is completed. Besides, a detailed analysis shows that despite the differences, the initialisation of the new product catalogue from the old list could be automated. A compensation policy is thus adopted and procedures defined to allow the creation of new products in the catalogue by reuse of the old one.

4.3 Fully flush a process with persistence of some information

Requirement of the engineer

The business process has been modified. The instance lifecycle is too much advanced or the cost of the migration is too high comparatively to the instance value to make the instance migrate. Furthermore no change force (i.e. no new internal or external exigencies, like a new law) forbids the continuation of the lifecycle according the As-Is model. Finally some information concerning these instances is wished to be kept in the To-Be model even after the end of the lifecycle in the As-Is. Indeed, while this information disappeared in the As-Is with the end of the instance lifecycle, they subsist in the To-Be. The historic of the As-Is instance thus migrates in the new system.

Situation

$$\text{As-Is} \neq \text{To-Be}$$

The As-Is part and its "corresponding" To-Be part are very different. (The intentions have been replaced, removed, merged or split).

Intention

Fully flush a process with persistence of some information

Heuristics

The As-Is and the To-Be models are relatively different at least concerning the end of the processes they model. The value of the instances does not justify a migration. The instances of the As-Is can thus end their lifecycle according to this model. Nevertheless, some information needs to be conserved.

The application of this pattern has several consequences:

- A part of the As-Is model corresponding to the end of the instance lifecycle needs to be added in the To-Be model.
- Only the instances that attain this part according to the As-Is model can continue according to it. Thus no instance having followed the To-Be model (completely or only a part after a migration) can end its life cycle following this added part.
- Some information of the instances being conserved, it is necessary to specify the manner they are kept.
- The information being kept can be considered as information of any other instance even these of the To-Be. Before their use, a migration can perhaps be needed.

Target

S_{cg1} = AddStrategy(J, Stop) $\wedge$ S_{cg1}.name = « by fully flushing the instances of the As-Is » (where J corresponds to the intention where the decision to fully flush is taken).

S_{cg2} = AddStrategy(Start, I) $\wedge$ S_{cg2}.name = « by keeping information of the flushed instances of the As-Is » (where I corresponds to the intention where comparable information concerning the To-Be instances are captured).

Example
In the case of room booking, the hotel owners asking for a smooth transition from a contract-based to a customer-centric room booking management, the room booking contracts in progress are let to terminate the way they started that means according the As-Is model. Nevertheless, the information concerning the clients of these contracts is conserved in order to be reused in the new system.

5 Lessons learnt and future work

Evolution is an important aspect of the software lifecycle. Therefore, mastering software evolution is a crucial issue that should be addressed in an appropriate way. We proposed a process to guide the elicitation of business change requirements by propagation of business evolution. We then outlined an approach to elicit new change requirements by studying the consequences of business change requirements on the running instances of the system. In this context we defined a set of modification policies and introduced the notion of compliance gap by linking them together. This link between the modification policy and the compliance gaps was still intuitive, our proposal is to formalise it with methodological patterns.

This paper identifies a collection of forty-two patterns and details three of them. Each of the patterns describes a problem, its context and a solution.

These patterns were defined based on industrial experiences in which an intention/strategy map formalism was used.

All the proposed patterns cannot be used in any situation. For example, some of the listed patterns like the "Fully flush the process with persistence of some information" pattern imply the coexistence of the old system and the new one. However, such a coexistence is not always possible or wanted in practice. Our proposal is then to use one of the other patterns if it is possible or revise the business change requirements if not.

Our agenda combines research in three complementary ways: (i) we want to evaluate our patterns. Indeed, patterns are defined to be generic, flexible, easy to use...We believe that further evaluation of our patterns is needed for example through empirical experiments with students as well as interview with experts from the industry or on additional case-studies. (ii) we believe that further work is also needed to complete our patterns description and find other ways to guide business gap identification; (iii) we believe that further analysis is needed to define rules identifying impacts of some criteria on the pattern parts.

References

[1] Alexander, C., Ishikawa, S., Silverstein, M., et al.: A Pattern Language. Oxford University Press, New York (1977)

[2] Bandinelli, S., Fuggetta, A., Ghezzi, C.: *Software Process Model Evolution in the SPADE Environment.* IEEE Transactions on Software Engineering, 19(12) pp.1128-1144, (1993).

[3] Beck, K.: Smalltalk, Best Practice Patterns. Volume 1, Coding. Prentice Hall, Englewood Cliffs, NJ. (1997)

[4] Buschmann, F., Meunier, R., Rohnert, et al.: Pattern-Oriented Software Architecture - A System of Patterns. John Wiley (1996)

[5] Coad, P.: Object-Oriented Patterns. Communications of the ACM, Vol. 35, No. 9 (1992). pp 152-159

[6] Conradi R., Fernström, C., Fuggetta A.: *A Conceptual Framework for Evolving Software Process. ACM SIGSOFT Software Engineering Notes*, 18(4): 26--34, October 1993.

[7] Coplien J., D. Schmidt (eds.), *Pattern Languages of Program Design*, Addison Wesley, Reading, MA, 1995.

[8] Deneckère, R.: *Approche d'extension de méthodes fondée sur l'utilisation de composants génériques*. PhD thesis, 2001

[9] Deneckère, R.: *Using meta-patterns to construct patterns*, Proceedings of the Conference on Object-Oriented Information Systems: OOIS'2002, Montpellier, France, September 2002.

[10] Fowler M., *Analysis Patterns: Reusable Object Models*, Addison-Wesley, 1997.

[11] Gamma E., R. Helm, R. Johnson, J. Vlissides, *Design Patterns: Elements of Reusable Object-Oriented Software*, Addison Wesley, Reading, MA, 1994

[12] G. Grosz, C. Rolland, S. Schwer, C. Souveyet, V. Plihon, C. Ben Achour, C. Gnaho. Modelling and Engineering the Requirement Engineering Process : an Overview of the Nature Approach. Requirements Engineering journal, Vol. (2), pp. 115-131, (ed. P. Loucopoulos, C. Potts), Springer Verlag. 1997.

[13] Jarke, M., Pohl, K.: *Requirements Engineering in 2001: Managing a Changing Reality*. IEEE Software Engineering Journal, pp. 257-266. November 1994.

[14] Joeris, G., Herzog, O.: Managing Evolving Workflow Specifications With Schema Versioning and Migration Rules. TZI Technical Report 15, University of Bremen, 1999

[15] Liu, C., Orlowska, M., H. Li.: *Automating Handover in Dynamic Workflow Environments*. Proceedings of 10th CAiSE, Pisa, Italy, 1998.

[16] C. Rolland, N. Prakash, A. Benjamen, *A Multi-Model View of process Modelling*, Requirements Engineering Journal, Vol 4, pp 169-187, 1999.

[17] C. Rolland, N. Prakash, *Matching ERP System Functionality to Customer Requirements*, Proceedings of the 5th International Symposium on Requirements Engineering (RE'01), Toronto, Canada, pp. 66-75, 2001.

[18] Rolland, C., Salinesi, C., Etien, A.: *Eliciting Gaps in Requirements Change*. To appear in Requirement Engineering Journal. 2003.

[19] S. Sadiq and M. Orlowska. Architectural Considerations in Systems Supporting Dynamic Workflow Modification. Proceedings of the workshop on Software Architectures for Business Process Management at CAiSE'99, Heidelberg, Germany. June14-18, 1999.

[20] Sadiq, S.: *Handling Dynamic Schema Change in Process Models*. Australian Database Conference, Canberra, Australia. Jan 27 - Feb 02, 2000.

[21] Salinesi, C., Presso, M. J. : *A Method to Analyse Changes in the Realisation of Business Intentions and Strategies for Information System Adaptation*. Proceedings of EDOC'02, Lausanne, Switzerland, September, 2002.

[22] Salinesi, C., Wäyrynen J.: A Methodological Framework for Understanding IS Adaptation through Enterprise Change. In Proceedings of OOIS'02, 8th International Conference on Object-Oriented Information Systems, Montpellier, France, September 2002

[23] Salinesi, C., Rolland, C.: F*itting Business Models to Systems Functionality Exploring the Fitness Relationship*. Proceedings of CAiSE'03, Velden, Austria, 16 - 20 June, 2003.

[24] C. Salinesi, A. Etien, *Compliance Gaps: a Requirements Elicitation Approach in the Context of System Evolution*, Proceedings of the 9th International Conference on Object-Oriented Information Systems (OOIS'03), Geneva, Switzerland, September 2003.

[25] Van der Aalst, W.: *Generic Workflow Models: How to Handle Dynamic Change and Capture Management Information*. In M. Lenzerini and U. Dayal, editors, Proceedings of the Fourth IFCIS International Conference on Cooperative Information Systems, pp. 115-126, Edinburgh, Scotland. September 1999.

Information Modelling and Knowledge Bases XVI
Y. Kiyoki et al. (Eds.)
IOS Press, 2005

An Approach to Guide Requirement Elicitation by Analysing the Causes and Consequences of Threats

Elena IVANKINA(1)(2), Camille SALINESI(2)
(1) Département Systèmes d'Information et de Décision - ESSEC
Avenue Bernard Hirsch B.P. 105
95021 Cergy-Pontoise Cedex France
(2) CRI, Université Paris 1 – Panthéon Sorbonne
90, rue de Tolbiac
75013 Paris
Elena.Ivankina@malix.univ-paris1.fr, camille@univ-paris1.fr

Abstract: Eliciting system requirements is not an easy task, especially if one considers the amount of problems that can arise once the system is at use. A difficulty to ensure requirements completeness is to foresee these problems and identify the requirements that can help to overcome them. Existing approaches are based on probabilistic prediction of usage or system-related problems. Very few of these approaches are able to take into account both the causes of the potential problems and the impacts that these problems can have on the system usage. This paper presents an approach that proposes to elicit requirements by analysing how such kinds of problems can threaten the adequacy of a system to its intended usage. The approach aims at improving requirements completeness by (i) guiding a systematic identification of threats, (ii) evaluating their cause and (iii) evaluating their consequences on the system. This approach extends the CREWS – *l'Ecritoire* method with techniques adapted from the risk engineering domain. CREWS-*l'Ecritoire* couples abstract goals with concrete scenarios to elicit requirements. It proposes a number of rules to model goals, author scenarios, and discovers new goals by analysing scenarios. Our proposal stands in a new kind of goal discovery rule that is based on threat discovery, cause and consequence analysis, and solution identification. The paper gives an overview of the proposed rule, then shows how it integrates into CREWS-*l'Ecritoire*, and provides an example of application.

1 Introduction

Requirement (in-)completeness is one of the well-known major issues of Requirement Engineering (RE) [23]. One of its particularly difficult aspects is to identify the requirements that deal with problems that arise in the usage of the system. Such threats to the adequate fitness of the system to its potential usages can be of many kinds. Their common point is that if they occur, the effects can be expensive and time consuming both in terms of business performance and to maintain the system. It is thus natural to try to predict potential problems, so as to totally prevent them from occurring, or at least if they occur, to prevent their negative effects on the business. Our intention is thus to predict and remove or decrease the effect of such threats before the system is used (when the cost of adaptation is still manageable) rather than at run time.

Several RE approaches are already developed to deal with threats in Information System (IS) design and use. Two particular kinds of threats are in particular identified in the RE literature: obstacles and conflicts [9], [8], [1]. Both concepts relate to the goals that users have in mind when they use the facilities offered by their IS. An obstacle is defined as a phenomenon that occurs in the IS environment and obstructs the achievement of a goal [1], [9]. A conflict is when the achievement of two different goals obstructs each other [16], [3].

This paper proposes, as [21], [22], an approach that considers obstacles, conflicts and other system menaces as fundamentally similar as they all correspond to the partial or total hindering of one or several goal(s). The generic concept that we use to refer to those is the one of threat. Threats are different from risks [2], [7] because they only concern problems with the realisation of a system usage goal. Risks are more general as they can compromise any IS project goal, including goals that do not directly relate to the system usage.

The approach proposed in this paper is structured as a pattern composed of three fragments: (i) one fragment to guide the discovery of threats, (ii) one fragment to analyse possible causes of these threats, and (iii) one fragment to identify ways to remove threats or alter their consequences and elicit the corresponding system requirements. Threat discovery is achieved by heuristic rules generalised from our experience with goal modelling [20]. The analysis of threats causes is achieved using cause and consequences analysis diagrams inspired from [6]. Requirements are elicited using a repository of design solutions gathered from the system engineering literature.

The proposed pattern will be the backbone of a new goal identification guiding rule to be introduced into the CREWS-*L'Ecritoire* requirements elicitation approach. The next section gives a brief introduction to CREWS-*L'Ecritoire*. Section 3 presents the three parts of the pattern. Section 4 illustrates the application of the pattern with an example drawn from the *Meeting Scheduler* case study [4]. The concluding section discusses how we intend to refine the pattern into enactable rules that can be implemented into the CREWS-*L'Ecritoire* prototype.

2 Requirements elicitation in CREWS-*L'Ecritoire*

The CREWS-*L'Ecritoire* approach guides the discovery/elicitation of requirements by exploiting a bi-directional coupling between goals and scenarios. This coupling allows movement from goals to scenarios and vice-versa:

- When a goal is discovered, a scenario is authored to illustrate it. In this sense, the goal-scenario coupling is exploited in the forward direction from goals to scenarios.
- Once a scenario has been authored, it is analysed to yield goals. This leads to goal discovery by moving along the goal-scenario relationship in the reverse direction.

The exact sequence of steps of the CREWS-*L'Ecritoire* requirements elicitation process is:

1. Initial Goal Identification

Repeat 2. Goal Analysis

3. Scenario Authoring

4. Goal Discovery

Until all goals have been elicited

It can be seen that goal discovery and scenario authoring are complementary steps. Requirements are incrementally elicited by repeating the goal-analysis, scenario-authoring, goal-discovery cycle. As mentioned in the introduction, the objective of the *L'Ecritoire* approach is to guide this process. For this purpose, each of the three steps of the cycle is supported by systematic mechanisms formalised into guiding rules. The next subsection defines the main concepts used in *L'Ecritoire*, the following subsection gives a brief overview of the goal discovery rules that already exist in the approach.

2.1 Concepts and terminology

In *L'Ecritoire*, requirements are modelled under the form of a collection of requirement chunks organised in a hierarchy. The requirement chunks collection contains goals, scenarios, and links between them as follows.

A *Requirement Chunk* (RC) is a <Goal, Scenario> pair. Since a goal is by nature intentional and a scenario is concrete, a requirement chunk is a possible way of achieving the goal.

Goals can be defined as "something that some stakeholder hopes to achieve in the future" [14]. In *L'Ecritoire*, a *goal* is expressed as a clause with a main verb and several parameters, where each parameter plays a different role with respect to the verb. Goals can be of three different kinds: those that the system users hope to achieve in their business, those that are expected to be achieved by system components, and those that can be seen as services offered by the system.

A *scenario* is "a possible behaviour limited to a set of purposeful interactions taking place among several agents" [19]. It is composed of one or more *actions*, an *action* being an interaction *from* one agent *to* another. The combination of actions in a scenario describes a unique path. A scenario is characterised by an initial and a final state. An *initial state* attached to a scenario defines a precondition for the scenario to be triggered. A *final state* defines a state reached at the end of the scenario. We distinguish between *normal* and *exceptional* scenarios. The former lead to the achievement of its associated goal whereas the latter fails in goal achievement.

Three levels of abstraction called *contextual*, *functional*, and *physical* are introduced by *L'Ecritoire* to organise requirements chunks. The contextual level identifies the services that a system should provide to an organisation and their rationale. The functional level focuses on the interactions between the system and its user to achieve the needed services. Finally, the physical level deals with the actual performance of the interactions. In addition to these three levels, *L'Ecritoire* offers to organise requirements chunks using links.

There are three types of relationships among requirement chunks namely, the composition, alternative, and refinement relationships. The first two of these lead to a horizontal AND/OR structure between requirement chunks. These are extensions of conventional AND/OR relationships between goals. AND relationships among requirement chunks link together those chunks that require each other to define a completely functioning system. Requirement chunks related through OR relationships represent alternative ways of fulfilling the same goal. The third kind of relationship relates requirement chunks at different levels of abstraction. The refinement relationship establishes a vertical link between requirement chunks.

2.2 The requirements elicitation process

The guidance mechanism for goal analysis is based on a linguistic analysis of goal statements. It includes, on the one hand, a general goal statement pattern adapted from the linguistic structure of goals defined by [15]. On the other hand, more precise patterns are proposed for different families of goals organised into a hierarchy. These patterns are used to enhance goal formulation, as well as to support creativity while analysing the initial goals that are discovered in the approach.

The mechanism for scenario authoring combines style/content guidelines and linguistic devices grounded on Fillmore's theory of natural language semantics [20]. The former advise scenario authors on how to write correct scenarios whereas the latter provides semi-automatic help to check, correct, conceptualise, and complete a scenario.

Finally, *L'Ecritoire* proposes enactable rules offering three different goal discovery strategies namely, refinement strategy, composition strategy, and alternative strategy. The first helps discovering goals at a lower level of abstraction than that of the analysed requirement chunk. The second discovers goals linked to the original one by an AND relationship. The last discovers alternatives. The principle of each kind of rule is to look for specific patterns in the analysed scenarios, such as the presence of constrains, the presence of specific kinds of actions, or of specific sequences of actions, etc. New goals can be discovered (hence the requirements collection completed) by repeatedly applying the rules on each requirement chunks. Requirements chunk can be analysed once they have been completed, i.e. once the scenarios that illustrate the corresponding goals have been correctly authored.

Contrary to the goal discovery rules that already exist in *L'Ecritoire*, the heuristic that we propose in this paper does not aim at completing the requirements hierarchy in a particular direction with respect to the AND/OR/Refine relationship of the requirement chunks hierarchy. On the contrary, our proposal is driven by the concept of threat which is not a built in concept of *L'Ecritoire*.

3 Overview of the proposed pattern

The pattern that we propose is composed of three parts: one that identifies threats; the second part that analyses their causes; and the third part that helps eliciting requirements to overcome the issue presented by the threats. Each of the three pattern parts is introduced at its turn in the three following sub-sections.

3.1 Threats identification

As defined in the introduction, a threat represents a situation in which the system or a part of it is put in danger. At the level of requirements, the system can be considered as potentially in danger when one of its goals is not (partially or at all) realized. Based on the investigation of examples, we found that the non-realization of a given goal can be detailed by: (i) the fact that the pre-conditions expressed in the scenario's initial states are not fulfilled, (ii) the fact that the post-conditions expressed in the scenario's final states are not fulfilled, or (iii) the goal is not exactly achieved as required. Therefore, a threat is defined by (i) the goal that is not achieved, and (ii) the details on what is not achieved in the goal

To guide the identification of threats, we propose to apply the 'negation' strategy. As shown in figure 1, the idea behind this strategy is to formulate the details on what is not achieved in a goal through the negation of the initial states of the corresponding scenario, of the corresponding final state and/or of the goal formulation itself. Starting from a single requirement chunk, analysing these three elements of details can produce different threats. The other way round, a single threat can be detailed by goal negation, initial state negation, and final state negation at the same time. Therefore the analysis of negation details can be redundant.

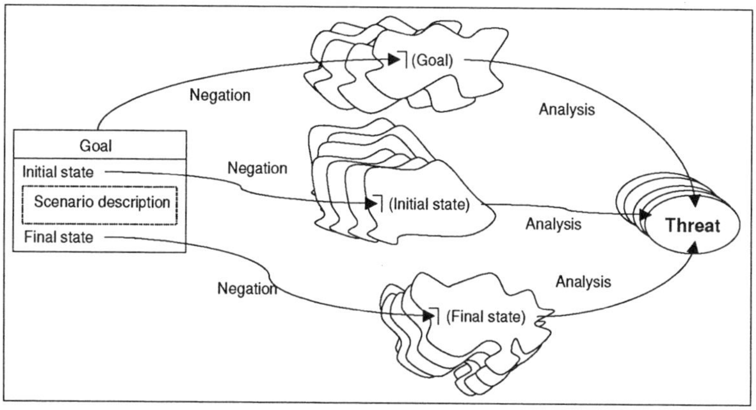

Figure 1. Negation Strategy

The negation strategy allows to define threats on a very general level. This may be sufficient to identify general solutions [12]. However, we believe that a more complete collection of goals and scenarios can be elicited if the identified threats are more detailed. Our proposal is to detail threats using a causal analysis technique.

3.2 Analysing the causes of threats

To analyse the causes of threats, we suggest to adapt the principles underlying Ishikawa's cause and effect diagrams [6]. This kind of diagram was elaborated by Kaoru Ishikawa for qualitative management and control in industrial environments. It is now used in various disciplines, in particular in the domain of risk management.

As figure 2 shows it, to construct a cause and effect diagram, it is necessary to start with a problem. Once the problem stated, four to six families of causes are identified (the number must remain reasonable, but there is no limitation). In the original approach four families of causes of industrial production problems are proposed: Materials, Machines, Manpower, and Methods. Each of these families is used as a guide to list the causes of the problem. For example, C4 and C5 are two manpower-related causes of the problem at hand. Once the list of causes completed, each individual cause can itself be taken as a problem and recursively analysed with the same diagram. This is for example the case of C6, which is detailed with C6.1, C6.2, C6.3 and C6.4. The result is a tree which root is the initial problem (the "effect"), and leaves the most detailed causes of this problem.

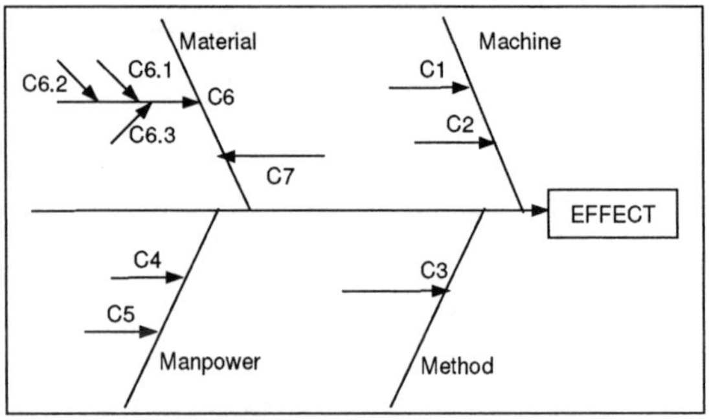

Figure 2: Ishikawa's cause and effect diagram

In our approach, the initial problems are formulated under the form of threats discovered by the negation strategy. Besides, we propose five families of causes adapted to the domain of Information Systems. These families where defined as a mix between Ishikawa's original families and the four worlds framework on Information Systems developed in the F^3 project [17]. This framework was chosen because its purpose was to tackle all perspectives on IS. According to the framework, there are four different perspectives: the subject, the usage, the system, and the development. As shown in the meta-model presented in Figure 3 (using the UML notation), the five families of causes of threats are: Users, Material, Equipment, Environment, and Method.

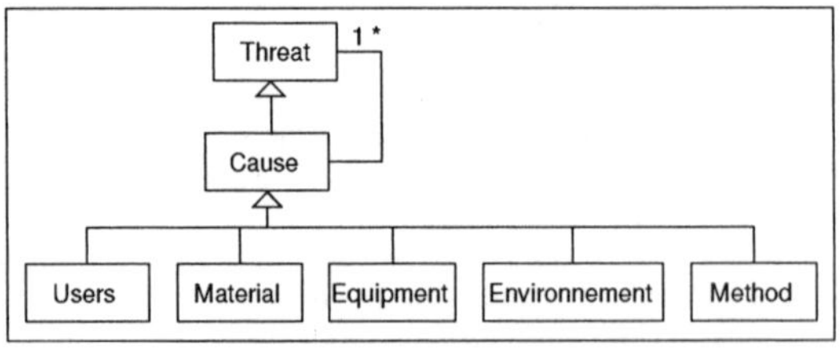

Figure 3 : Threat meta-model

- The causes of threats that relate to *material* can be found in the soft layer of the system. They can be met when the system does not implement correctly its specifications. The most typical examples of such an issue are: implementation bugs, incompatible data, information that do not match their specification, overflows, impossible operations, etc.
- Causes that relate to the *user* family are those that are met when human agents that access the IS have an unexpected behaviour. The issue can be organisational (e.g. when a customer doesn't take contact with the appropriate business actor), as well as at the individual level (e.g. there can be a skill issue, an incorrect perception of the system, a physical difficulty to use the system, malicious intentions such as hacking or intrusion, etc).
- The *equipment* family of causes corresponds to hardware issues. They can result from a problem of physical availability, incompatibility between devices, ageing of system components, etc.
- The *method* family covers the causes of threats that take their origins in the system design, in the IS project organisation, or in the engineering methods used in the project.

Typical examples are incomplete business rules, missing functions, untested usage cases, incomplete requirement specifications, lack of communication between project participants, lack of involvement, inadequate analysis strategies, etc.

- The last family relates to the *environment* of the system. This includes the physical and organisational environment of the user (e.g. work conditions, noise, etc), as well as the more general environment of the system (e.g. electro-magnetic interferences, heat, communication issues with other systems, etc)

These families of causes are not mutually exclusive. For example, a malicious use of the system can be achieved through an unforeseen device; the cause of the issue is threefold : (i) the user is a hacker (ii) the security requirements were incomplete, and (iii) the system is physically compatible with other devices than the ones for which it was designed. As this example shows, the diagram and the proposed families of causes help finding large numbers of causes of threats. This way, the causes of each threat that was identified can be analysed on a very detailed level. However, too much detail can also bring unnecessary complexity. Let us recall that in *L'Ecritoire*, requirement chunks can be described at different levels of abstraction/detail. Therefore, issues can also be found at different levels, and their causes also analysed on different levels. It is recommended to analyse the cause of threat at the same level as the level at which the threatened goal was defined.

3.3 The strategies of treatment of threats

Once the causes of a threat analysed, it is possible to look for adequate solutions. We believe that there is a number of solutions that can be described on a generic level, and reused from one threat to the other. Our proposal stands in a collection of threat treatment strategies that we identified through a series of interviews with IS designers and developers, and based on our own experience. Each treatment strategy proposes a parry to different kinds of threats. All strategies were organised in a refinement hierarchy. As shows figure 4, there are four general ways to treat threats: to ignore them, to reduce them, and to kill them.

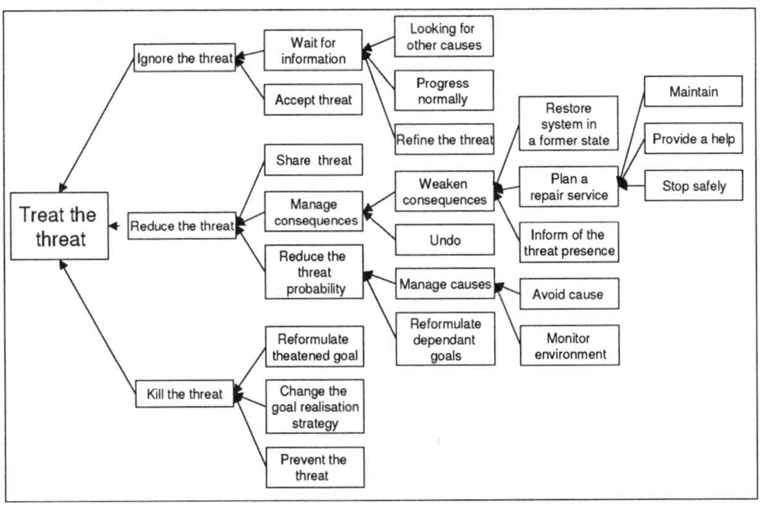

Figure 4: Threat management strategies

The three high level strategies propose to treat the causes of threat and parry them as follows :

- Ignoring the threat consists in accepting it, either definitively, or temporarily, e.g. the time to have more information. In this case, the system is designed to dynamically decide on the treatment based on the information that it is able to gather on the threat. If the information needed to make the decision is not complete, the system can wait and look for other causes of the threat, progress normally until the information is gathered, or refine the threat.
- The second kind of treatment consists in designing the system so that it reduces the threat. Several strategies are proposed to do so, sharing, reduction and management of consequences. The consequences of a threat can be managed by the system in several ways : the system can be designed to inform users of the threat, to stop safely, to provide the user with help, and to provide maintenance features. Each of these goals manage the consequences of the threat by weakening it. An alternative is to undo the consequences of the threat. Besides, a threat can be reduced by decreasing its probability, i.e. by affecting to the system the responsibility of acting so that at least some of the causes of the threat are not met.
- The third strategy consists in eliminating the threat by acting on the system goal that it affects. In this kind of solution, the system is re-designed, either by reformulating the affected goal, or by changing the way the goal is achieved by the system (in *L'Ecritoire,* identified by the strategy and described by the scenario). A third solution consists in affecting to the system the responsibility of preventing the threat from occurring.

It can be seen from above that each strategy is not applicable in all situations. Therefore, we propose to describe them with : (i) the kind of threat on which they apply, (ii) conditions of application, (iii) arguments (pros and cons) of selection, and (iv) effects on the requirement chunks hierarchy. All these descriptions are not provided in this paper for the sake of space. We are exploring criteria to select treatment strategies e.g. according to the level of gravity of the threats to which they correspond. Besides, the proposed hierarchy of treatments is not rigid. It can be enriched by adding new treatment strategies and linking them to the existing ones using refinement. We are also aware that the proposed classification is not unique. Other ways of organising our collection of treatment strategies could possibly allow the identification of more strategies, and more arguments to choose among them.

4 Example of application

The example taken in this section is that of the Meeting Scheduler [4]. The overall purpose of this system is to assist in the organization of a meeting with a number of participants. An initiator proposes to organize a meeting with participants. The participants and the meeting initiator use the meeting scheduler to decide on the date and location of the meeting.

4.1 Threat identification

As shown in Figure 5, there is already a number of requirement chunks that can be elicited using *L'Ecritoire*. Each requirement chunk is ranked in one of the three pre-defined levels of abstraction, and linked to the other requirement chunks with alternative, complementary and refinement relationships. The resulting requirement chunk hierarchy contains a number of alternative goals on the highest level. The one that are studied in the following tells in which part of the business process the meeting scheduler shall be involved : "Gather (a group of participants)$_{object}$ (at a given date)$_{time}$ (in a given place) $_{location}$ (for the initiator) $_{beneficiary}$ (using the meeting scheduler) $_{means}$ (by a centralized organization) $_{maner}$". The alternative goals show that the meeting scheduler could also be designed for other purposes. The figure shows that this goal is illustrated in the requirement chunk with a scenario described as a flow of user-system interactions. These interactions were refined and analysed as complementary service goals at the functional level.

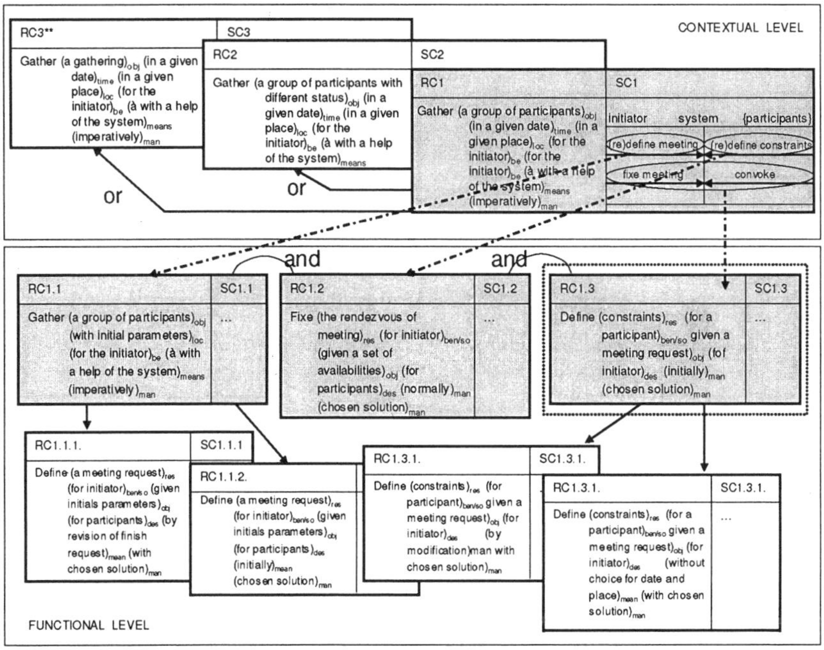

Figure 5: Requirement chunks hierarchy generated by *L'Ecritoire* for the meeting scheduler example

Each service goal at the functional level is at its turn described with a scenario and the accompanying initial states and final states. For example, the meeting scheduler system shall allow to "Define constraints from a participant given a meeting request for the initiator, etc" as described in the following table.

Table 1: Requirement Chunk N°1.3

Initial state	The user is connected as a participant The participant has identified a meeting The system is ready
Scenario description	1. The system informs the participant of the meeting request 2. If the meeting definition allows it 3. The participant defines his/her preferences 4. If the status of the participant allows it 5. The participant defines his/her constraints 6. If the role of the participant and the demand of the initiator authorizes it 7. The participant defines its equipment needs
Final state	The constraints of the participant are confirmed The system is ready

The negation strategy can be applied on this requirement chunk. For example, the negations of the initial states and final states can have different interpretations:

- ¬"The user is connected as participant"
  - "The user is not connected as a participant"
  - "The connected user is not a participant"
- ¬ "The participant has identified a meeting"
  - "The participant has not identified a meeting"
- ¬ "The system is ready" (as an initial state)
  - "System is not ready"
- ¬ "The constraints of the participant are confirmed "
  - "The constraints of the participant are not confirmed"
- ¬ "The system is ready" (as a final state)
  - "The system is partially ready"

Each of them can generate a threat. Similarly, the negation strategy can be applied on the goal statement itself. This generates new threats as well as threats that could have already been identified by applying the negation strategy on the initial and final states. The potential threats in this set can have different interpretation, depending to context, or system specificities; the statement of the generated threats must then be refined. Besides, there are some unrealistic threats that must be eliminated. Examples of threats identified using the negation strategy, after rephrasing and removal of unrealistic threats are:

Th1: The participant uses the 'constraint definition' system service to define other things than constraints (i.e. not constraints)

Th2 : The participant defines other participant's constraints

Th3: The participant uses the 'constraint definition' system service to define constraints for somebody else than the initiator (i.e. not for the initiator)

Th4: The participant uses the 'constraint definition' system service to define constraints that have already been defined (i.e. not initially)

Th5: The user accesses the 'constraint definition' system service without being connected as a participant.

Th6: The 'constraint definition' system service is used by a non-participant user

Th7: The participant uses the 'constraint definition' system service without having initially identified a meeting

Th8: The participant uses the 'constraint definition' system service whereas the system is not ready

Th9: The participant uses the 'constraint definition' system service, but the constraints are not confirmed, and leaves the system in a partially or not ready state.

The cause of each of these threats can be analysed separately, and new requirements elicited to treat them. The following two sections respectively deal with the analysis of causes and treatments of Th7, namely "The participant uses the 'constraint definition' system service without having initially identified a meeting".

4.2 Causes of threat Th7

The causes of threat *Th7* can be found in each of the five families (users, material, equipment, environment, method). For example

- In the *Equipment* family the cause of the threat can stand in an incompatibility between the kind of terminal used and the meeting identification facility (e.g. typically when the system is accessed with a PDA or a mobile phone).
- In the *Material* family, the problem can be that the information needed to identify meetings is not visible. Another example of material cause for *Th7* is when the meeting selection criteria are defined in a language which alphabet cannot be displayed by the system.
- In the *User* family, the problem can be that the participant is not able to use the interface to select the meeting, or that the user is not connected as a participant. Another example of user causes for this issue is when the language used to present meeting selection criteria is not understood by the participant.
- Different environmental causes can be found for threat *Th7*.For example, the fact that the participant is distant and its meeting selection has not been received by the meeting scheduler in due time is one of them.
- From the Method point of view, the former examples of *Th7* causes can be identified by an inadequate specification of multinational requirements.

Note that some of these causes had themselves already been identified as threats (e.g. Th5 is a *User* cause of TH7) Besides, the causes are complementary: altogether the convergence of these examples of causes have the a single effect (meeting is not identified). Figure 6 shows other examples of causes for this threat.

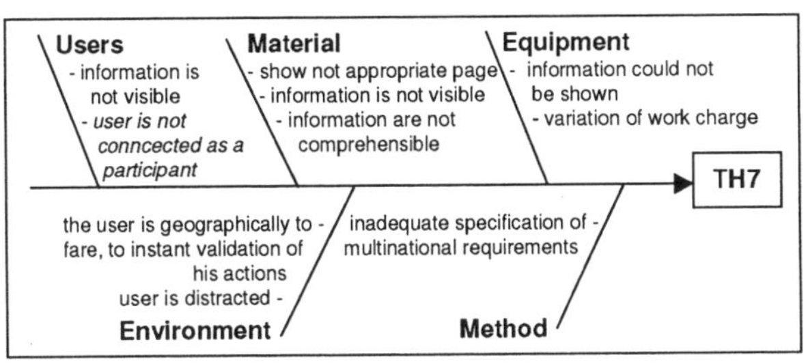

Figure 6: Causes of threat *Th7*

Each of the causes identified in Figure 6 was in turn analysed as a threat. For example Figure 7 shows the causes that can be found for threat *Th5*, namely "The user accesses the 'constraint definition' system service without being connected as a participant".

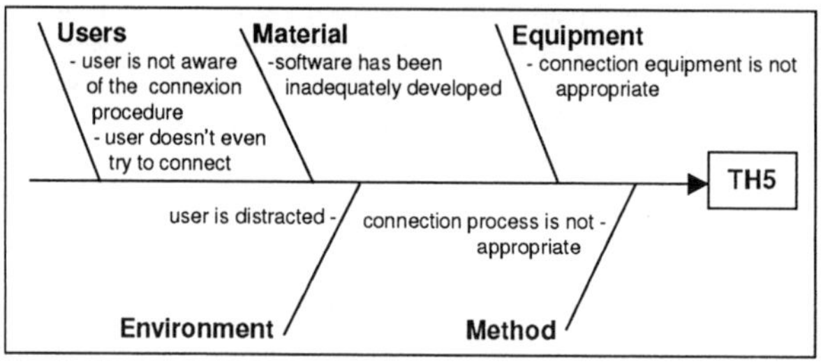

Figure 7: Causes of threat *Th5*

4.3 *Treatment strategies for threat Th5*

Each cause was separately studied to identify their importance in generating the threats, and to identify the level of gravity of the threats when they are due to these causes. As shown below, this intuitive way of reasoning lead to selecting one or several treatment strategies for each cause when this was found necessary.

Figure 7 shows that *Th5* can occur when the user is not aware of the connexion procedure. This can occur when the user uses the system for the first time, or when the connexion procedure has changed. This is a very likely situation as the meeting scheduler system is intended to be frequently used by occasional (end hence inexperienced) users. The impact on the initial goal (to define the participant's constraints) is high, and can even have consequences on the overall organisation of the meeting (e.g. if the user plays an important role in the meeting). The threat cannot be ignored or killed. Its consequences can however be weakened by providing a help to users and the threat probability can be reduced by monitoring user connexions and helping users when they fail to connect to the system. There are also different moments when the connexion help can be provided to users : with the meeting invitation, at the occasion of a first connexion, when the connexion fails, or when the connexion procedure is changed. The following goals could thus be added in the requirement chunks hierarchy :

- Provide (connexion help)$_{object}$ (to users)$_{destination}$ (at the moment of the meeting invitation)$_{time}$
- Provide (connexion help)$_{object}$ (to users)$_{destination}$ (at the occasion of a first connexion)$_{time}$
- Provide (connexion help)$_{object}$ (to users)$_{destination}$ (when the connexion procedure is changed)$_{time}$

and

- Provide (connexion help)$_{object}$ (to users)$_{destination}$ (when the connexion fails)$_{time}$

together with

- Monitor (connexion failures)$_{object}$ (from users)$_{source}$

Similarly, Figure 7 shows that the reason for Th5 could be that the user doesn't even try to connect. This situation is likely to occur when the user has not received the invitation to the meeting, when he/she doesn't have access to the system (e.g. he/she is abroad for a while), or when the system is down. The degree of importance of the threat depends directly on the duration for which the user doesn't access the system. If this is for a few minutes, there should be no particular treatment. If this is for such a long time that the meeting cannot be organised before the date at which it is supposed to occur, then the impact is high and an adequate treatment should be foreseen. The potential consequences of the threat can first be weakened by informing the user. In this case, the added goal is :

- Inform (the user)$_{destination}$ (of the need to connect)$_{object}$ (when constraints waiting time is exceeded)$_{time}$

or at the extreme by excluding the user from the meeting:

- Exclude (the user)$_{object}$ (from the meeting)$_{source}$ (when waiting time for meeting organisation is exceeded)$_{time}$

Another cause for this threat is when the software has been inadequately developed. The connexion requirement is identified and implemented in the system, but it does not function correctly. The consequence of the threat in this case is extreme as it impacts the ability of the system to let people organise meeting at all. There is no way to let the system overcome the threat by itself. However, it can still monitor user connexions and inform the development team when it finds that no connexion succeeds when meetings are organised. It is then up to the development team to explore the issue and correct the system. The new goal of the system is thus to inform developers of the threat :

- Inform (development team)$_{destination}$ (of inability for users to connect)$_{object}$ (when no succeeding user connexion is observed)$_{time}$

Other goals can be found by analysing each of the other causes of threat Th5 and of the other identified threats. Our proposal is not to systematically adopt each of them. Indeed, some of them are redundant, and they do not have the same level of priority. Therefore, we propose to let stakeholders choose in the list of generated goals those that are the most important. Each selected goal must then be integrated in the requirements hierarchy and concretised with a scenario. Pending goals can be recorded in *L'Ecritoire* and treated in a later stage of the requirements elicitation process, as proposed in [19].

5 Conclusion and future work

This paper proposes to extend the CREWS-*L'Ecritoire* requirements elicitation approach with a requirement elicitation rule based on threat analysis. The proposed rule is presented under the form of a methodological pattern decomposed in three parts: one to identify threats, one to refine their understanding, and one to identify how the system can treat the identified threats.

Each part of the pattern is guided with a specific approach. We propose to guide threat identification by a heuristic called 'negation strategy' which applies on requirements expressed as system goals. The second part of the pattern is guided by a pre-structured cause-effect diagram inspired from Ishikawa's diagrams. The discovery of new requirements is guided by the selection of a treatment strategy from an organised collection of pre-defined strategies. The elicited requirements are presented under the form of goals. These can be integrated and linked to other goals in CREWS-*L'Ecritoire*'s requirement chunk hierarchy.

The application of our methodological pattern is illustrated with the example of the meeting scheduler. Additional elements should be added to complete our pattern and provide the level of guidance expected in a CREWS-*L'Ecritoire* rule. For example :

- More heuristics could be proposed to identify threats. Such heuristics could be found from industrial projects, or adapted from other works.

- The analysis of threat causes could be further guided by improving the structure of cause-effect trees. For example sub-families of causes or generic collections of threats causes could be proposed. Cause analysis could also be structured, e.g. to express that it is the conjunction of causes that generate the threat.
- The ontology of treatments could be further extended. The collection of treatments could also be classified in a different way to facilitate their selection.
- Pre-defined analysis of generic threats could be proposed to accelerate cause analysis; this would require to define finer-grain families of threat causes.
- Treatment strategies could systematically be linked to threats to facilitate their selection. Threats could also be further qualified to enrich selection criteria.
- Once treatment strategies are identified, the stakeholder should select among them which one is a requirements. This selection could be guided, e.g. using bundling, prioritisation, etc
- The purpose of CREWS-L'Ecritoire goal discovery rules, is not just to guide the identification of new goals, but also to guide their integration in the requirement chunk hierarchy. We believe that in addition to treatment strategies, links with the existing requirements should systematically be identified to guide this integration.

Another important axis of our work will be to evaluate our approach. One particular concern is the validity of our five families of causes of threats. We are currently interviewing industrials to evaluate their effectiveness. The proposal could be further refined or re-designed depending on the comments that will be made. Empirical evaluations (to evaluate the efficiency of our approach) will be undertaken in a later stage of our research program. Another goal that we intend to achieve later on in our research program is the generalisation of our approach so as to integrate it in other goal-based requirements engineering approaches such as KAOS, I*, or extended Use Case modelling techniques.

References

[1] Annie I. Antòn, Colin Potts, Kenji Takahanshi, *Inquiry Based Requirements Analysis*, 1994 IEEE

[2] Boehm B., Ross R., *Theory-W Software Project Management: Principles et Examples*, IEEE Transactions on Software Engineering, vol.15, pp.902-916, 1989

[3] Easterbrook S., Nuseibeh B., Cramer J., Finkelstein A., *Co-ordinating Conflicting ViewPoints by Managing Inconsistency*, School of Cognitive and Computing Science – University of Sussex, Department of Computing – Imperial College, 1994

[4] Martin S. Feather, Stephen Fickas , Anthony Finkelstein, and Axel van Lamsweerde *Requirements & Specification Exemplars* Automated Software Engineering, Kluwer Pubs., Vol. 4 No. 4, 1997.

[5] Julia Galliers, Shailey Minocha and Alistair G. Sutcliffe CREWS Report 98-34; *A Causal Modèle of Human Error for Safety Critical User Interface Design* submitted to Special Issue on Interface Issues and Designs for Safety-Critical Interactive Systems of the ACM Transactions on Computer - Human Interaction. London 1998

[6] Kaoru Ishikawa, Guide to Quality control, Asian Productivity Organisation 1976

[7] Jyrki Kontio, *CS-TR-3782, UMIACS-TR-97-38, The Riskit Method for Software Risk Management, version 1.00*, Institute for Advanced Computer Studies and Department of Computer Science, University of Maryland

[8] Axel van Lamesweerde, Darimont R., Letier E., *Managing Conflicts in Goal-Driven Requirements Engineering*, IEEE Transactions in Software Engineering, Special Issue on Managing Inconsistency in Software Development, 1998

[9] Axel van Lamsweerde and Emmanuel Letier, *Integrating Obstacles in Goal-Driven Requirements Engineering*, Proc. ICSE'98 – 20th Intl. Conference on Software Engineering, Kyoto, April 1998, Vol, 53-64

[10] Axel van Lamsweerde, Robert Darimont, Philippe Massonet, *Goal-Directed Elaboration of Requirements for a Meeting Schedulers*: Problems and Lessons Learnt, IEEE 1995

[11] A. van Lamsweerde and E. Letier, *Handling Obstacles in Goal- Oriented Requirement Engineering*, IEEE-TSE, Special Issue on Exception Handling, 2000.

[12] Neil Maiden, Shailey Minocha, Michele Ryan, Keith Hutchings, Keith Manning; A Co-operative Scenario-based Approach to Acquisition and Validation of System Requirements : How Exceptions can help!,

CREWS Report 97-01 appeared in 'Proceedings of the Workshop on Human Error and Systems Development', 19-22 nd March, 1997, Glasgow University, Scotland, United Kingdom, organised by Dr. Chris Johnson

[13] Jolita Ralyté, CREWS Report Series 99 – 12, *Reusing Scenario Based Approaches In Requirements Engineering Methods* : CREWS Method Base, Proceedings of REP'99, 1st International Workshop on the Requirements Engineering Process, Florence, Italy, 2nd - 3rd Septembre 1999

[14] V. Plihon, J. Ralyté, A. Benjamen, N.A.M. Maiden, A. Sutcliffe, E. Dubois, P. Heymans *A Reuse-Oriented Approach for the Construction of Scenario Based Methods*, London 1998

[15] N. Prat. *Goal Formalisation and Classification for Requirements Engineering.* in 3rd International Workshop on Requirements Engineering: Foundation for Software Quality (REFSQ'97), Barcelone, Espagne, June 1997

[16] Robinson W.N., Volcov S., *Conflict Oriented Requirements Restructuring*, Working Paper CIS-96-15, 1996

[17] Colette Rolland, Camille Ben Achour - Salinesi, Corine Cauvet, Jolita Ralyté, Aistair Sutcliffe, Neil A.M. Maiden, Matthias Jarke, Peter Haumer, Klaus Pohl, Eris Dubois, Patrick Heymans. A Proposal for a Scenario Classification Framework. *Requirements Engineering Journal*, Vol.3, N°1, pp.23-47, 1998. Also available as CREWS report N°96-01 at URL http://sunsite.informatik.rwth-aachen.de/CREWS

[18] W. N. Robinsion, S. D. Pawlowski, V. Volkov *Requirements Interaction Management*, Atlanta, http://cis.gsu.edu/~wrobinso, GSU CIS Working Paper 99-7 August 30, 1999

[19] Colette Rolland, Carine Souveyet, Camille Ben Achour *Guiding Goal Modelling Using Scenarios*, CREWS Report Series 98-27, TSE special issue on scenario management, 1998

[20] C. Ben Achour- Salinesi, C. Rolland, N.A.M. Maiden, C. Souveyet. *Guiding Use Case Authoring : Results of an Empirical Study*. Proceedings of RE'99, fourth IEEE International Symposium on Requirements Engineering, University of Limerick, Ireland. June 1999

[21] Alistair Sutcliffe, Julia Galliers and Shailey Minocha, *Human Errors and System Requirements*, CREWS Report 98-36, International Symposium on Requirements Engineering, RE'1999. London 1999

[22] Alistair Sutcliffe and Shailey Minocha, *Analyzing Socio-Technical System Requirements*, CREWS Report 98-37, International Symposium on Requirements Engineering, RE'1999.

[23] Didar Zowghi, Vincenzo Gervasi. *On the Interplay between Consistency, Completeness, and Correctness in Requirements Evolution*. The Information and Software Technology Journal, Elsevier, Volume 45, Issue 14, pp993-1009, 2003

Information Modelling and Knowledge Bases XVI
Y. Kiyoki et al. (Eds.)
IOS Press, 2005

Application Development Based On Database Components

Bernhard Thalheim
Computer Science Institute, Christian-Albrechts University Kiel,
Olshausenstrasse 40, 24098 Kiel, Germany

Abstract. Large database applications often have a very complex structuring that complicate maintenance, extension, querying, programming. Due to this complexity systems become unmaintenable. We observe, however, that large database applications often use an implicit structuring into connected components. We propose to initially use this internal structuring for application development. The application architecture is based on database components. Database components can be composed to an application system. This paper shows how components may be developed, composed and applied.

1 Introduction

Complex Applications Result in Large Schemata.

Monographs and database course books usually base explanations on small or 'toy' examples. Reality is, however, completely different. Database schemata tend to be large, not surveyable, incomprehensible and partially inconsistent due to application, the database development life cycle and due to the number of team members involved at different time intervals. Thus, consistent management of the database schema might become a nightmare and may lead to legacy problems. The size of the schemata may be very large.

It is a common observation that large database schemata are error-prone, difficult to maintain and to extend and not surveyable. Moreover, development of retrieval and operation facilities requires highest professional skills in abstraction, memorization and programming. Such schemata reach sizes of more than 1000 attribute, entity and relationship types. Since they are not comprehensible any change to the schema is performed by extending the schema and thus making it even more complex. Database designers and programmers are not able to capture the schema.

Application schemata could be simpler only to a certain extent if software engineering approaches are applied. The repetition and redundancy in schemata is also caused by

- different usage of similar types of the schema,
- minor and small differences of the types structure in application views and

- semantic differences of variants of types.

Therefore, we need approaches which allow to reason on repeating structures inside schemata, on semantic differences and differences in usage of objects.

Large schemata also suffer from the *deficiency of variation detection: The same or similar content is often repeated in a schema without noticing it.*

Techniques to Decrease Complexity in Applications.

Large database schemata can be drastically simplified if techniques of *modular modeling* such as *modular design by units*[9] are used. It is an abstraction technique based on principles of hiding and encapsulation. Design by units allows to consider parts of the schema in a separate fashion. The parts are connected via types which function similar to bridges.

Data warehousing and user views are often based on *snowflake or star schemata*. The intuition behind such schemata is often hidden. Star and snowflake schemata are easier to understand, to query, to survey and to maintain. At the same time, these structures are of high redundancy and restricted modeling power. For instance, the central type in a star or snowflake schema is a relationship type which has attributes that use only numerical types. We may wonder, however, why we need to apply these restrictions and why we should not use this approach in general.

Co-design [9] of database applications aims in consistent development of all facets of database applications: structuring of the database by schema types and static integrity constraints, behavior modeling by specification of functionality and dynamic integrity constraints and interactivity modeling by assigning views to activities of actors in the corresponding dialogue steps. Co-design, thus, is based on the specification of the the database schema, functions, views and dialogue steps. At the same time, various abstraction layers are separated such as the conceptual layer, requirements acquisition layer and implementation layer.

Software becomes surveyable, extensible and maintainable if a clear separation of concerns and application parts is applied. In this case, a skeleton of the application structure is developed. This skeleton separates parts or services. Parts are connected through interfaces. Based on this *architecture*, an application can be developed part by part.

We *combine* modularity, star structuring, co-design, and architecture development to a novel framework based on components. Such combination seems to be not feasible. We discover, however, that we may integrate all these approaches by using a component-based approach. This skeleton can be refined during evolution of the schema. Then, each component is developed step by step. Structuring in component-based co-design is based on two constructs:

Components: Components are the main building blocks. They are used for structuring of the main data. The association among components is based on 'connector' types (called hinge or bridge types) that enable in associating the components in a variable fashion.

Skeleton-based construction: Components are assembled together by application of connector types. These connector types are usually relationship types.

An e-Government Application.

Modern parliamentary work has to face some problems due to the complexity of decision-making processes. Citizens as well as politicians are often lost in the variety of topics that are discussed; they are lost in complex and dynamic political structures and hidden responsibilities. The SeSAM e-Democracy portal provides an easy-to-use collaboration and communication platform for politicians and interested citizens. Information is presented always up-to-date in all aspects with respect to the users specific needs and wishes while the system itself is easy to maintain and adaptable to local peculiarities.

The information space of e-Government applications is rather complex. In Figure 1 we sketch this space.

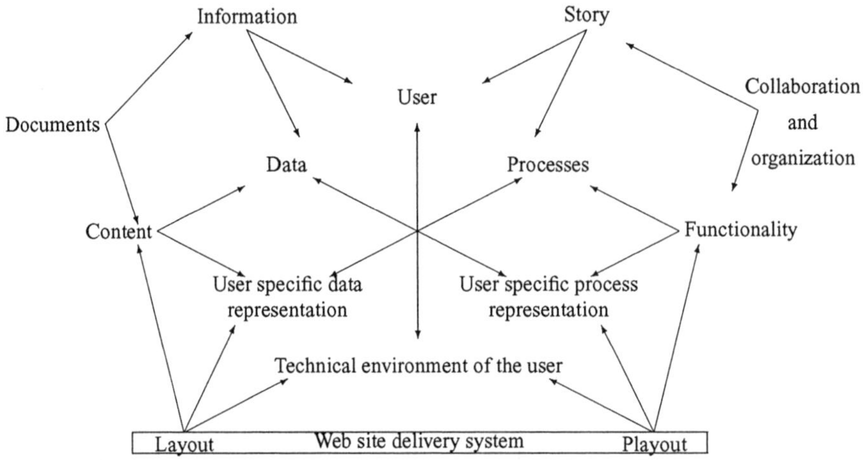

Figure 1: Dimensions of the e-Government System Space

Parliamentary systems provide information about persons, parties, parliamentary groups, and committees involved in the decision-making process, offer information about actual meetings, supports orchestration and syndication of complex documents (with older versions, annotations, protocols, or references to other related documents), and assists in collaboration, supports message exchange, suggestion making, sending criticism or opinions to responsible persons or groups. Such systems tend to be very large, become unmaintenable, are not extensible, and are finally not supporting but disturbing work.

We have analyzed systems available on the market and developed a database structure supporting the system. The outcome has been shocking. The initial schema used more than 800 relation types. Most of the relations are extensively interrelated to others. Extensibility, maintenance, development costs, integrity and semantics maintenance, change in technology, change of requirements becomes a nightmare for such systems. Experience gained in other projects have shown that the final system triples the complexity observed already for the in initial system.

For this reason, we decided to apply a new technique: **component-based system development**. Based on this technique, the SeSAM system system has been developed, is installed and used and can be step-by-step extended. The overall size of the final system is not smaller than the size of a monolithic system. But the system becomes maintainable, extensible, developable by smaller teams, and allows change of requirements and in technology.

Survey on the Paper.

We introduce the theoretical concepts of component-based co-design of database applications in the next section. Section 3 uses these constructs for development of methods for component integration that is non-invasive. We applied our approach to component construction in a number of projects, e.g., to an internet site that supports communication of parliamentarians in their daily work with documents. In the last section we demonstrate the power of our approach by discussing the document management in the SeSAM system.

2 Database Components and Construction of Schemes

Database Schemes in a Nutshell.

We use the extended ER model for representation of structuring and behavior generalizing the approach of [6]. The extended ER model (HERM) [9] has a generic algebra and logic, i.e., the algebra of derivable operations and the fragment of (hierarchical) predicate logic may be derived from the HERM algebra whenever the structure of the database is given.

A database type $\mathfrak{S} = (S, O, \Sigma)$ is given by

- a structure S defined by a type expression defined over the set of basic types B, a set of labels L and the constructors product (tuple), set and bag, i.e. an expression defined by the recursive type equality
  $t \;=\; B \,|\, t \times \ldots \times t \,|\, \{t\} \,|\, [t] \,|\, l : t \quad ,$
- a set of operations defined in the ER algebra and limited to S, and
- a set of (static and dynamic) integrity constraints defined in the hierarchical predicate logic with the base predicate P_S.

Objects of the database type $\mathfrak{S}^C$ are S-structured. Classes $\mathfrak{S}^C$ are sets of objects for which the set of static integrity constraints is valid.

Operations can be classified into "retrieval" operations enabling in generating values from the class $\mathfrak{S}^C$ and "modification" operations allowing to change the objects in the class $\mathfrak{S}^C$ if static and dynamic integrity constraints are not invalidated.

A database schema $\mathfrak{D} = (\mathfrak{S}_1, ..., \mathfrak{S}_m, \Sigma_G)$ is defined by

- a list of different database types and
- a set of global integrity constraints.

The HERM algebra can be used to define (parameterized) views $\mathfrak{V} = (V, O_V)$ on a schema $\mathfrak{D}$ via

- an (parameterized) algebraic expression V on $\mathfrak{D}$ and
- a set of (parameterized) operations of the HERM algebra applicable to V.

The view operations may be classified too into retrieval operations O_V^R and modification operations O_V^M. Based on this classification we derive an *output view* O^V of $\mathfrak{V}$ and an *input view* I^V of $\mathfrak{V}$.

In a similar way (but outside the scope of this paper) we may define transactions, interfaces, interactivity, recovery, etc.

Obviously, I^V and O^V are typed based on the type system. Data warehouse design is mainly view design [9].

Database Components and Component Algebra.
A **database component** is *database scheme that has an import and an export interface for connecting it to other components by standardized interface techniques.* Components are defined in a data warehouse setting. They consist of input elements, output elements and have a database structuring. Components may be considered as input-output machines that are extended by the set of all states S^C of the database with a set of corresponding input views $I^\mathcal{V}$ and a set of corresponding output views $O^\mathcal{V}$. Input and output of components is based on channels K. The structuring is specified by S_K. The structuring of channels is described by the function $type : C \rightarrow \mathcal{V}$ for the view schemata $\mathcal{V}$. Views are used for collaboration of components with the environment via data exchange. In general, the input and output sets may be considered as abstract words from M^* or as words on the database structuring.

A database component $\mathcal{K} = (S_K, I_K^\mathcal{V}, O_K^\mathcal{V}, S_K^C, \Delta_K)$ is specified by

(static) schema S_K describing the database schema of $\mathcal{K}$,

syntactic interface providing names (structures, functions) with parameters and database structure for S_K^C and $I_K^\mathcal{V}, O_K^\mathcal{V}$,

behavior relating the $I^\mathcal{V}, O^\mathcal{V}$ (view) channels
$\Delta_K : (S_K^C \times (I_K^\mathcal{V} \rightarrow M^*)) \rightarrow \mathcal{P}(S_K^C \times (O_K^\mathcal{V} \rightarrow M^*))$.

Components can be associated to each other. The association is restricted to domain-compatible input or output schemata which are free of name conflicts.

Components $\mathcal{K}_1 = (S_1, I_1^\mathcal{V}, O_1^\mathcal{V}, S_1^C, \Delta_1)$ and $\mathcal{K}_2 = (S_2, I_2^\mathcal{V}, O_2^\mathcal{V}, S_2^C, \Delta_2)$ are free of name conflicts if the set of attribute, entity and relationship type names are disjoint.

Channels C_1 and C_2 of components $\mathcal{K}_1 = (S_1, I_1^\mathcal{V}, O_1^\mathcal{V}, S_1^C, \Delta_1)$ and $\mathcal{K}_2 = (S_2, I_2^\mathcal{V}, O_2^\mathcal{V}, S_2^C, \Delta_2)$ are called *domain-compatible* if $dom(type(C_1)) = dom(type(C_2))$.

An output O_1^V of the component $\mathcal{K}_1$ is *domain-compatible* with an input I_2^V of the component $\mathcal{K}_2$ if $dom(type(O_1^V)) \subseteq dom(type(I_2^V))$

Component operations such as merge, fork, transmission are definable via application of superposition operations [4, 5]: Identification of channels, permutation of channels, renaming of channels, introduction of fictitious channels, and parallel composition with feedback displayed in Figure 2.

Thus, a component schema is usually characterized by a kernel entity type used for storing basic data, by a number of dimensions that are usually based on subtypes of the entity type `Document` such as `ApplicableFormat` and `LawsForSignature`, and on subtypes which are used for additional properties such as *Contract* and *Usage/Utilization* in Figure

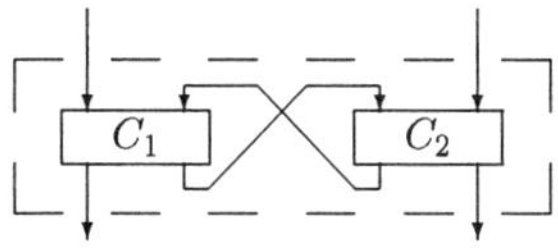

Figure 2: The Composition of Database Components

5. These additional properties are clustered according to their occurrence for the things under consideration. Furthermore, documents are classified by a set of *meta-properties*. Finally, documents may have their life and usage cycle, e.g., *language versions*. Therefore, we observe that the schema is in our case a schema with four dimensions: subtypes, additional characterization, versions and meta-characterizations.

The star schema is the main component schema used for construction. A star schema for a database type C_0 is defined by

- the (full) (HERM) schema $\mathcal{S} = (C_0, C_1, ..., C_n)$ covering all types on which C_0 has been defined,
- the subset of *strong types* $C_1, ..., C_k$ forming a set of keys $K_1, ..., K_s$ for C_0, i.e., $\cup_{i=1}^{s} K_i = \{C_1, ..., C_k\}$ and $K_i \rightarrow C_0$, $C_0 \rightarrow K_i$ for $1 \leq i \leq s$ and $card(C_0, C_i) = (1, n)$ for $(1 \leq i \leq k)$.
- the extension types $C_{k+1}, ..., C_m$ satisfying the (general) cardinality constraint $card(C_0, C_j) = (0, 1)$ for $((k+1) \leq i \leq n)$.

The extension types may form their own $(0, 1)$ specialization tree (hierarchical inclusion dependency set). The cardinality constraints for extension types are partial functional dependencies.

There are various variants for representation of a star schemata:

- Representation based on an entity type with attributes $C_1, ..., C_k$ and $C_{k+1}, ..., C_l$ and specializations forming a specialization tree $C_{l+1}, ..., C_n$.
- Representation based on a relationship type C_0 with components $C_1, ..., C_k$, with attributes $C_{k+1}, ..., C_l$ and specializations forming a specialization tree $C_{l+1}, ..., C_n$. In this case, C_0 is a *pivot element* [1] in the schema.
- Representation by be based on a hybrid form combining the two above.

Star schemata may occur in various variants within the same conceptual schema. Therefore, we need variants of the same schema for integration into the schema. We distinguish the following variants:

Integration and representation variants: For representation and for integration we can define views on the star type schema with the restriction of invariance of identifiability through one of its keys. Views define 'context' conditions for usage of elements of the star schema.

Versions: Objects defined on the star schema may be a replaced later by objects that display the actual use, e.g., *Documents* are obtained and stored in the *Archive*.

Variants replacing the entire type another through renaming or substitution of elements.

History variants: Temporality can be explicitly recorded by adding a history dimension, i.e., for recording of instantiation, run, usage at present or in the past, and archiving.

Lifespan variants of objects and their properties may be explicitly stored. The lifespan of products in the acquisition process can be based on the *Raw Document - Operational Document - Blueprint - Submission - Approved Document - Archived Document* cycle.

3 Non-Invasive Conceptual Construction Based on Components

Construction Requirements.

Component construction is based on a general component architecture or a skeleton. Each component is developed in separate. The advantage of the strict separation is an increase of modularization, parameterizability and conformance to standards.

We derive now a none-invasive construction approach which does not change components used for construction. Due to this restriction we gain a number of properties such as *adaptativity, seemless gluing, extensibility, aspect separation, scalability*, and *metamodeling* and *abstraction*.

Components and Harnesses.

The construction is based on harnesses and the application skeleton. The skeleton is a special form of a meta-schema architecture. It consists of a set of components and a set of harnesses for superposition operations. Harnesses are similar to wiring harnesses used in electrotechnics. A harness consists of a set of input-output channels that can be used to combine wrapped components.

Given a sets of components $\mathfrak{K} = \{\mathcal{K}_1, ..., \mathcal{K}_m\}$ and labels $\mathfrak{L} = \{L_1, ..., L_n\}$ with $n \geq m$. Given furthermore a total function $\tau : \mathfrak{L} \rightarrow \mathfrak{K}$ used for assigning roles to components in harnesses. The triple $(\mathfrak{K}, \mathfrak{L}, \tau)$ is called harness skeleton $\mathfrak{H}$. The arity of the skeleton is n.

The skeleton is graphically represented by doubly rounded boxes. Components are graphically represented by rounded boxes. The construction may lead to complex components called *units*.

The example in Figure 3 has been used in one of our projects. Parliamentarians and inhabitants are combined into a component *Users*. We may use a large variety of positions. A user may use a certain service through some devices. Appointments are based on the usage of services. Tools vary depending on services and on equipment. The final schema contains more than 2.500 attribute, entity, cluster and relationship types. The skeleton of the application is rather simple.

Harness Filters.

Components may be associated in a variety of ways. In the application in Figure 3 the usage of services depends on the properties of parties, the tools they may use, and the services provided. Services, parties, and tools have their own dimensionality. If we use the classical approach to schema development each subtype may cause the introduction of a new usage type. The schema explodes due to the introduction of a large variety of usage type. To overcome this difficulty we introduce filters.

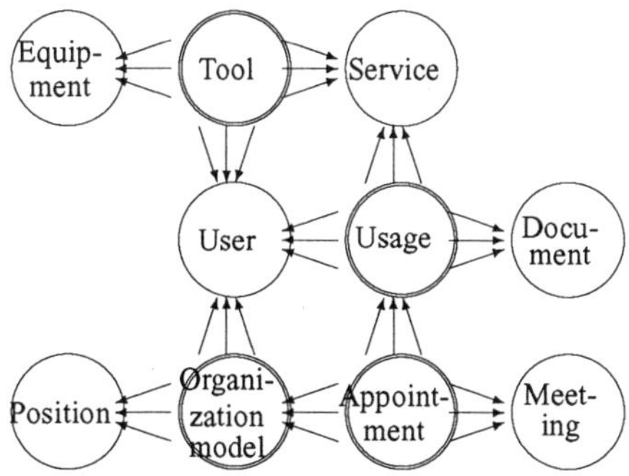

Figure 3: Skeleton of a Schema for *e-Government Service* Applications

Given component schemata of an n-ary harness skeleton. A **filter** of an n-ary harness is an n-ary relation defined of the multi-dimensional structure of the components, i.e. on the views defined for the components.

Filters may be represented either graphically or in a tabular form. In our example, we obtain the following filter. Components are already presented in Figure 3. We develop a number of services which might be used depending on the role, rights, and positions of the users. For

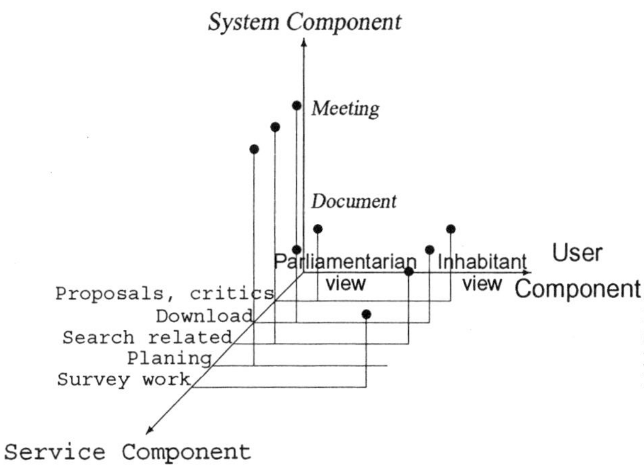

	Parliam.	Inhab.
Sur	*Doc*	*Doc*
Pla	*Meet,Doc*	*Doc*
SRe	*Meet*	*Doc*
Load	*Meet*	-
PrCr	-	*Doc*

instance, the parliamentarian is interested in search of related documents in the role of an inhabitant and in search of related meetings.

The implementation of filters is rather straightforward. Each harness has a filter. Since views are defined together with their identification mechanism, an n-ary harness may be represented by an $(n+1)$-ary relationship type associating the components with their roles and extended by the filter.

A **harness** consists of the harness skeleton $\mathfrak{H} = (\mathfrak{K}, \mathfrak{L}, \tau)$ and the harness filter $\mathfrak{F} = \{(L_i, \mathcal{V}^{L_i}) \mid 1 \leq i \leq n,\ L_i \in \mathfrak{L},\ \mathcal{V}^{L_i} \subseteq \mathcal{V}_{\tau(L_i)}\}$ for a set of wrapped components $(\mathcal{K}_i, \mathcal{V}_i)$.

Operators Used For Non-Invasive Schema Construction.
In [13] a number of composition operators for construction of entity and relationship types has been introduced: constructor-based composition, bulk composition, lifespan composition (architecture-based composition, evolution composition, circulation composition, incremental composition, network composition, loop composition), and context composition.

We generalize now these composition operators to component-based schema construction.

Constructor harnesses are based on composition operations such as *product*, *nest*, *disjoint union*, *difference* and *set* operators.

Bulk harnesses allow to bound components, types or classes which share the same skeleton. Two harness skeletons $\mathfrak{H}_1 = (\mathfrak{K}_1, \mathfrak{L}_1\tau_1)$ and $\mathfrak{H}_2 = (\mathfrak{K}_2, \mathfrak{L}_2\tau_2)$ are called unifiable if they are defined over the same set of components, $|\mathfrak{L}_1| = |\mathfrak{L}_1| = n$, and there exists a permutation ρ on $\{1, ..., n\}$ such that $\mathcal{K}_{\tau_1(i)} = \mathcal{K}_{\tau_2(\rho(i))}$. The bulk harness of unifiable harnesses $\mathfrak{H}_1, ..., \mathfrak{H}_p$ is constructed by renaming the labels L_j of each harness $\mathfrak{H}_i$ to $L_{i,j}$ and combining the label functions τ_i.

Application-separating harnesses: An enterprize is usually split into departments or units which run their own applications and use their own data. Sharing of data is provided by specific harnesses.

Distribution-based harnesses: Data, functions and control may be distributed. The exchange is provided through specific combinations which might either be based on exchange components that are connected to the sites by harnesses or be based on combination harnesses.

Application-separation-based harnesses have been widely used for complex structuring. The architecture of SAP R/3 often has been displayed in the form of a waffle. For this reason, we prefer to call this composition *waffle composition* or architecture composition.

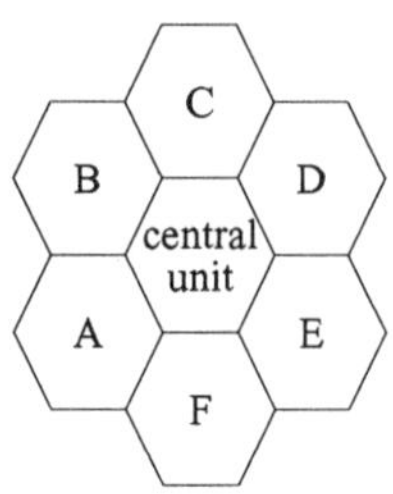

Figure 4: The Waffle Architecture Composition

4 Using the Power of Component Construction for Management of Administrative Documents

The SeSAM System.
The SeSAM system (system for simple storage, retrieval and modification of internet content) has been been developed as an extension of the internet platform www.cottbus.de. It

supports the work of parliamentarians, their internal and external communication and their work with documents. Information is displayed to visitors of the website depending on their role, the content available, the task currently under consideration, the associations within the content set, and the technical environment. Content of the website is determined by the schedule of parliamentary sessions, on groups in the parliamentary, on the state of the document processing, and discussion forums currently open.

SeSAM is a content management system. It supports the export and the import of documents, news, and contributions. Data are sorted into folders and projects. Documents have a number of versions. SeSAM has sub-systems for accounting, configuration, archiving, and oral communication. Within this paper we concentrate on the document management component of the system.

SeSAM components have been also integrated into the inhabitant information service of www.cottbus.de. Inhabitants are supported by download facilities for documents, by intelligent search facilities for document retrieval depending on their daily needs, by communication facilities for direct communication with the city administration and by adaptation of the presentation to their current technical environment such as channel, display equipment and personal profile.

Document Characterization.

Documents in administrations are very complex. They evolve over lifetime. We can distinguish different phases:

Raw document: Raw documents or frames or prototypes are document forms with mostly empty data, with emphasis and default values. During call of a document by a department, the document is instantiated by the default values associated with the department or unit calling the document. Instantiation can be also performed by computing queries on the document base, especially on archived documents. Components or attributes to be instantiated during the call of the raw document are called configuration parameters.

Operational document: The entry of data into the document may be a rather complex workflow with several stages, different associated actors, various responsibilities and different stages of preparations.

Blueprint: The document is almost ready. The content which is required for a filled document exists. The documents needs a number of approvals in order to be submitted.

Submission: The document to be submitted has the complete content to be required by the business processes and rules. It is used in the working process. The document may be changed. In this case, the document is pushed to the state blueprint.

Archived document: The document or parts of the documents or views on the document are stored in the archive. Archived documents have a summary or docket[8]. They cannot be changed. They can be retrieved.

These lifespan phases are supported by the lifespan composition of document star schemes and use incremental composition.

Documents may have a complex structuring which is partially displayed in Figure 5. We, however, assume:

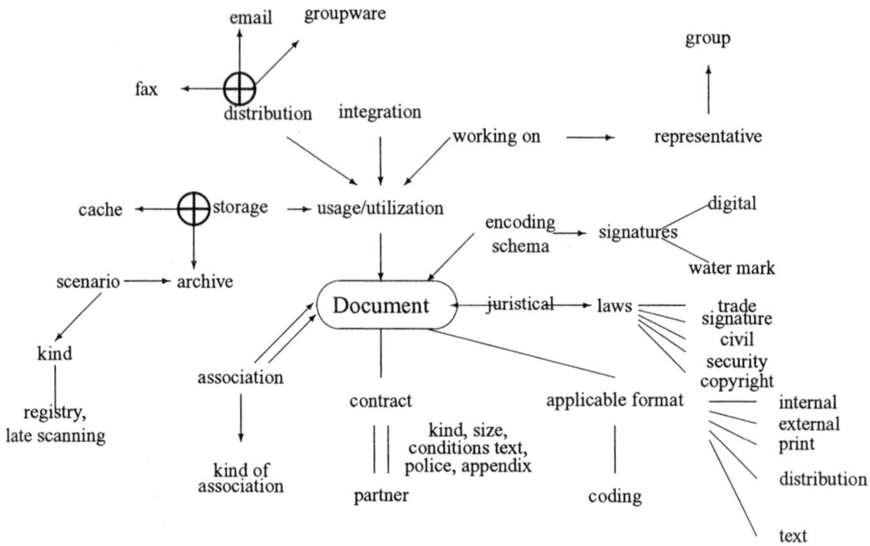

Figure 5: Combining the Star Structure and the Context of Documents

Inductive construction: *Each document can be constructed on the basis of simpler documents and base documents by application of constructors.*
This assumption restricts the application areas. However, based on this assumption we are able to compose the document step by step. In this case, structural recursion can be applied to construction of necessary functionality.

The semantics changes dynamically. Thus, we are not able to specify the entire document by means of static integrity constraints. In order to model the change of semantics we introduce the concept of states. We associate documents with their state thus providing a version control. The actual document is generated by unfolding the versions.

Version control may be rather complex. We can use the concepts of the life space of a document which is similar to the concept of the story space:

Life flow of operational documents is defined by a rooted graph $\mathcal{G} = (N, E)$ which nodes reflect the variety of states a document may have and which edges represent the actions causing a change of the document.

Nodes have a unique identity and represent the delta of the change to the document imposed by the action which is applied to the predecessor.

Directed edges are associating two nodes. They have a label that is used for representation of action kind, action time, actor, role, rights etc.

Construction of documents is based on the concept of snapshots and on cuts or stratums in the life flow graph: At a given moment of time t and for a given document the graph is captured by *unwinding* the document graph:

- Those nodes N_t are captured which time stamp is smaller than the given point of time.
- The subgraph $\mathcal{G}_t = (N_t, E \cap (N_t \times N_t))$ is used by iterative construction of the document beginning from the root of the graph.

Construction may lead to documents which have a number of optional components. The graph $\mathcal{G}_t$ is thus used to generate a snapshot D_t of the document D.

Further, given a group of users U. A view of the document is based on the subgraph $\mathcal{G}_{t,U}$ which uses only those edges which visibility rights belong to the group U or are assigned to '*public*'. Only those nodes are used for the delta capturing which are reachable from the root.

The construction mechanism may also be used for keeping the *current version* of a document. The current version describes the latest state.

Supporting Functionality for Documents.
The construction is performed by the

ICE: Information and Content Extractor

which has been developed for runtime and context-sensitive adaptation of the database design systems RADD and ID^2 to the designers [9]. A similar application is the adaptation of web search tools. The ICE is based on the following concepts:

Built-in abstraction for data for coarsing/zooming, indexing, viewing, and aggregation of documents enable in surveying and landscaping documents.

Farms of views can be inductively and incrementally generated with or without materialization and with or without recharging after modifications.

Just-in-time views on the basis of materialized or prefetched views support quick response by right document, at the right place, at the right time, in the right condition, and at the right cost.

Parametric views have parameters. A special group of parameters are the context parameters which enable in adaptation to the technical context (infrastructure, channel), to the task-ruled context (workflow, history), and to the actor context (profile, role).

Runtime versioning supports the management of views currently in use. These views disappear ('fade') whenever they are not needed. The handling is based on garbage collection procedures.

We notice that ICE has been also used in one of our industrial project for the generation of *bulk workflows*. This concept is very similar to the discussed above document concept. It is a useful concept for e-Learning sites. Bulk workflows support the typical parallel workflow in offices:

- Work is carried out in a high parallel form.
- The flow of tasks requires feedback and support of waiting queues.

- Work is often controlled from outside.

Bulk workflows are based on

unfolding of general workflows with a generic runtime workflow,

potentially high parallelism of workflows with intermediate states, to-do lists and resource sharing,

multiple-choice workflows for roles, free choice and bundling of tasks, and

transactions-based meta-workflows with management of resources and roles.

The operations we use for the change in documents are based on paragraph anchors. Paragraph anchors are automatically generated together with the operations applied to the document. We use the following elementary operations:

Insertion of new components c into the document at a point p assigned. Given a document identifier d, a node n in the life flow, and a moment of time. Then, an actor a which has the right r^* to change the document in a role r causes a complex insert at a the point p:
$insert(d, n, c, p, t) = insert(\mathcal{G}, \{n'(new, c), (n, l(add, t, a, r, r^*, ...), n')\});$
$update(\mathcal{G}, n, add(p))$.

Deletion of anchored paragraphs p from the document identified by d in a node n in the life flow at a moment of time t by an actor which has the right r^* to change the document in a role r is defined by:
$delete(d, n, p, t) = insert(\mathcal{G}, \{n'(new, \lambda), (n, l(erase, t, a, r, r^*, ...), n')\});$
$update(\mathcal{G}, n, add(p))$.

Updating of anchored paragraphs p by a component c in the document identified by d in a node n in the life flow at a moment of time t by an actor which has the right r^* to change the document in a role r is defined by:
$update(d, n, c, p, t) = insert(\mathcal{G}, \{n'(new, c), (n, l(replace, t, a, r, r^*, ...), n')\});$
$update(\mathcal{G}, n, add(p))$.

A modification of the document is a sequence of elementary operations. The modification forms a sub-graph. If the modification is a sequence we can condense the path to one edge between the end node and the starting node.

Strongly sequenced documents allow a simpler handling. If only one actor is acting at a certain point of time and all actions are only applicable to the most recent state then the life flow is a queue of actions applied. In this case we may store the most recent version of the document and invert the life flow to a history flow. We find, however, that due to a variety of rights and roles online documents cannot be modeled by strongly sequenced documents.

The *life flow of operational documents* corresponds to *construction of documents*. There is a large number of document management system [3] such as EDMSuite (IBM), Domino.Doc (Lotos), ImagePlus. Different standardization committees such as the ODMA (open document management API) and DMA (document management alliance) have been working on a standard. The problem of all these proposals is, however, that none of them was able to capture the complex structuring of a documents life flow. Our approach allows the integration of life flow, content of documents and workflow.

Varying Documents for Play-Out.
Documents have a large variety of representation views:

Actor view: An actor or a group of actors have their own views based on the access rights that are valid for the actor.

Role view: A document can be used in a number of roles. These roles are supported by a variety of partial views on the document.

Function view: Documents are often instruments for complex business processes. The specific content necessary for the document in the business process is generated as a function view.

Task view: Documents are used to support tasks. These tasks require the insertion of certain data into the document. The task view consists thus of a representation similar to a printed form.

Node view: The responsibility of the content for a document is often distributed among a set of users. A document may have parts for which a user plays the role master. The copies on other nodes have the role slave. Slaves do not have the right for updates or other modifications.

Documents can be represented by snowflakes. They have a variety of representations.

Configuration of a document: As already described the construction of documents can be based on the life flow. The snapshot at a certain moment of time displays the state of the document.

Run-time adaptation depending on the state: Concepts used for the representation of documents are adhesive and cohesive to each other in a different manner. If a component currently under consideration is less cohesive to some of the components than these components may be cut out during consideration of the component currently under consideration.

This representation structuring of documents leads to the
conception of recursive folders of documents.
If the structure is defined on the basis of a star sub-schema and the life flow is representable by a snowflake then the variety of views is representable by a multi-list. The different branches of the multi-list are opened at different moments of time.

Summarizing we can model the document by a multi-dimensional structuring:

Intext: The intext of a document follows the internal logical structure of the document itself, is used for the main characterization of the document, and carries the specialization hierarchies.

Context-dependent representation: The context of the document is a manifold of associations to other documents, the temporal and spatial relations, the application context, the interface context, and the meta-characterizations of the document.

Version under consideration: The document version, the archive version, the log version and the view versions of the document are combined in the version space of the document.

Life flow of the document: The evolution of the document and the changes imposed by a variety of actors are represented by the life flow of the document.

The four dimensions are used for run-time configuration of documents.

Documents are classified by meta-parameters. These meta-parameters are *dimensions*. One of these meta-parameters is the applicable overlay structure:

- The presentation forms are represented by parameters enabling in adaptation of presentations. Web applications use the full-screen modus of representation although most of users try to use web browsers in a half-screen modus. We may have also small information containers [2]. Therefore, the representation of documents is supported by functions for generating display options, for decomposing documents and thus generating junks of the same document, for adaptation to the size of the window, for enumeration, for cutting pages of the document, for creating page turners and passages.

- User-defined overlay structures use *anchors* provided by the document structure. The anchoring is usually hidden from the user. The anchors may be used to create user-specific bookmarking. Bookmarks may be of arbitrary graph structure. A variety of ordering and presentation functions can be applied. Anchors have their own name space which consists of the identification of the document and the internal anchor name. These anchors can be combined with the index or abstraction facilities.

  Specific overlay information is made in form of *annotations*. Annotations can be related to the content or to the functionality.

  Thumb nails are specific bookmarks. They support scanning and zapping through documents.

- Documents are often formulars. Therefore, a user-defined version is instantiating some sections, paragraphs or forms provided by the document. Since these instantiations relate the document also to other data sources, an update functions is included into the form insert procedures. Forms may be also added to the document as well as specific appendices. These functions must be reversible. Forms may be included into the representation or may not.

- Documents may be combined or associated by the user. The *bundling mechanism* enables in combining sets of documents by the user. This mechanism generalizes the folder concept. We do not store, however, the same document in different folders but mirror the documents in a master-slave mode. Whenever, we need the full document the master document is retrieved. Whenever a change is made then the master document is used. Whenever we need a user-defined document then the master document is used together with the slave that is bound to the user or to the workflow.

- Documents are characterized by keywords, topics or ontology items. The characterization may be applicable to entire documents or to anchors in documents. The ontology characterization of documents supports intelligent search especially associative search.
- Documents often use specific menu functions. These functions are added to the header of the document. They enable in variable display, in applying the provided functionality and in export of the content.
- Finally, dockets[8] of documents are specific overlay structures. Dockets provide information
  - on the content (*abstracts* or *summaries*),
  - on the delivery instruction,
  - on the parameters of functions for opening the document (opening with(out) zooming, breath, size, activation modus for multimedia components etc.)
  - on the tight association to other documents (versions, releases etc.),
  - on the meta-information such as resources, restriction, copyright, roles, distribution policy etc.
  - on the content providers, content reviewers and review evaluators with quality control policies,
  - on applicable workflows and the current status of completion and
  - on the receipt of the document which enable in tracing the document life cycle.

The overlay structuring of documents is supported by the onion approach [14]. Onion generation is based on layering that used for generation of website functionality and content. The onion approach nicely fits with the translational approach. It is our aim to generate consistent sets of associated XML documents. Let $\mathcal{X}$ be the set of all XML documents under consideration. Further given a generation algorithm G applicable to XML documents that allows to generate new XML documents on the basis of the given ones. Let us denote by $[\mathcal{X}]_G$ the transitive closure of the generation algorithm applied to $\mathcal{X}$. A set $\mathcal{X}$ of XML documents is called consistent according to G if all inner references in $\mathcal{X}$ belong to $[\mathcal{X}]_G$, i.e., no dangling inner references are in the set $\mathcal{X}$. In this case $\mathcal{X}$ is called XML suite.

Document systems should be supported by a specific data warehouse architecture:

Play-out servers present, store and protect released content. The play-out of documents depends on their usage. Typical widely used documents are documents used in logistics:

- Bills have their own numbering and their own format. They serve also as an contract of carriage between shipper and carrier.
- Certificate on the content and the origin of the contents are used for statistical research, and for accessing duties, particularly under trade agreements.
- Invoices declare against which payment is made. They are used for clearing documents.
- Dock receipts are issued by the forwarder on experter's behalf. They include shipment description, physical details, and shipping information.

- Bills of lading are used as contracts between carrier and shipper, spell out legal responsibilities and liability limits for all parties to the shipment.
- Packing lists provide details on the packing procedure of the container.
- Sight, time drafts instruct the buyer's bank to collect payment.

Production servers have controlled access to documents and host dockets.

Specific docket servers manage trusted content exchange between the servers.

Generic docket servers communicate and encapsulate value-adding services.

5 Conclusion and Outlook

Component construction has been widely used in other engineering areas. Database modeling is still based on handicraft approaches, i.e., each new application is developed from scratch or uses solutions which are again based on handicraft. Database systems become now part of middleware solutions. Thus, a plug-in approach must be developed that allows a stepwise integration into existing infrastructure. Component construction supports a plug-in approach.

This paper has introduced a general approach to component construction. We observe a number of advantages of component construction such as: simpler sub-schemata, simpler combination, simpler integrity maintenance, modularity and extensibility. Component construction does not lead, in general, to optimal schemata defined on the number of types. Handicraft approaches may lead to such schemata. However, the advantages will play an important role whenever the application is complex and the database schema becomes large.

We defined harnesses and harness construction. This mechanism allows to construct information systems based on components which are associated to each other by hyper-edge wires of harnesses. We have applied this approach in the SeSAM project. The application led to simpler schema construction and led to a schema that is easy to extend, to query and to maintain.

Component construction may be extended by construction with component functionality. The development of a theory of component functionality is our main research direction in future.

References

[1] J. Biskup and T. Polle. Decomposition of database classes under path functional dependencies and onto contraints. In *Proc. FoIKS'2000*, LNCS 1762, pages 31–49. Springer, 2000, 2000.

[2] T. Feyer, K.-D. Schewe, and B. Thalheim. Conceptual design and development of information services. In *Proc. ER'98, LNCS 1507, Springer, 1998*, pages 7–20. Springer, Berlin, 1998.

[3] U. Kampffmeyer, B. Merkel, and C. Feldmann. *Document management.* Teubner, Wiesbaden, 1998.

[4] V. B. Kudrjavcev. *Functional systems.* Moscov Lomonossov University Press, Moscov, 1982. (in Russian).

[5] A. I. Malzew. *Algebraic systems.* Nauka, Moscow, 1970.

[6] J. Paredaens, P. De Bra, M. Gyssens, and D. Van Gucht. *The structure of the relational database model.* Springer, Berlin, 1989.

[7] K.-D. Schewe and B. Thalheim. Readings in object-oriented databases. Reprint, BTU-Cottbus, accessible through http://www.is.informatik.uni-kiel.de/~thalheim, Collection of papers by C. Beeri, K.-D. Schewe, J.-W. Schmidt, D. Stemple, B. Thalheim, I. Wetzel, 1998.

[8] J. W. Schmidt and H.-W. Schering. Dockets: a model for adding vaulue to content. In *Proc. ER'99, LNCS 1728*, pages 248–262. Springer, Berlin, 1999.

[9] B. Thalheim. *Entity-relationship modeling – Foundations of database technology*. Springer, Berlin, 2000. See also http://www.is.informatik.uni-kiel.de/~thalheim/HERM.htm.

[10] B. Thalheim. Readings in fundamentals of interaction in information systems. Reprint, BTU-Cottbus, accessible through http://www.is.informatik.uni-kiel.de/~thalheim, Collection of papers by C. Binder, W. Clauß, A. Düsterhöft, T. Feyer, T. Gutacker, B. Heinze, J. Lewerenz, M. Roll, B. Schewe, K.-D. Schewe, K. Seelig, S. Srinivasa, B. Thalheim, 2000.

[11] B. Thalheim. Component construction of database schemes. In *Proc. ER'02, LNCS 2503*, pages 20–34, 2002.

[12] B. Thalheim. Co-design of structuring, funtionality, distribution, and interactivity of large information systems. Coumputer Science Reports 15/03, Cottbus University of Technology, Computer Science Institute, 2003.

[13] B. Thalheim. Database component ware. In *Proc. ADC'2003, Journal on Research and Practice in Information Technology, vol 17*, pages 1–13, 2003.

[14] B. Thalheim and A. Düsterhöft. Sitelang: Conceptual modeling of internet sites. In *Proc. ER'01*, volume 2224 of *LNCS*, pages 179–192. Springer, 2001.

Remark: *It is not my aim to discuss related database and internet technology. Our group uses the website modeling language SiteLang [14], the power of advanced object-oriented approaches [7], the theory of interaction [10], and the co-design approach [9, 12].*

I thank the EJC'2004 reviewers of the paper for their substantial suggestions for improvement of this paper.

Information Modelling and Knowledge Bases XVI
Y. Kiyoki et al. (Eds.)
IOS Press, 2005

A Dynamic Retrieval Space Creation Method for Semantic Information Retrieval

Xing CHEN* and Yasushi KIYOKI**

**Department of Information & Computer Sciences*
Kanagawa Institute of Technology
1030 Simo-Ogino, Atsugi-shi, Kanagawa 243-0292, Japan
chen@ic.kanagawa-it.ac.jp

***Department of Environmental Information*
Keio University
Fujisawa, Kanagawa 252-8520, Japan
kiyoki@sfc.keio.ac.jp

Abstract Semantic search for databases and knowledge bases is becoming very important for realizing a new information resource environment over world wide networks. The vector space model for implementing semantic search gives great attention in this field. The vector space model is advantageous for computing semantic relationships between data. In order to realize semantic search, a vector space is created by extracting meanings of terms statically. Semantic search is performed by mapping retrieval candidates and queries onto the retrieval space and calculating the correlations between retrieval candidates and queries.

We have noticed that the semantic of a retrieval candidate may be different among individual users. Furthermore, even for the same user, the semantic of a same retrieval candidate may be different in his/her different queries. Therefore, it is required to extract the meaning of the retrieval candidates and queries dynamically in the query processing.

In this paper, we present a dynamic semantic space creation method for semantic search for document data. In this method, the meanings of a query and retrieval candidates are dynamically extracted by a learning mechanism during the users' retrieval actions. The retrieval space is created dynamically during query processing. The most important feature of our method is that the retrieval spaces, which are mostly adapted to the meanings of the queries, are created during query processing. The retrieval results which are mostly closely to the meanings of the queries are given to users.

1. Introduction

The vector space model is known well as an advantageous approach for realizing semantic search. In the vector space model, the semantic retrieval is performed by mapping retrieval candidates and queries onto an orthogonal vector space according to the contents of retrieval candidates and queries. The correlations between queries and retrieval candidates on the retrieval space are calculated. Retrieval candidates are ranked according to their correlations to the queries as the retrieval results.

The orthogonal vector space can be created by applying one of the following two important methods, Latent Semantic Indexing (LSI) [4, 10, 11, 12] and the Mathematical Model of Meaning (MMM) [2, 3, 5, 6, 7, 8, 9]. In both of the methods, the meanings of terms appearing in retrieval candidates are calculated by a mathematical method called SVD.

In LSI, the SVD calculation is applied directly to retrieval candidates and the orthogonal space is created according to the correlations of terms appearing in the retrieval candidates. That is, the meanings of terms are defined according to the appearance of terms in the retrieval candidates.

In MMM, a dictionary, for example, an English-English dictionary is referred to create a semantic vector space. The orthogonal space is created by applying the SVD calculation to calculate the correlations of basic words defined in the dictionary. In this way, the meanings of terms are defined according to the definitions for the terms in the dictionary.

We have presented another method to extract meanings of terms through users' query processing [1]. In the paper [1], it is basically declared that each term includes different meanings. That is, different users define a same term into different meanings in their queries and even a same user defines a same term into different meanings in his/her different queries. Therefore, a basic issue is how to recognize the meaning of a query, that is, the user's intention. To recognize the meaning of a query, we consider that the meaning of a query can be recognized by using user's feedbacks. If a user selects some items which are well represent his/her query intention from retrieval results, the selected results are recorded as the user's feedback. In the method, queries and retrieval candidates are vectorized based on words. Queries' vectors are adjusted according to users' feedbacks to well represent the users' intentions in the subsequent retrieval processes. We refer this adjusting process to as the "learning mechanism" [1].

In our method, a vector space is created by some previously prepared documents, which we refer to as "sample documents." Those sample documents are classified into several clusters. Some common terms, which frequently appear in the documents in the same cluster and rarely appear in the documents of the other clusters, are extracted from the sample documents. We refer these extracted common terms to as "keyword terms".

A retrieval space is constructed based on the keyword terms. Documents are mapped onto the retrieval space according to the keyword terms appearing in them. When a query is given, the document clusters, which are most related to the query, are selected and a subspace related to those clusters is selected from the entire retrieval space. Documents on the subspace are ranked according to the values of their norms on the subspace. Because the components of vectors on the retrieval space are directly expressed by words, our method makes it possible to perform the dynamic meaning recognition by using users' feedbacks during the query process.

When an orthogonal space is created for the semantic information retrieval, a term is represented as a vector on the retrieval space. We refer to the vector as the term vector. If the meanings of two terms are similar, these two term vectors are close to each other on the space. Queries and Retrieval candidates are mapped on the retrieval space and represented as vectors according to the term vectors. When the retrieval space is created statically, the term vectors are fixed on the space. If the meaning of a query is changed during the query processing, the term vectors have to be modified to the position at which the meanings of the terms are best represented. However, the term vectors will not be automatically changed as the changing of the query vectors. Because the term vectors are not automatically changed as the changing of the query vectors, the vectors of the retrieval candidates will not be changed to the positions at which the correlations between the retrieval candidates and the queries will be best represented.

If the retrieval space is reconstructed as the changing of the meanings of the queries, all the retrieval candidate vectors will be changed during the reconstruction. In this paper, we present a new mechanism for the dynamic retrieval space creation. This mechanism is applied to the retrieval space created by the methods that we have proposed in [1]. In this mechanism, the operations for combining the existing axes of the retrieval space and creating the new axes of the retrieval space are performed. That is, the dimension of the retrieval space is dynamically decreased and increased according to the users' queries. A retrieval space that is mostly suitable to a user's query is dynamically created during the query processing. In this paper, we also present a fast retrieval candidate mapping method to map retrieval candidates to the dynamically created retrieval space.

In the following, we briefly review the method for the retrieval space creation method represented in [1] in Section 2. In Section 3, we present our new mechanism for the dynamic retrieval space creation. Experimental results are shown in Section 4. In Section 5, we summarize the proposed mechanism.

2. Retrieval space creation and document mapping based on the document classification

In this section, we briefly review the retrieval space creation, document mapping and the semantic retrieval methods that we have proposed in [1]. Our mechanism for fast retrieval space creation is applied to the retrieval space created by these methods.

2.1 The retrieval space construction method

We create a retrieval space by classifying a pre-selected sample document set into several clusters. Each of the clusters has a feature that some common terms frequently appear among the documents in the same cluster but rarely appear in the documents of the other clusters. The common terms which frequently appear in a cluster C_i are referred to as C_i's keyword terms. We use a term set K_i to represent the keyword terms of C_i,

$$K_i = \{t_1, t_2 \cdots t_a\},$$

where, t_1, t_2 ... t_a are the keyword terms.

By using the keyword term sets and the sample document clusters, we construct a matrix as shown in Fig. 1, which is referred to as *K-C* matrix. Fig. 1 shows that the sample documents are classified into q clusters. In the *K-C* matrix, each of the rows corresponds to a keyword set K_i, and each of the columns corresponds to a cluster. The ij^{th} entry of the matrix is the number of the keyword terms in the set K_i appearing in the cluster C_j. Because keyword terms in the set K_i only appear in the cluster C_i, therefore, if i is not equal to j, $i \neq j$, the value of the ij^{th} entry is 0. The value of the ii^{th} entry is the number of the elements of the set K_i, $|K_i|$. As shown in Fig. 1, the *K-C* matrix is a diagonal matrix.

As shown in Fig. 1, each cluster is represented as a q dimensional vector. We use $\mathbf{C}_i$ to represent the vector of the cluster C_i. We use a unit vector $\mathbf{c}_i$, where its norm is 1 ($|\mathbf{c}_i|=1$), to represent the cluster vector $\mathbf{C}_i$ as $\mathbf{C}_i=|K_i|\mathbf{c}_i$. When the sample documents are classified into q clusters, we obtain q cluster vectors. Therefore, we define q unit vectors $\mathbf{c}_1,\mathbf{c}_2,\ldots,\mathbf{c}_q$, to represent the q cluster vectors. We refer the vector space constructed by the q unit vectors to as the "*retrieval space*". Because the inner product of two different unit vector is 0, $(\mathbf{c}_i \cdot \mathbf{c}_j)=0,\ i \neq j$, and there are q unit vectors, the retrieval space is a q dimensional orthogonal space.

K \ C	C_1	…	C_i	…	C_q
K_1	$\|K_1\|$	0	0	0	0
…	0	…	0	0	0
K_i	0	0	$\|K_i\|$	0	0
…	0	0	0	…	0
K_q	0	0	0	0	$\|K_q\|$

Fig. 1 The *K-C* matrix which is the diagonal matrix

2.2 Document mapping

A document d_j, is vectorized according to the keyword terms. The count of the occurrences of the C_i's keyword terms in the document d_j is defined to $e_{i,j}$. We set the value of $e_{i,j}$ based on the following rule: *when a keyword term in the keyword set K_i appearing in the document d_j, it is counted only once even if it appears in the document many times* . If the number of the keyword sets is q, the document d_j is vectorized to a q dimensional vector. We represent the vector of the document d_j as $\mathbf{d}_j$:

$$\mathbf{d}_j = \begin{bmatrix} e_{1,j} \\ e_{2,j} \\ \vdots \\ e_{q,j} \end{bmatrix}$$

We use v_t to represent the counted value of a term t. In the following, we use the expression $\sum_{t \in K_i}\{v_t\}$ to represent the sum of $v_{t_1}+v_{t_2}+\cdots+v_{t_a}$, where, t_1, t_2 ... t_a are the elements of the keyword term set K_i:

$$\sum_{t\in K_i}\{v_t\}=\{v_{t_1}+v_{t_2}+\cdots+v_{t_a}|t_1\in K_i, t_2\in K_i,\cdots,t_a\in K_i\}.$$

In this way, the calculation for $e_{i,j}$ is represented by the following formula:

$$e_{i,j}=\sum_{t\in K_i}\{v_t \mid if\ t\in d_j\ v_t=1\ else\ v_t=0\},$$

where, K_i is the set of the C_i's keyword terms and v_t is the counted value of one of the C_i's keyword term t in the document d_j. If the document d_j contains the keyword term t, v_t is set to 1. If the document d_j does not contain the keyword term t, v_t is set to 0.

With the definition of the retrieval space, we express the document vector $\mathbf{d}_j$ as

$$\mathbf{d}_j=\sum_{i=1}^{q}e_{i,j}\mathbf{c}_i .$$

In this way, documents are mapped onto the q dimensional retrieval space.

2.3 The semantic document retrieval

Document retrieval is performed through two steps. The first step is to select a subspace by a given query. The second step is to rank the documents on the subspace.

Subspaces are selected from the retrieval space based on queries. The subspace is a v-dimensional space which is a part of the q-dimensional retrieval space, where v is smaller than q. The v-dimensional subspace correlates to v clusters. The subspace is selected by the following steps.

(1) When a query Q is given, the pattern matching method is used to search the documents which contain the same terms in the query. Documents that have the same terms as those in the query are extracted.

(2) From all the component items of the selected document vector, the cluster vector $\mathbf{c}_i$ is extracted where the related component item $|e_{i,j}\mathbf{c}_i|$ has the maximum value.

$$|e_{i,j}\mathbf{c}_i|=MAX(e_{1,j},e_{2,j},\cdots,e_{q,j})$$

We use a vector $\mathbf{q}$ referred to as the query vector to represent the extracted clusters. The extracted cluster vector $\mathbf{c}_i$ is added to the vector $\mathbf{q}$,

$\mathbf{q}=\mathbf{q}+\mathbf{c}_i$; where the initial value of $\mathbf{q}$ is $\mathbf{0}$.

(3) A retrieval subspace S corresponding to the query is selected from the entire retrieval space by calculating the inner product of $\mathbf{c}_i$ and $\mathbf{q}$. If the value of the inner product ($\mathbf{c}_i\cdot\mathbf{q}$) is greater than or equal to a threshold ε, which is referred to as the subspace selection threshold, $\mathbf{c}_i$ is added to S.

These steps (2) and (3) are repeated for each extracted document vector.

When the subspace S is obtained, the document vector on the subspace is projected and represented as

$$\mathbf{d}_j = \sum_{i=1}^{q} \{e_{i,j}\mathbf{c}_i | \mathbf{c}_i \in S\}.$$

In the second step, the documents are ranked by calculating the norms of the document vectors on the selected subspace:

$$|\mathbf{d}_j| = \left| \sum_{i=1}^{q} \{e_{i,j}\mathbf{c}_i | \mathbf{c}_i \in S\} \right|.$$

In our model, a learning mechanism is applied to improve the retrieval accuracy. The learning mechanism implements the retrieval subspace reselection based on users' feedbacks. In our learning mechanism, users are required to indicate the documents which are appropriate to their retrieval requirements from the retrieval results. When documents are indicated, the query vector **q** is modified under the following steps:

(1) From all the component items of an indicated document vector, the cluster vector $\mathbf{c}_i$ is extracted where the related component item $|e_{i,j}\mathbf{c}_i|$ has the maximum value. The extracted cluster vector $\mathbf{c}_i$ is added to the vector **q** to modify it as

$$\mathbf{q}=\mathbf{q}+\mathbf{c}_i.$$

The initial value of the vector **q** is the value used in the previous subspace selection.

(2) The retrieval subspace S is set to as an empty set. Then, the inner product ($\mathbf{c}_i \cdot \mathbf{q}$) of $\mathbf{c}_i$ and **q** is calculated. After the inner product of the total extracted cluster vectors is calculated, the cluster vectors, whose inner product values are greater or equal than a threshold ε, are added to the subspace set S.

Based on these steps, when a cluster is repeatedly selected, the query vector becomes closer and closer to the vector of the cluster. By setting an appropriate threshold, we can obtain the subspace that is closest to the requirement of the query.

3. Dynamic retrieval space reconstruction

As mentioned in Section 2, for a given query, a query vector is created. A subspace is selected according to the query vector from the retrieval space and the retrieval candidates on the subspace are ranked based on their norms as the retrieval result. The norms of the retrieval candidates on the subspace represent correlations of the retrieval candidates and the given query. In order to select the appropriate subspace, the learning mechanism is used to modify the query vector. When all the retrieval candidates are appropriately mapped onto the subspace, the retrieval candidates that are mostly close to the meaning of the query appear in the retrieval result. However, if there are some retrieval candidates that are not appropriately mapped on the retrieval space, they will not be on the selected subspace as shown in Fig.2.

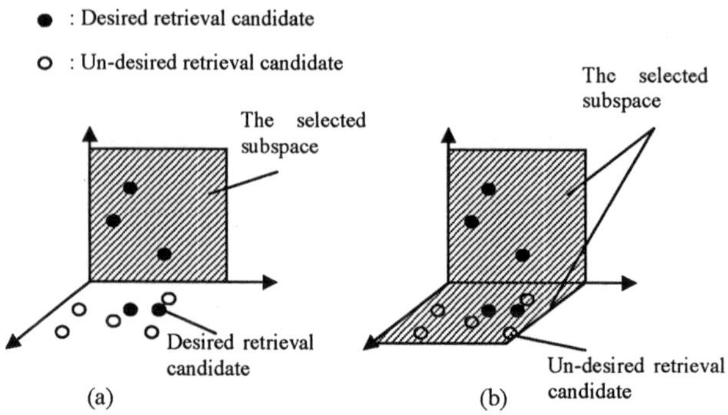

Fig.2 Retrieval candidates and the selected subspace

In the case that some of the retrieval candidates are not mapped on the selected subspace, all the desired retrieval candidates do not appear in the retrieval result. Fig.2 shows an example. Fig.2 (a) shows that some of the desired retrieval candidates are not on the selected subspace. When another subspace is reselected as shown in Fig.2 (b), all the desired retrieval candidates are on the selected subspace. However, some of the un-desired retrieval candidates are also on the subspace and they will appear in the retrieval result.

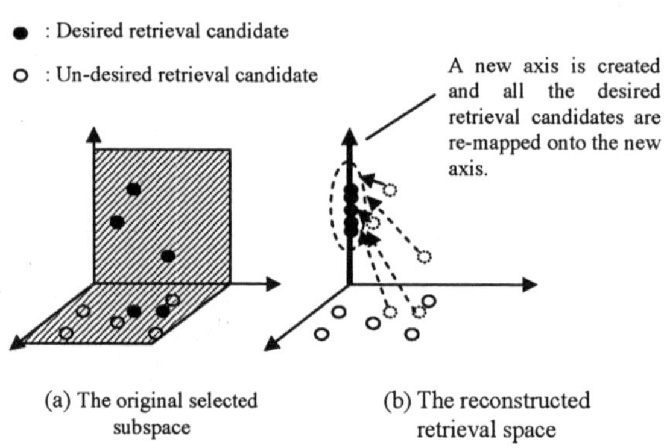

Fig.3 The reconstruction of the retrieval space

We solve this problem by reconstructing the retrieval space. Fig.3 illustrates the basic idea. In our method, a new axis is created and all the desired retrieval candidates are re-mapped onto the new axis. In Fig.3 (a), it is shown that the desired retrieval candidates are distributed among three axes of the original retrieval space. In Fig.3 (b), it is shown that a new axis is created and all the desired retrieval candidates are re-mapped onto the new axis.

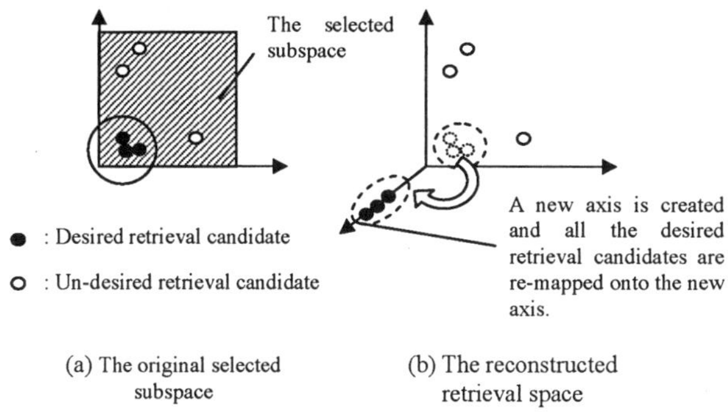

Fig.4 The reconstruction of the retrieval space

Another problem is that the desired retrieval candidates are ranked near the bottom in the retrieval result. This problem is caused when the desired retrieval candidates are mapped near the original point on the selected subspace as shown in Fig.4 (a). We solve this problem by creating a new axis on the retrieval space. The desired retrieval candidates are re-mapped onto the new axis as shown in Fig.4 (b). Therefore, when new axis is selected as the subspace (a one-dimensional subspace), only the desired retrieval candidates appear in the retrieval result.

For realizing this retrieval space reconstruction, we define two basic operations: (1) the combining operation and (2) the decomposing operation, as shown in Fig.5. The combining operation combines the axes on the selected axes into one axis, as shown in Fig.5 (a). The decomposing operation creates a new axis from the selected subspace, as shown in Fig.5 (b).

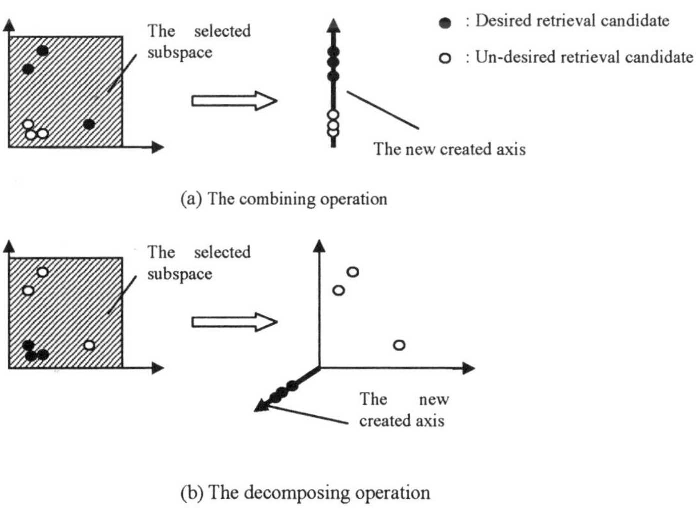

Fig.5 The combining and the decomposing operations

With these two operations, the reconstruction of the retrieval space shown in Fig.3 can be decomposed into three operation steps. The first step is the combining operation. This operation combines the three axes on the selected subspace into one axis as shown in Fig.6 (b). The second step is the decomposing operation. In this step, the decomposing operation creates a new axis and re-maps the un-desired retrieval candidates onto the new axis as shown in Fig.6 (c). The third step is also a decomposing operation. In this step, the decomposing operation creates another new axis and re-maps the retrieval candidates again as shown in Fig.6 (d).

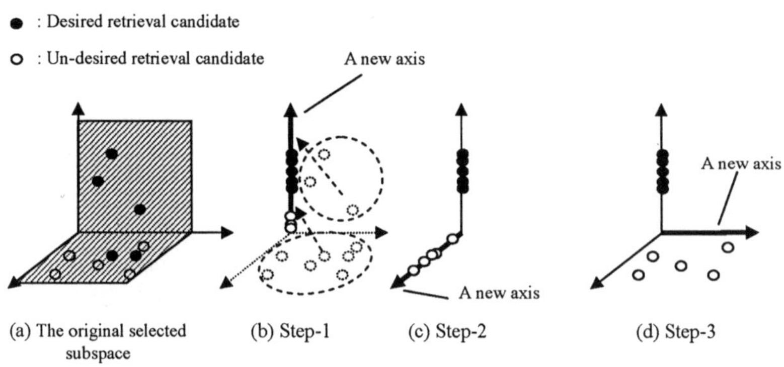

Fig.6 The reconstruction of the retrieval space

In the following, we describe details on the combining and the decomposing operations.

3.1 The combining operation

During query processing, a subspace is selected from the retrieval space, according to the query vector, and the retrieval candidates on the subspace are ranked according to their norms as the retrieval result. In the following, we refer to the selected subspace as the *CS space*. If there are some retrieval candidates which are not mapped on the *CS* space, the combining operation is performed to re-map those retrieval candidates to the *CS* space. In the following, we refer to the space as the *DS space*, on which some expected retrieval candidates exist which are needed to be re-mapped to the *CS* space.

In the combining operation, the axes on the *CS* space are combined into one new axis. The axes on *DS* space are deleted. In Fig.7, it is shown that there are two axes on the space which will be combined into one axis. It is also shown that there are some desired retrieval candidates on the *DS* space. There is only one axis on the *DS* space. This axis will be deleted. And then, the retrieval candidates on the *DS* space are re-mapped onto the new axis.

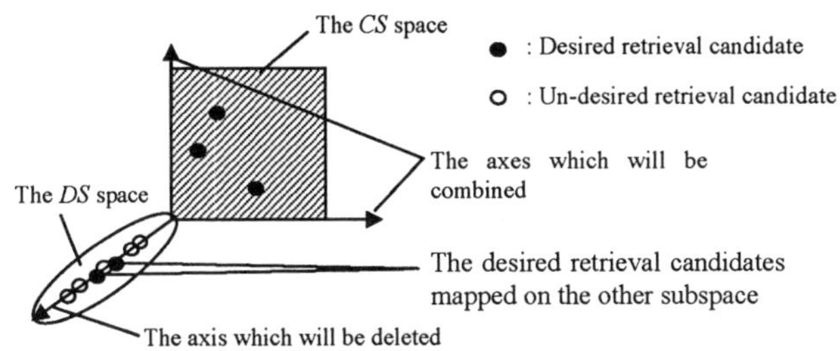

Fig.7 The *CS* space and the *DS* space

As mentioned in Section 2, the retrieval space is created from the *K-C* matrix. Each axis of the space corresponds to each column of the matrix. When the *K-C* matrix has q columns, a q dimensional retrieval space is created.

The *K-C* matrix is a diagonal matrix. The value of an entity on the diagonal line is the norm of the relative keyword set. When two axes are combined into one axis, the combining operation combines the two relative entities of the axes on the matrix. For example, for a two given axes $\mathbf{c}_i$ and $\mathbf{c}_j$, $(i \neq j)$, the entity $|K_i|$ at the ith row and i^{th} column in the *K-C* matrix corresponds to the axis $\mathbf{c}_i$ and the entity $|K_j|$ at the jth row and j^{th} column corresponds to the axis $\mathbf{c}_j$. When the axes $\mathbf{c}_i$ and $\mathbf{c}_j$ are needed to be combined, the combining operation combines the ith row and the jth row into one row and the ith column and the jth column into one column as shown in Fig.8.

We use '$\oplus$' to represent the combining operation. As shown in Fig.8, when the two rows corresponding to K_i and K_j are combined, it is represented as $K_i \oplus K_j$. The combining operation of C_i and C_j is represented as $C_i \oplus C_j$.

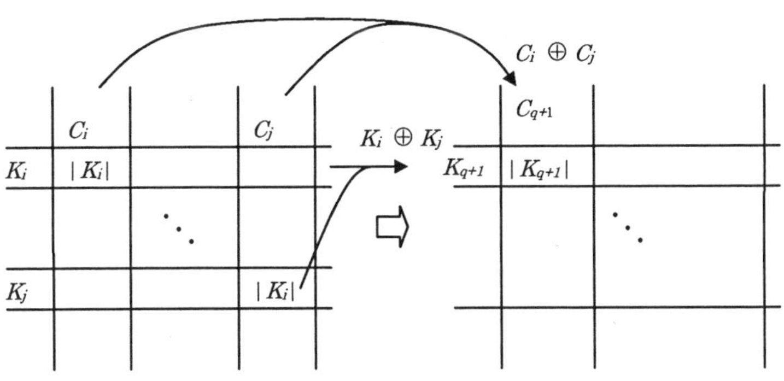

Fig.8 The combining operation

When a new axis is created by the combining operation, a new row and a new column are created in the *K-C* matrix. We use a new subscript number $q+1$ to represent the new row and the new column. The new row corresponds to a new created keyword set. Thus, the new row is represented as K_{q+1} and the new column is represented as C_{q+1}. We use a unit vector $\mathbf{c}_{q+1}$, where its norm is 1 ($|\mathbf{c}_{q+1}|=1$), to represent the new created axis. Thus, the vector of the new column $\mathbf{C}_{q+1}$ can be represented as,

$$\mathbf{C}_{q+1}=|\mathrm{K}_{q+1}|\mathbf{c}_{q+1}.$$

When two axes are combined by the combining operation, those two axes are removed from the retrieval space and a new combined axis is added into the space. When more than two axes are needed to be combined into one axis, we combine two axes into a new axis at first. After that, we combine the rest of the axes one by one with the new axis

In the following, we give the calculation method to obtain the value of $|K_{q+1}|$. The value of $|K_{q+1}|$ is the number of the elements in the keyword set K_{q+1}. The calculation is performed in two steps.

In the first step, the axes on the *DS* space are combined by combining their relative entities on the *K-C* matrix. When the two axes $\mathbf{c}_i$ and $\mathbf{c}_j$ are given, their corresponding entities in the *K-C* matrix are $|K_i|$ and $|K_j|$. The union operation '$\cup$' is performed to combine the keyword sets K_i and K_j:

$$K_{q+1}=K_i\cup K_j$$

In the second step, the entities of the *K-C* matrix corresponding to the axes on the *DS* space are deleted. New keyword terms are extracted from the desired retrieval candidates that are mapped on the *DS* space. The extracted keyword terms are added into the keyword set K_{q+1}.

The above two steps can be expressed by the following formula:

$$K_{q+1}=\{t_i \mid t_i\in K_i\cup K_j \textit{ or } t_i \textit{ is a new extracted keyword term}\}$$

If there are more than two axes to be combined, the combining operation firstly combines two axes into a new axis. After that, it combines the rest of the axes one by one with the new axis. This operation is repeatedly performed until all the axes are combined.

We summarize the combining operation into the following steps:

(1) Two axes $\mathbf{c}_i$ and $\mathbf{c}_j$ are randomly selected from all the axes that are needed to be combined. The axis $\mathbf{c}_i$ is set as the new axis $\mathbf{c}_{q+1}$. After that, the axis $\mathbf{c}_i$ is removed from the retrieval space.
(2) The two axes $\mathbf{c}_{q+1}$ and $\mathbf{c}_j$ are combined into one new axis $\mathbf{c}_{q+1}$ and $\mathbf{c}_j$ is removed from the retrieval space.
(3) If all axes are combined, the combining operation is terminated. Elsewhere, one axis $\mathbf{c}_j$ is randomly selected from the rest axes which are needed to be combined and step-2 and step-3 are repeatedly executed.

When the new retrieval space is created, a retrieval candidate vector $\mathbf{d}_j$ is re-mapped onto the new retrieval space,

$$\mathbf{d}_j=\sum_{k=1}^{q+1}\{e_{k,j}\mathbf{c}_k \mid \mathbf{c}_k \notin \textit{the removed axes}\}.$$

3.2 The decomposition operation

The decomposition operation is performed by adding new keyword term sets into the *K-C* matrix. The keyword terms are extracted from the retrieval candidates which appear in the retrieval results but are ranked around the bottom in the retrieval results. The *K-C* matrix is extended with new rows and columns when the decomposition operation is performed. The *K-C* matrix is still a diagonal matrix after the extension. We define a new subscript number $q+1$ to represent the new row and the new column. Thus, the new row is represented as K_{q+1} and the new column is represented as C_{q+1}. The value of the entity of the q+1th row and the q+1th column is the number of the elements in the set K_{q+1}. We use a unit vector $\mathbf{c}_{q+1}$, where its norm is 1 ($|\mathbf{c}_{q+1}|=1$), to represent the created axis. Therefore, the new created column is represented as a vector $\mathbf{C}_{q+1}$,

$$\mathbf{C}_{q+1}=|\mathrm{K}_{q+1}|\mathbf{c}_{q+1}.$$

If p new entries are added into the *K-C* matrix, the *K-C* matrix is extended with p new rows and p new columns. The extended *K-C* matrix is also a diagonal matrix and the values of the entities on the diagonal line of the *K-C* matrix are $|K_1|$, $|K_2|$, ..., $|K_{q+p}|$. As the extension of the *K-C* matrix, the dimension of retrieval space is extended from its original q dimension into $q+p$ dimension. We use $\mathbf{c}_{q+k}$ $(1 \le k \le p)$ to represent a new extended axis. Thus, an extended column can be expressed as a vector $\mathbf{C}_{q+k}$,

$$\mathbf{C}_{q+k}=|\mathrm{K}_{q+k}|\mathbf{c}_{q+k},$$

When p new axes are added into the retrieval space, a retrieval candidate $\mathbf{d}_j$ is vectorized as

$$\mathbf{d}_j = \sum_{i=1}^{q+p} e_{i,j}\mathbf{c}_i$$

As the original values of the entities $|K_1|$, $|K_2|$, ..., $|K_q|$ of the *K-C* matrix are not changed, the components of all the existing retrieval candidate vectors on the original q dimensional space are not changed. Therefore, we can decompose a retrieval candidate vector $\mathbf{d}_j$ on the extended retrieval space into two parts, which are the components of the vector on the original axes of the retrieval space and the components of the vector on the new added axes:

$$\begin{aligned}\mathbf{d}_j &= \sum_{i=1}^{q+p} e_{i,j}\mathbf{c}_i \\ &= \sum_{i=1}^{q} e_{i,j}\mathbf{c}_i + \sum_{i=q+1}^{q+p} e_{i,j}\mathbf{c}_i ,\end{aligned}$$

In this formula, the components of the vector on the original axes are expressed in the first part as

$$\sum_{i=1}^{q} e_{i,j}\mathbf{c}_i .$$

As the components on the original axes are not changed, the first part of the expression for the vector $\mathbf{d}_j$ is not needed to be recalculated.

The components on the new added axes are expressed in the second part as

$$\sum_{i=q+1}^{q+p} e_{i,j}\mathbf{c}_i .$$

As the second part of the expression for the vector $\mathbf{d}_j$ correlates to the new added axes, the values of the vector $\mathbf{d}_j$ in the second part of the expression must be calculated. If a large number of retrieval candidates exist, long calculation time is consumed for the calculation. In the following, we present our method for reducing this calculation overhead.

As mentioned previously, we create a new axis to re-map the desired retrieval candidates which are ranked near and around the bottom in the retrieval result. Therefore, only the retrieval candidates which are mapped near and around the original point on the selected subspace are needed to be re-mapped to the new created axes. If the selected subspace is defined as S, the re-mapping operation is can be expressed by the following formula:

$$\mathbf{d}_j = \begin{cases} \sum_{i=1}^{q}\{e_{i,j}\mathbf{c}_i \mid \mathbf{c}_i \in S\}, & \left|\sum_{i=1}^{q}\{e_{i,j}\mathbf{c}_i \mid \mathbf{c}_i \in S\}\right| > \varepsilon \\ \sum_{i=1}^{q}\{e_{i,j}\mathbf{c}_i \mid \mathbf{c}_i \in S\} + \sum_{i=q+1}^{q+p}\{e_{i,j}\mathbf{c}_i \mid \mathbf{c}_i \in S\}, & \left|\sum_{i=1}^{q}\{e_{i,j}\mathbf{c}_i \mid \mathbf{c}_i \in S\}\right| \le \varepsilon \end{cases},$$

where ε is the threshold which defines the retrieval candidates needed to be re-mapped.

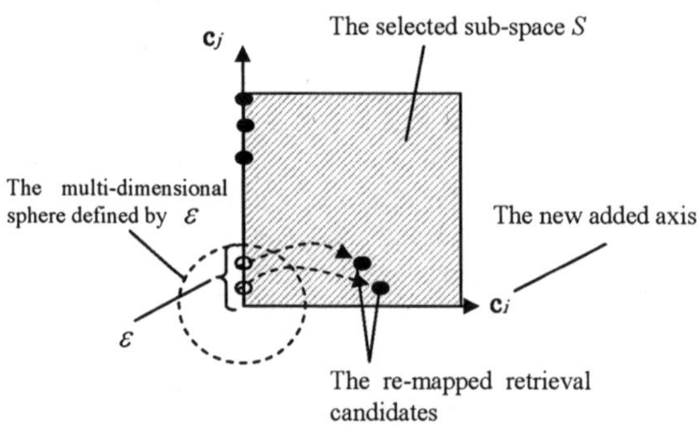

Fig. 9 The illustration of the re-mapping operation

The re-mapping operation is illustrated in Fig.9. The threshold ε defines a multi-dimensional sphere on the selected subspace S. In Fig.9, the sphere is represented as a circle by the dot line. The retrieval candidates in the sphere are re-mapped. In Fig.9, two retrieval candidates are in the sphere, so they are re-mapped to the new added axis. In the formula of the re-mapping operation, the retrieval candidate in the sphere is represented as

$$\left|\sum_{i=1}^{q}\{e_{i,j}\mathbf{c}_i \mid \mathbf{c}_i \in S\}\right| \le \varepsilon ,$$

Because only the retrieval candidates inside the sphere are re-mapped, the calculation overhead for the re-mapping operation is greatly reduced.

3.3 The deleted axes re-creating in the combining operation

In the combining operation, after the axes on the *CS* space are combined into one new axis and the axes on the *DS* space are deleted, the retrieval candidates are divided into two groups on the new created axis as shown in Fig.10 (a). One group contains the desired retrieval candidates and the other group contains the un-desired retrieval candidates. The undesired retrieval candidates are mapped near and around the original point on the new created axis as shown in Fig.10 (a). In order to re-map the retrieval candidates which are mapped near and around the original point, other new axes are created. After the new axes are created, the retrieval candidates mapped near the original point are re-mapped to the subspace created by the new axes as shown in Fig.10 (b).

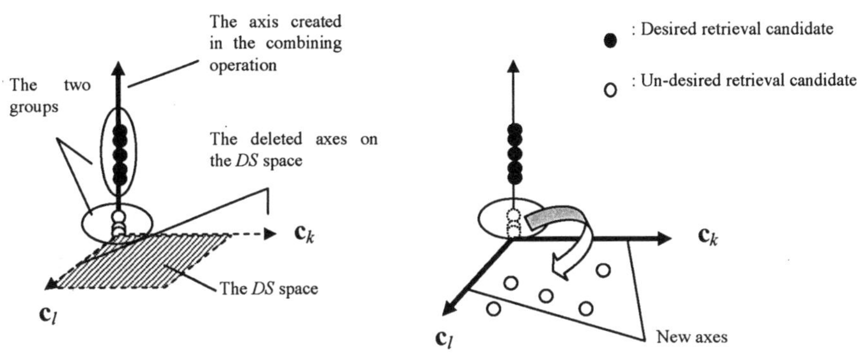

(a) A new axis is created and the axes on the *DS* space are deleted

(b) The deleted axes on the DS space are re-created

Fig.10 The deleted axes re-creation

The keyword terms are extracted from the retrieval candidates that are mapped near the original point. The keyword term extraction is performed based on the original clusters corresponding to the deleted axes. For example, in Fig.10, the axes $\mathbf{c}_k$ and $\mathbf{c}_l$ are deleted. These two axes correspond to the cluster C_k and C_l respectively. Therefore, new keyword sets K_k and K_l are created based on the cluster C_k and C_l, respectively. The new created keyword sets are used to replace the original existing keyword sets. In the example of Fig.10, the new created keyword set K_k is used to replace the original keyword set K_k, and the new keyword set K_l is used to replace the original keyword set K_l. The correlated entities in the *K-C* matrix are replaced at the same time. That is, the entity at the kth row and k^{th} column is replaced by $|K_k|$, which is the number of the elements of the new keyword set K_k, and the entity at the lth row and l^{th} column is replaced by $|K_l|$ which is the number of the elements of the new keyword set K_l.

When the above operations are performed, the un-desired retrieval candidates are re-mapped back onto the *DS* space. The desired retrieval candidates are removed from the *DS* space to the new axes created by the combining operation.

4. Experiment

We have prepared retrieval candidates under 10 categories, C_1, C_2, ..., C_{10}, as shown in Fig.11. From these retrieval candidates, we select 100 retrieval candidates, d_1, d_2, ..., d_{100}, as the training sample to create a retrieval space. After the retrieval space is created, we map other 100 retrieval candidates, d_{101}, d_{102}, ..., and d_{200}, onto the retrieval space.

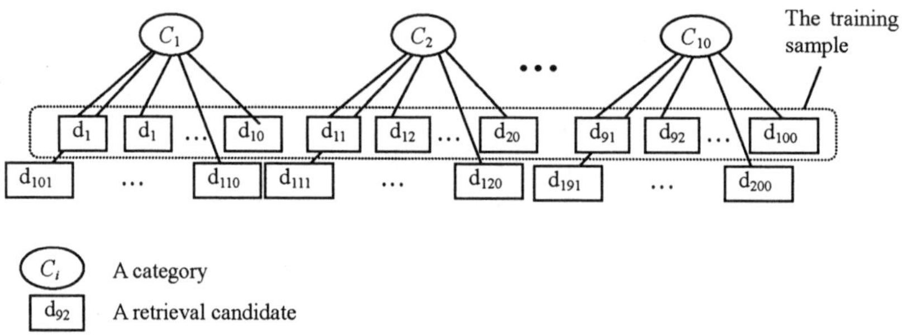

Fig.11 The retrieval candidates prepared for the experiment

As shown in Fig.11, for each category, we use 10 retrieval candidates as the training sample to extract the keyword terms. As there are 10 categories, 10 keyword term sets are created. A 10×10 *K-C* matrix is created based on the 10 keyword term sets and the 10 categories. Each row of the matrix corresponds to a keyword term set and each column of the matrix corresponds to a category. Because the *K-C* matrix has 10 rows and 10 columns, a 10 dimensional retrieval space is created.

Each retrieval candidate is expected to be mapped onto the appropriate axis. For example, the retrieval candidate d_{106} belongs to the category C_1. The axis that correlates to the category is $\mathbf{c}_1$. Therefore, the appropriate axis for the retrieval candidate d_{106} is $\mathbf{c}_1$. In the experiment, the retrieval candidate d_{106} is vectorized as

$$\mathrm{d}_{106}=3\mathbf{c}_1+0\mathbf{c}_2+0\mathbf{c}_3+0\mathbf{c}_4+0\mathbf{c}_5+0\mathbf{c}_6+0\mathbf{c}_7+0\mathbf{c}_8+0\mathbf{c}_9+0\mathbf{c}_{10} \\ =3\mathbf{c}_1 .$$

That is, the retrieval candidate d_{106} is mapped onto the expected axis $\mathbf{c}_1$.

In all the 200 retrieval candidates, 116 retrieval candidates are mapped onto a single axis, 48 retrieval candidates are mapped onto two axes and 20 retrieval candidates are mapped onto three axes. There is a retrieval candidate that is mapped onto six axes, which is the maximum number.

In the experiment, retrieval is performed by changing the selected subspace for a given query. First, a subspace that contains only a single axis is selected. Then, another axis is added to the selected subspace. Finally, the selected subspace is extended to 6 axes. That is, for the given query, 6 subspaces are used in the experiment. The number of the axes in each subspace is different. The number of the axes contained in each subspace increases from 1 to 6. The retrieval candidates are ranked according to the norms of their corresponding vectors on each subspace. The retrieval candidate that has the largest norm is ranked at the top. The average precision rate and the recall rate are represented in Fig.12.

As shown in Fig.12, when the selected subspace contains a single axis, both the recall rate and the precision rate are near to 0.6 because nearly 40% of the retrieval candidates are the un-desired retrieval candidates and the desired retrieval candidates are distributed on the other subspaces. When the axis that contains the desired retrieval candidates is added onto the selected subspace, the recall rate grows up but the precision rate decreases. When the recall rate reaches around 1.0, the precision rate decreases to around 0.2.

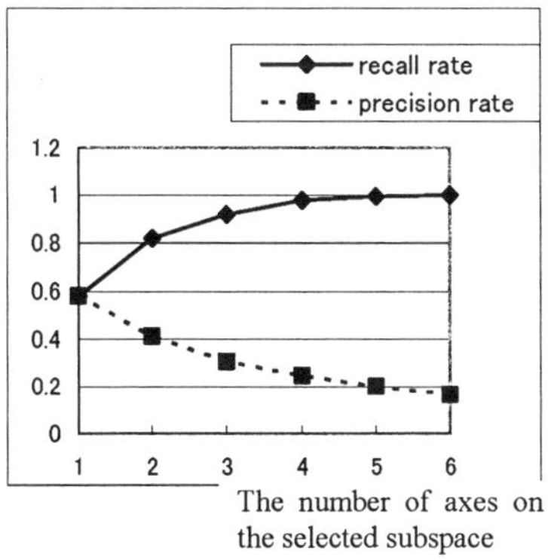

Fig.12 The average recall rate and the precision rate

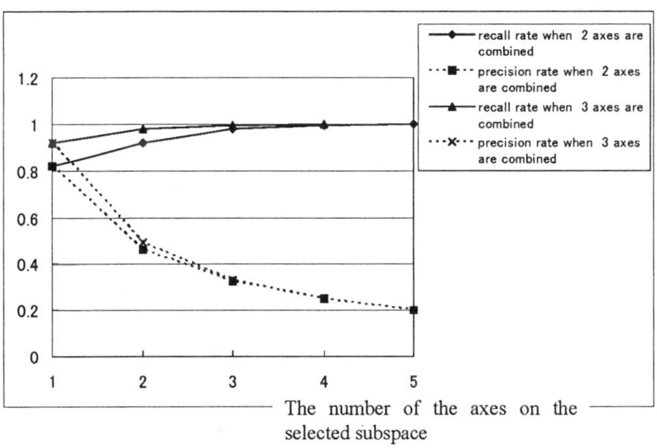

Fig.13 The relationship of the precision rate and the recall rate with the number of the combined axes

When the combining operation is performed, both the recall rate and the precision rate are improved as shown Fig.13. In the experiment, two axes which contain 58% and 24% of desired retrieval candidates respectively are combined into a new single axis. In the query execution after this combining, both the precision rate and the recall rate are improved. Comparing with Fig.12, the precision rate and the recall rate are improved from 58% to

82% in the case that two axes are combined. When three axes are combined, the precision rate and the recall rate are further improved to 92%.

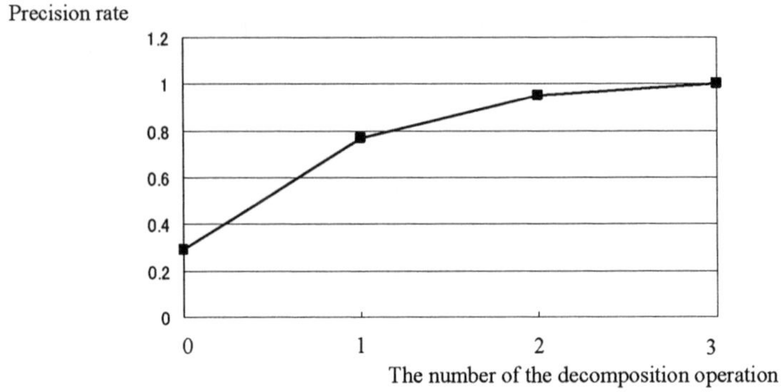

Fig.14 The relationship of the precision rate and the number of the decomposition operation

We use the worst retrieval candidate distributing case for verifying the efficiency of the decomposing operation. In this case, there are 20 desired candidates. The desired retrieval candidates are not ranked at the top on the selected subspace. Before the decomposition operation is applied, to obtain the total 20 desired retrieval candidates, 48 un-desired retrieval candidates are included in the retrieval result. In this case, the recall rate is 1.0 and the precision rate is 0.29. After the decomposition operation is performed once, the number of the un-desired retrieval candidates is reduced to 6 when the total desired retrieval candidates are presented in the retrieval result. The precision rate is improved from 0.29 to 0.77. When the decomposition operation is performed again, the number of the un-desired retrieval candidates is reduced to 1. The precision rate is improved to 0.95.

5. Conclusion

In this paper, we have presented new methods to dynamically reconstruct the retrieval space for the semantic information retrieval. We have introduced two retrieval space reconstructing operations, the combing operation and the decomposition operation. When the desired retrieval candidates and the un-desired retrieval candidates are mixed on a selected subspace, the combining operation combines the axes on the subspace and re-maps the desired retrieval candidates on a single axis. Based on this operation, both the precision rate and the recall rate of the retrieval results are improved. The decomposition operation is performed when the desired retrieval candidates are ranked in the bottom positions and the un-desired retrieval candidates are ranked in the top positions. The decomposition operation creates new axes and re-maps the retrieval candidates onto the new axes. Based on this operation, the desired retrieval candidates are re-mapped onto a new subspace and the retrieval candidates are ranked on the new subspace. As only the desired retrieval candidates are re-mapped on the new subspace, both the precision rate and the recall rate of the retrieval results are improved. Our experimental results clarify the feasibility and effectiveness of our methods.

As our future work, we extend our semantic information retrieval methods to apply them to several important multimedia database fields, such as image and music databases.

References

[1] Chen, X. and Kiyoki, Y., "*A query-meaning recognition method with a learning mechanism for document information retrieval,*" Proceedings of the 13th European-Japanese Conference on Information Modeling and Knowledge Bases, pp. 47-62, 2003.

[2] Chen, X., Kiyoki, Y. and Kitagawa, T., "*A multi-language oriented intelligent information retrieval system utilizing a semantic associative search method,*" Proceedings of the 17th IASTED International Conference on Applied Informatics, pp.135-140, 1999.

[3] Chen, X., Kiyoki, Y. and Kitagawa, T., "*A semantic metadata-translation method for multilingual cross-language information retrieval,*" Information Modelling and Knowledge Bases XII, Vol. 67, pp.299-315, 2001.

[4] Deerwester, S., Dumais, S., Furnas, G.W., Landauer, T.K. and Harshman, R., "*Indexing by latent semantic analysis,*" Journal of the American Society for Information Science, Vol. 41, No. 6, pp.391-407, 1990.

[5] Kitagawa, T. and Kiyoki, Y., "*A mathematical model of meaning and its application to multidatabase systems,*" Proc. 3rd IEEE International Workshop on Research Issues on Data Engineering: Interoperability in Multidatabase Systems, pp.130-135, April 1993.

[6] Kiyoki, Y. and Kitagawa, T., "*A semantic associative search method for knowledge acquisition,*" Information Modelling and Knowledge Bases (IOS Press), Vol. VI, pp.121-130, 1995.

[7] Kiyoki, Y., Kitagawa, T. and Hitomi, Y., "*A fundamental framework for realizing semantic interoperability in a multidatabase environment,*" International Journal of Integrated Computer-Aided Engineering, Vol.2, No.1(Special Issue on Multidatabase and Interoperable Systems), pp.3-20, John Wiley & Sons, Jan. 1995.

[8] Kiyoki, Y., Kitagawa, T. and Hayama, T., "*A metadatabase system for semantic image search by a mathematical model of meaning,*" ACM SIGMOD Record, Vol.23, No. 4, pp.34-41, Dec. 1994.

[9] Kiyoki, Y, Chen, X. and Kitagawa, T., "*A WWW intelligent information retrieval system utilizing a semantic associative search method,*" APWeb'98, 1st Asia Pacific Web Conference on Web Technologies and Applications, pp. 93-102, 1998.

[10] Kolda, T. and O'Leary, D.P., "*A semi-discrete matrix decomposition for latent semantic indexing in information retrieval,*" ACM Transactions on Information Systems, Vol. 16, pp.322-346, 1998.

[11] Dumais S. D, "*Using latent semantic indexing (LSI) for information retrieval, information filtering and other things,*" Cognitive Technology Workshop, April 1997.

[12] Mori T., Kokubu T. and Tanaka T., "*Cross-lingual information retrieval based on LSI with multiple word spaces,*" Proc. NTCIR Workshop 2 Meeting on Evaluation of Chinese & Japanese Text Retrieval and Text Summarization, pp.5-67-5-74, Tokyo, Japan, March. 2001.

Information Modelling and Knowledge Bases XVI
Y. Kiyoki et al. (Eds.)
IOS Press, 2005

A System for Tracing Implicit Relationships in a Web of Contexts

Mina AKAISHI*, Nicolas SPYRATOS†,
Makoto OHIGASHI‡ and Yuzuru TANAKA‡

*RCAST, University of Tokyo,
4-6-1 Komaba Meguro-ku Tokyo 153-8904, JAPAN
†Laboratoire de Recherche en Informatique, Université de Paris-Sud
LRI-Bat 490, 91405 Orsay Cedex, France
‡Meme Media Laboratory,
Hokkaido University, Sapporo 060-8628 JAPAN
E-mail: *mina@ai.rcast.u-tokyo.ac.jp, spyratos@lri.fr, ohigashi|tanaka@meme.hokudai.ac.jp*

Abstract. [1]
We propose a framework for discovering implicit relationships between descriptors through objects in large information bases. Our approach assumes that the contents of the information base are organized into manageable subsets called contexts, each user setting up his(her) own web of contexts. While setting up a web of contexts, a user may create implicit relationships between descriptors and objects (relationships of which the user may not even be aware). We provide a framework for discovering such implicit relationships in a web of user-defined contexts; this is done based on extended notions of synonymy/homonymy. We also describe an implementation of the proposed framework based on the IntelligentBox system, a component-ware system developed at the Meme Media Laboratory.

1 Introduction

The rapid proliferation of information sources in recent years and the advent of the Internet have created a world wide web of interconnected information resources. Today, the Web represents the largest collection of information resources that an individual has ever been able to access - and it is continuously growing at accelerating paces.

Several kinds of tools have been developed in recent years to help individual users to access Web resources. The so-called search engines are the most popular among these tools, as they allow users to access the resources they index using a very simple search mechanism, namely keywords or combinations of keywords. However, the answers obtained from search engines usually contain large amounts of information that is not related to what the user had in mind when asking the question. This problem comes from the fact that any given search engine indexes resources that cater to the interests of very large and highly heterogeneous populations of users. So, on the one hand we have a large collection of information resources (hereafter referred to as the *information base*), and on the other hand a very large and highly heterogeneous population of users.

[1]Work by SPYRATOS was partially supported by the DELOS EU Network ofExcellence in Digital Libraries, Contract No G038-507618

To cope with this situation, in previous work [4, 1, 3], we introduced a personalization mechanism that allows each user to define his (her) own sub-collections of resources of interest and interconnect them for later (and more convenient) use. Such "customized" sub-collections is what we called in [7, 8, 9] user contexts, or simply *contexts*. Intuitively, a context is a set of resources (hereafter referred to as *objects*) within which each resource is associated with two items: a set of *descriptors* and a set of *references* to other contexts. More formally, a context is a set of triples of the form: $\langle object, descriptors, references \rangle$.

We assume that the set of objects, the set of contexts and the set of descriptors are mutually disjoint sets. As a consequence, a context cannot belong to the contents of any other context. However, a context can be referred to from within another context.

It is important to note that our context mechanism is not concerned with the content or internal structure of the objects: as long as the objects are identifiable, they can belong to the contents of any context. Actually, the information base itself is considered as a context - the context containing all the objects of the information base.

It is also important to note that our context mechanism has a minimal interference with the normal operation of the information base: it can be implemented on top of the system managing the information base, at a minimal cost.

There are two main reasons for using a web of contexts on top of an existing information base:

(a) to allow users to define their own, personalized sub-collections, and to describe objects using familiar names (local to the sub-collection);

(b) to speed-up searching for desired objects, by traversing appropriate contexts.

Indeed, when a user sets up a web of contexts, he or she does so in order to better conceptualize the information of interest, and also to facilitate access to that information at some later point in time.

However, in doing so, the user may create new relationships between names and objects, through the reference paths traversing the contexts. As contexts represent conceptual units, usually built up separately, these new relationships may have been created subconsciously, so the user may not even be aware of their existence. As a consequence, showing to the user such unexpected (and sometimes surprising) relationships, may prove useful to the user in several ways, including re-thinking the web of contexts already created.

In previous work [2, 5, 4, 1, 3], we have presented a mechanism for context management, in a web of inter-related contexts, i.e., a set of tools for creation or deletion of a context, insertion/deletion/modification of a triple in a context, search for objects of interest in a context, and traversal of contexts in search of a context of interest.

The present paper builds upon our previous work and proposes a framework for discovering *implicit* relationships between descriptors through the objects they describe. Such implicit relationships are usually created during context design time, and the designer may not even be aware of their existence. The main contribution of this paper is to provide a framework for the discovery of relationships between names and objects in a web of user-defined contexts. This is done based on extended notions of synonymy/homonymy using name and object paths.

The remainder of this paper is organized as follows. In section 2 we recall the definition of context that we use, as well as some auxiliary concepts. In section 3 we define the notions of name extension and object intension on which the tracing of implicit relationships between names and objects is based. In section 4 we describe an implementation of the proposed framework based on the IntelligentBox system, a

Perfumes, Cosmetics	ground floor
Women's Clothing	1st floor
Men's Clothing	2nd floor
Children's Clothing	3rd floor
Sports, Casual	4th floor
Furniture	5th floor
Restaurant	6th floor

(a)

< line# 1,	{ Perfumes, Cosmetics },	0 >
< line# 2,	{ Women's Clothing },	1 >
< line# 3,	{ Men's Clothing },	2 >
< line# 4,	{ Children's Clothing },	3 >
< line# 5,	{ Sports, Casual },	4 >
< line# 6,	{ Furniture },	5 >
< line# 7,	{ Restaurant },	6 >

(b)

Figure 1: An example of contexts.

component-ware system developed at the Meme Media Laboratory [6]. Finally, in section 5 we offer some concluding remarks.

2 Context

In this section, we define the notion of context that we use and the query language for context traversal. Our definitions are inspired from the notion of context presented in [7, 8, 9], with some extensions.

2.1 Example of Context

Perhaps the simplest example of context is the directory of a department store. In most department stores, this directory is usually located on the ground floor (facing the entering customers) and looks like the table of Figure 1(a). A directory, consists of a number of lines and each line contains one or more keywords, or *descriptors*, describing items for sale in the store plus the floor on which the items can be found. For example, in the directory of table 1(a), the first line contains the keywords "Perfumes" and "Cosmetics" plus the indication "ground floor" meaning that such items are found on the ground floor. Similarly, the second line indicates that women's clothing can be found on the first floor, and so on.

We can view the department store directory as a context whose contents are triples of the form: $\langle$*line-number*, *descriptors*, *floor-number*$\rangle$, as shown in Figure 1(b). Under this view, the line number in the triple is the object, the keywords are the descriptors of that object, and the floor number is the identifier of just another context, i.e., of a floor context. Each floor context, in turn, is again a set of triples, where each triple consists of a floor sector number, a set of descriptors describing the items available in that floor sector and a sub-sector number (assuming the sector is divided into sub-sectors, otherwise the sub-sector number is set to Nil, indicating that the floor sector is not divided into sub-sectors).

A customer entering the department store looks at (or "accesses") the directory, and then examines one by one the triples of the directory. Aided by the descriptors the customer selects a floor number, i.e., a new context. Once on the selected floor, the customer repeats the same steps (but this time looking at the floor directory) in order to find the exact sector on the floor where the items of interest are located. Note however that the items themselves are *not* part of the sector context. In other words, the context mechanism just helps you find *where* the items of interest are located, based on their descriptors. Incidentally, when the items of interest are information items, such as electronic documents, then their descriptors are usually referred to as *metadata* - as opposed to the actual data contained in the information item.

2.2 The notion of Context

A context consists of an identifier c plus a *content*. The content of c is a set of triples of the form, ⟨*object-descriptors*, *object*, *object-references*⟩, where *object-descriptors* is a set of descriptions of the object content and *object-references* is a set of other contexts (eventually empty).

We note that, contrary to [7, 8, 9], we allow our contexts to have multiple references. As we shall see, this feature allows greater flexibility and provides for higher expressive power. We also note that an object descriptor can be a piece of text, an image, a sound score, or anything that helps describe the content of the object; it can even be an "object viewer"[4, 1, 3], i.e., a tool allowing the user to just have a look at the contents of the object.

In some application environments the object descriptors are just keywords, and such keywords may come from a controlled vocabulary. Furthermore, such a vocabulary may be structured by a subsumption relation. For example, if the objects are the books of a computer science library then their descriptors most likely will be keywords from the ACM Computing Classification System (ACM, 1999, http://www.acm.org/class/). However, for the purposes of this paper, it is immaterial whether descriptor definitions follow given rules or not.

What is important to keep in mind is that descriptors and references are context dependent: an object can belong to different contexts and may have different names and/or different references in each context. This feature is useful when we want to view an object from different perspectives.

It is important to note that a context is a "light-weight" abstraction mechanism, different from the usual abstraction mechanisms (such as classification, attribution and inheritance). However, it can be used orthogonally together with those mechanisms. We stress the fact that searching for desired objects is based solely on object descriptors, i.e., without consideration of content or internal object structure. This is precisely the reason why we called our context mechanism a "light-weight" abstraction mechanism.

2.3 Accessing information through paths

Accessing information in a web of contexts often involves navigating from one object to another by following references, i.e., by traversing contexts. From an object within a given context, we can reach any object that belongs to one of its references and, recursively, any object that lies on a path. A path is a sequence of pairs of the form (c_i, t_i), $i = 1, ..., n$, where
(i) c_i is a context and $t_i = \langle d_i, o_i, r_i \rangle$ is a triple in the content of c_i, and
(ii) c_{i+1} is in r_i, for $i = 1, ..., n$-1.
We also define the concepts of descriptor path and object path as follows :

- a descriptor path is a dot-separated sequence of descriptors $d_1...d_k$, where d_i is a descriptor in triple t_i, $i = 1, ...,k$.
- an object path is a dot-separated sequence of objects $o_1...o_k$, where o_i is an object in triple t_i, $i = 1, ...,k$.

Paths form the basis for reaching objects in a context navigating through the references of objects.

3 Implicit Relationships

As we mentioned earlier, each context is a conceptual unit, probably set up separately by its designer then connected through references to other contexts, in a web of contexts. In this section, we describe the framework for discovering *implicit* relationships in a web of contexts.

3.1 Implicit Relationships between descriptors through objects

In any specific context, each descriptor describes one or more objects in that context, and each object has one or more descriptors describing it . For example, in a library context, the term **Programming** may describe one or more different books concerning programming, and a book concerning programming may also have the term **Theory** associated to it (e.g., if it's a book concerning the theory of programming).

We define the *extension* of a descriptor to be the set of all objects with which the descriptor is associated, and the *intension* of an object to be the set of all descriptors with which the object is associated. More formally we have:

Definition 1

- Let c be a context, and let d be a descriptor in c.
  The *extension* of d in c, denoted by $E_c(d)$, is defined as follows:

$$E_c(d) = \{o|(d, o, r) \text{ is in } c\}$$

- Let c be a context, and let o be an object in c.
  The *intension* of o in c, denoted by $I_c(o)$, is defined as follows:

$$I_c(o) = \{d|(d, o, r) \text{ is in } c\}$$

Clearly, E_c and I_c can be seen as set-valued functions, the former receiving a descriptor as input and returning a set of objects, while the latter receiving an object as input and returning a set of descriptors.

We note that all objects in $E_c(d)$ are related by the fact that they all share the descriptor d, i.e. they are *homonyms*, and all descriptors in $I_c(o)$ are related by the fact that they all describe the same object o, i.e. they are *synonyms*. However, these relationships are *explicit*, in the sense that it is the context designer that has explicitly defined them, at context design time. By "composing" these explicit relationships one can define *implicit* relationships as well, either between descriptors or between objects.

Indeed, the above definitions are given with respect to a single context c. However, we can extend them easily to a web of contexts, by replacing the set of descriptors appearing in c with the set of descriptors appearing in the whole web of contexts, and the set of objects appearing in c with the set of objects appearing in the whole web of contexts.

First, let us call *environment* of a descriptor d any context in which d appears, and let us denote by $Env(d)$ the set of all environments of d. Then, we can generalize the notion of extension over the whole web of contexts as follows:

$$E(d) = \bigcup_{c \in Env(d)} E_c(d).$$

c_{p1}: Profile

KITANO Takeshi	o_1	$\{c_{p1}\}$
Family	o_{10}	$\{c_{10}\}$
Movies	o_{11}	$\{c_{m1}\ldots\}$
-	-	-

c_{m1}: ZATOICHI

ZATOICHI	o_{m1}	-
Director	o_1	$\{c_{p1}\}$
Scriptwriter	o_1	$\{c_{p1}\}$
Actor, Zatoichi	o_1	$\{c_1, c_2\}$
-	-	-

c_{p2}: Profile

KUROSAWA Akira	o_2	$\{c_{p2}\}$
Family	o_{20}	$\{c_{20}\}$
Movies	o_{21}	$\{c_{m2}\ldots\}$
-	-	-

c_{m2}: SEVEN SAMURAI

SEVEN SAMURAI	o_{m2}	-
Director	o_2	$\{c_{p2}\}$
Scriptwriter	o_2	$\{c_{p2}\}$
Actor, Kikuchiyo	o_3	$\{c_{p3}\}$
-	-	-

Figure 2: The examples of a web of contexts.

Next, let us call *facet* of an object o, any context in which o appears, and let us denote by $Fac(o)$ the set of all facets of o. Then, we can generalize the notion of intension over the whole web of contexts as follows:

$$I(o) = \bigcup_{c \in Fac(o)} I_c(o).$$

E and I can be seen as set-valued functions extending the functions E_c and I_c (seen earlier) over the whole web of contexts. The implicit relationships between descriptors that we have in mind are obtained by composing the functions E and I to obtain a monotonic operator over sets of descriptors. This operator is denoted by IE and it is defined as follows: for every set of descriptors D, $IE(D) = I(E(D))$.

Clearly, the function IE is monotonic with respect to set inclusion, so its repeated application attains a fixpoint, say after n_0 steps (assuming the set of all descriptors is finite): $(IE)^{n_0+1}(D) = (IE)^{n_0}(D)$. We define two descriptors d and d' to be equivalent, denoted $d \approx d'$, if $(IE)^{n_0}(d) = (IE)^{n_0}(d')$.

Clearly, the relation $\approx$ is an equivalence relation over the set of all descriptors, and its equivalence classes describe all relationships (explicit and implicit) that exist among descriptors. The following algorithm computes the equivalence classes of descriptors under the relation $\approx$:

1. Pick a descriptor not yet "visited".
2. Apply E to find a set of objects, then apply I to the result to find a set of descriptors.
3. Repeat 2 until fixpoint.
4. Repeat 1-3 until all descriptors are visited.

In what follows, we present a number of examples showing the usefulness of the relation $\approx$ between descriptors.

Figure 2 shows a simple example of a web of contexts. The context c_{p1} is a context identifier, and *"profile"* is the name of the context. *"KITANO"* is a name of object

o_1, and refers to a set of context $\{c_{p1}\}$. In the same context, the object o_{11} in named *"Movies"* and refers a set of contexts that includes the context c_{m1}. That context c_{m1} contains information about the movie *"ZATOICHI"*.

Let us see now if there is an implicit relation between *"KITANO"* and *"KUROSAWA"* through these contexts.

$E(\text{"}KITANO\text{"}) = \{o_1\}$
$IE(\text{"}KITANO\text{"}) = \{\text{"}KITANO\text{"}, \text{"}Takeshi\text{"}, \text{"}Director\text{"}, \text{"}Scriptwriter\text{"},$
$\text{"}Actor\text{"}, \text{"}Zatoichi\text{"}\}$
$E(IE(\text{"}KITANO\text{"})) = \{o_1, o_2\}$
$(IE)^2(\text{"}KITANO\text{"}) = \{\text{"}KITANO\text{"}, \text{"}Takeshi\text{"}, \text{"}Director\text{"}, \text{"}Scriptwriter\text{"},$
$\text{"}Actor\text{"}, \text{"}Zatoichi\text{"}, \text{"}KUROSAWA\text{"}, \text{"}Akira\text{"}\}$

$E(\text{"}KUROSAWA\text{"}) = \{o_2\}$
$IE(\text{"}KUROSAWA\text{"}) = \{\text{"}KUROSAWA\text{"}, \text{"}Akira\text{"}, \text{"}Director\text{"}, \text{"}Scriptwriter\text{"}\}$
$E(IE(\text{"}KUROSAWA\text{"})) = \{o_2, o_1\}$
$(IE)^2(\text{"}KUROSAWA\text{"}) = \{\text{"}KUROSAWA\text{"}, \text{"}Akira\text{"}, \text{"}Director\text{"}, \text{"}Scriptwriter\text{"},$
$\text{"}KITANO\text{"}, \text{"}Takeshi\text{"}, \text{"}Actor\text{"}, \text{"}Zatoichi\text{"}\}$

$(IE)^2(\text{"}KITANO\text{"}) = (IE)^2(\text{"}KUROSAWA\text{"})$, therefore, *"KITANO"* $\approx$ *"KUROSAWA"*. The descriptors *"KITANO"* and *"KUROSAWA"* are equivalent. This means that a set of implicit relation paths, or *IR-path*, exists between these two descriptors:

$IR\text{-}paths(\text{"}KITANO\text{"}, \text{"}KUROSAWA\text{"})$
$= \{\text{"}KITANO\text{"}.o_1.\text{"}Director\text{"}.o_2.\text{"}KUROSAWA\text{"},$
$\text{"}KITANO\text{"}.o_1.\text{"}Scriptwriter\text{"}.o_2.\text{"}KUROSAWA\text{"}\}$

The first path means that object o_1 has *"KITANO"* as a descriptor and *"Director"* as another descriptor. Futhermore, *"Director"* is also the descriptor of object o_2 whose *"KUROSAWA"* is another descriptor. So this path suggests the relation that both *"KITANO"* and *"KUROSAWA"* are directors of movies. In the same way, the second path means that they are also scriptwriters of movies.

4 CIA architecture

In this section we describe an implementation of the proposed framework based on the IntelligentBox system, a component-ware system developed at the Meme Media Laboratory [6].

In [5], we have proposed a basic architecture called a CIA (Context based Information Access space). The basic components of such a space are what we call media objects that are software entities defined in the IntelligentBox system. We use two kinds of media objects, a context space and an object space.

Based on the basic CIA architecture, we propose now an extension for the implementation of the functions E and I (Extension and Intension). This extension does not break the basic CIA architecture.

In what follows, we first recall the basic components of a CIA, and then describe the extension implementing the functions E and I.

4.1 Context Space and Object Space

A context c is implemented as a context space, denoted $space(c)$, and each triple t in the contents of c is implemented as an object space, denoted $space(t, c)$. For each triple t in the contents of c, the object space $space(t, c)$ is bound to the context space $space(c)$. Roughly speaking, a context space consists of the object spaces into which its objects are mapped. Note however that an object space is bound to one and only one context space, thus if a triple t belongs to two different contexts, say c and c', it will be mapped into two different object spaces, namely, $space(t, c)$ and $space(t, c')$, respectively. The object space $space(t, c)$ will be bound to the context space $space(c)$, while the object space $space(t, c')$ will be bound to the context space $space(c')$.

In a context space, each object space of a triple models the descriptors and the references of the triple, and moreover it provides a *viewer* to the content of the object in that triple. As we shall see later, the viewer simply allows inspecting the content of the object. We recall that the content of an object is *not* part of the context mechanism, i.e., the contents of objects are developed independently, and only their identifiers can appear in a context.

Figure 3 shows the component structure corresponding to a context space. A Context Space is a virtual space that includes a set of pointers to object spaces. A pointer to an object space is called an object space port (OSP) and a pointer to context space is called a context space port (CSP). The port is an entrance to access a target space.

A CSP box holds the context-id c. An OSP Generator is constructed by the DRRU (Database Record Reification Unit). The detail of the DRRU is described in the section 4.2. The OSP Generator receives the context-id c and reifies the object spaces that correspond to the contents of context c. An object space is a 3D visualization of a triplet. The left part of Figure 3 shows the component structure corresponding to each object space. Each OSP of an OSP generator box is connected to a virtual 3D space, called an object space.

In an object space, there are object descriptors, an object viewer and a set of references that correspond to the triplet data in the context. An object descriptor can be a piece of text, an image, a music score and so on. Appropriate media object(s) are used as Object Descriptors. How to construct each object descriptor is up to the designer of an object descriptor. An object viewer is also an appropriate media object to represent an object. How to construct each object viewer is up to the designer of the object viewer. A reference is a set of pointers to context spaces, they are represented by CSPs.

4.2 Information Access Module

Figure 4 shows the access mechanism to information bases from an IntelligentBox environment. We provide a database access interface and predefined template query as 3D components. This module is called as a Database Record Reification Unit (DRRU). It consists of three basic components. A Query Definition Box receives some arguments as inputs, and then generates a query statement. A Database Proxy Box works as an interface between a database system and the IntelligentBox system. It evaluates a given query. A Data Manager Box reifies database records as Boxes that structure is predefined as the template Box(es). Arbitrary Boxes can be pre-registered as a template Box. When the Data Manager Box receives the collection data, each copy of the template Box is instantiated with the value of the corresponding element of the collection. The

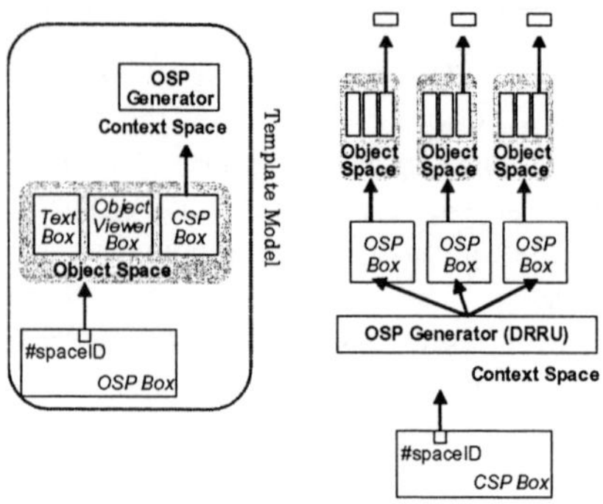

Figure 3: The components structure for generating the Objet and Context Spaces.

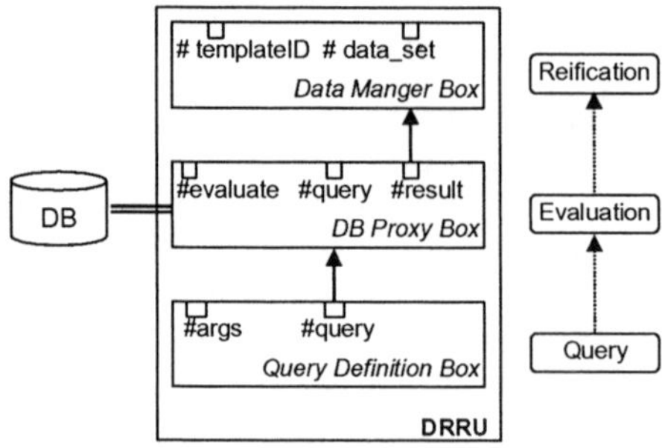

Figure 4: A Database Record Reification Unit.

Data Set Manager Box distributes each element of the collection to each Box, which becomes an instantiation of the template model.

4.3 Implementing the functions E and I

Figure 5 shows the component structure of the extended object viewer that enables to navigate a user to another information space according to the function Extension and Intension. The LOD (Level of Detail) box changes the level of detail depending on the distance of a user's view point and the location of an object. When a user is close to the object o, a $DRRU_i$ (Database Record Reification Unit for intension of the object) retrieves the all descriptors of the object o. The result, that is intension of the object o, appears as the media objects that are pre-defined as the template model. The template model for $DRRU_i$ is shown at the right of Figure 5. The basis of the template model is a Text box. Each Text box represents each descriptor. Moreover, a $DRRU_e$ (DRRU for Extension of the term) is attached on the Text Box through the LOD box. The $DRRU_e$ generates object spaces that correspond to Extension of descriptor text. For such reason, the template model for a $DRRU_e$ is an object space that component structure is shown in Figure 5.

This extension provides users ways between Extension of terms and Intension of objects. It supports to find the implicit relations through a set of terms and a set of objects.

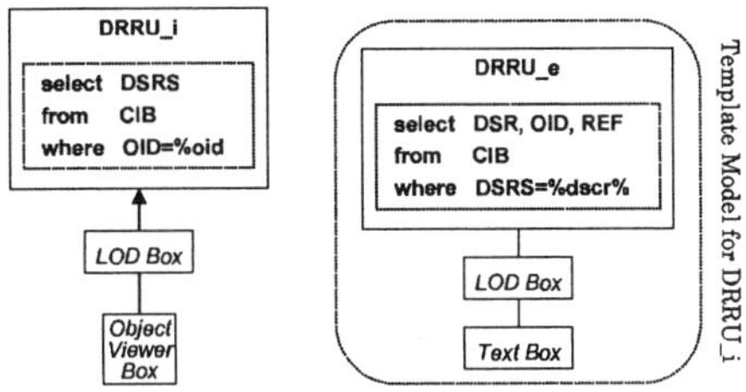

Figure 5: The extended structure of an object viewer for generating the *Extension* and *Intension* Spaces.

5 Concluding Remarks

We have defined the implicit relationship between descriptors through objects in a web of contexts in large information bases. We have also proposed a framework for discovering implicit relationships. This framework provides a user to discover relationships of which the user may not even be aware. It also provides a mechanism for context management in a web of inter-related contexts, i.e., a set of tools for creating or deleting a context, inserting or deleting the objects in a context, searching for objects of interest in a context, and traversal of contexts in search of a context of interest.

We have implemented these concepts in the form of what we call Information Access Space, based on the IntelligentBox System. As it stands now, our prototype allows for context traversal and querying of a collection of contexts.

References

[1] Mina AKAISHI, Makoto OHIGASHI, Nicolas SPYRATOS, and Yuzuru TANAKA. Information access space framework over contextualized information base. In *Seventh International Conference on Information Visualisation (IV 2003)*, pages 414–420, 2003.

[2] Mina AKAISHI, Nicolas SPYRATOS, and Yuzuru TANAKA. A component-based application framework for context-driven information access. In *The 11th European-Japanese Conference on Information Modeling and Knowledge Bases*, pages 242–253, 2001.

[3] Mina AKAISHI, Nicolas SPYRATOS, and Yuzuru TANAKA. Construction of information access space based on a contextualized information base. In *7th World Multiconference on Systemics, Cybernetics and Informatics (SCI 2003)*, pages 24–29, 2003.

[4] Mina AKAISHI, Nicolas SPYRATOS, and Yuzuru TANAKA. Contextual search in large collections of information resources. In *13th European-Japanese Conference (EJC 2003)*, pages 313–320, 2003.

[5] Mina AKAISHI, Hiroyuki YAMAMOTO, Makoto OHIGASHI, Nicolas SPYRATOS, and Yuzuru TANAKA. 3D visual construction of a context-based information access space. In *The 12th European-Japanese Conference on Information Modeling and Knowledge Bases*, pages 24–37, 2002.

[6] Y. Okada and Y Tanaka. IntelligentBox: A constructive visual software development system for interactive 3D graphic applications. In *Proc. Of Computer Animation '95*, pages 114–125, 1994.

[7] M. Teodorakis, A. Analyti, P. Constantopoulos, and N Spyratos. Context in information bases. In *Proc, of the 3rd Int. Conference on Cooperative Information Systems (coopIS '98)*, pages 260–270, 1998.

[8] M. Theodorakis, A. Analyti, P. Constantopoulos, and N. Spyratos. Contextualization as an abstraction mechanism for conceptual modelling. In *Proc. of the 18th Intern. Conference on Conceptual Modelling (ER '99)*, pages 475–489, 1999.

[9] M. Theodorakis, A. Analyti, P. Constantopoulos, and N Spyratos. Querying contextualized information bases. In *Proc. of the 24th Intern. Conference on Information and Communication Technologies and Programming (ICT&P '99)*, pages 260–270, 1999.

A Filtering Tool for WWW Search Engines based on Semantic Relation between Input Keywords

Akiyo Matsumoto ∗, Tatsuhiro Konishi ∗, Akira Takagi †, Teruo Koyama ‡, Yoshio Miyake §, Yukihiro Itoh ∗
∗ *Shizuoka University 3-5-1 Johoku, Hamamatsu-shi, Shizuoka, 432-8011 Japan*
† *CSK Corporation 2-6-1 Nishi-Shinjuku, Shinjuku-ku, Tokyo, 163-0227 Japan*
‡ *National Institute of Informatics 2-1-2 Hitotsubashi, Chiyoda-ku, Tokyo, 101-8430 Japan*
§ *Chukyo University 101 Tokodachi, Kaizu-cho, Toyota-shi, Aichi , 470-0393 Japan*

Abstract. In this paper, we propose a filtering tool for WWW search engines. Because ordinary search engines select candidate pages based on appearance frequency of keywords, the candidate pages may contain every keyword. However, there are some cases that each keyword appears in individual context and they have no semantic relation. Such pages tend to be useless. We try to improve precision of a search engine by checking whether input keywords have a semantic relation in the candidate pages or not. Semantic relations between keywords are represented by using a structure, such as sentence, table, itemized list and so on. In this paper we deal with sentence and table as a structure to represent semantic relations. We propose a method to detect semantic relations between input keywords represented in the structure of sentence and table. We also show an experimental evaluation of our method.

1 Introduction

In this paper, we discuss a filtering tool for WWW search engine. Many search engines select suitable WWW pages for users' requirements by using statistical methods. In general, search engines can find the pages which contain all input keywords, but they do not check the semantic relations between keywords. We think we cannot achieve satisfied level by only such methods and we should deal with a semantic factor of WWW pages. Thus we discuss how to measure the strength of semantic relations among keywords in retrieval conditions and try to adopt the measure to filtering WWW pages.

The fundamental strategy of search engines is to select the pages which suit the retrieval conditions input by users. Search engines also evaluate the importance of pages and filter them by using various factors such as the degree of "freshness", "population", and "reference" as well as "similarity", but the main factor by which almost all search engines select candidates for required pages is based on the degree of similarity between the retrieval conditions and documents.

As a method of measuring the degree of similarity between the retrieval conditions and documents, the "vector space model" and the "probabilistic model" are used widely [1]. In the both models, two kinds of statistical values called tf and idf are used in order to define similarities between the retrieval conditions and documents. The term frequency (tf) is the occurrence frequency of the term in the text and is normally reflective of term importance. The inverse document frequency (idf) is a factor which enhances the terms which appear in fewer documents, while downgrading the terms occurring in many documents [2]. In the vector space model, the weight of each terms in each document and a retrieval condition is calculated to define M-dimensional vectors corresponding to each document and the retrieval condition respectively (here, M is the total number of terms). The weight is calculated based on tf and idf. The degree of similarity between a retrieval condition and a document is defined as an angle which two vectors of the retrieval condition and the document make. In the probabilistic model, probabilities that each document suits with retrieval conditions by using tf and idf. However, there are some problems caused by the tf/idf method. The method based on tf/idf is based on the assumption that important terms showing a theme of a document are used in the document repeatedly and the terms appearing repeatedly in a document but unimportant are appeared in the other many documents. If every document is long enough such as newspaper articles, the assumption of tf is effective, since important words will appear in various sentences in the document. But, if the target documents to be searched are not so long even important terms are not used obviously much times than the other terms. Then tf of important terms and unimportant terms do not have some deference. In such cases, term importance is decided only with idf. Then accuracy of retrieval will be going down [3].

We think that there is a limit in such a statistical technique and we should treat structures representing meaning of an input retrieve condition or a page, in order to compute the degree of conformity. When two or more words are input as keywords in retrieval conditions, some semantic relations must exist among the keywords. The purpose of this research is to improve retrieval accuracy using the "relation". Most users of Web search engine tend to use only a few keywords. According to Kazama et al [4], in most cases, the number of keywords is one or two (1.42 on average) in query logs of Japanese search engine "ODIN". But, in case of only one keyword, the purpose of retrieval is very obscurity, moreover it is impossible to extract "relation" among keywords. In case of three or more keywords, it is fundamentally possible to process as combination of two words. And so, we focus the target of this research on two-keywords-queries.

In related work, Mine et al [5, 6] dealt with a dependency structure between input keywords. A dependency structure is a typical structure representing semantic relations among words. This system receives a query by a natural language, analyses syntactic structure of a query sentence and creates a frame which represents a dependency of a verb and a noun. Then it extracts the sentence containing the pair of the words from each document, transforms each sentence into the frame structure, and judges the degree of conformity by comparing the frames with the frame of the query. This method requires that the semantic relationship among words must be represented as a sentence. Therefore, their technique is applicable to documents which consist only of sentences. However it cannot be applied to documents containing various structures other than a sentence. Generally, semantic relations between keywords are represented in various structures such as a table, a title and its contents, and so on.

In order to develop a WWW search engine which has an ability to handle semantic relations among keywords, various structures representing semantic relations can be dealt with. As a first step, we paid attention to a sentence structure and a table structure as major structures representing semantic relations between keywords in a HTML document and construct meta-search engine for WWW.

In section 2 and 3, we discuss how to detect semantic relations between keywords represented by sentence structures and how to detect the relations represented by using table structures respectively. In section 4 we show the configuration of our system and in section 5 we report on an experimental evaluation of our system.

2 Judgment Using Dependency Structures of Sentences

2.1 Basic Idea

In this research, we measure the strength of semantic relations of keywords by using dependency structures of sentences. Our meta-search engine outputs preferentially the pages in which two keywords have a strong semantic relation.

Table 1: Example of input sentence and its parse tree

Ex.1	Ninki no-toPC no uriage rankingu. (The sales ranking of popular notebook PCs.)	Ex.2	Ninki no-toPC ke-su wo hanbaicyu-. (Popular notebook PC cases on sale.)
	rankingu (ranking) ⊢ uriage (sales) ⊢ no-toPC[no] (of notebook PCs) ⊢ ninki (popular)		hanbaicyu- (on sale) ⊢ ke-su[wo] (cases) ⊢ no-toPC (notebook PC) ⊢ ninki (popular)

See the two example sentence in Table 1. Both sentences include the words "popular" and "notebook PC" but have different dependency structures. The semantic relation between the two words must be strong when "popular" modifies "notebook PC" directly (Ex.1), while the semantic relation must be weaker when "popular" modifies not "notebook PC" but "cases" (Ex.2). In order to measure the strength of semantic relations between "adjective showing a state of an entity" (popular) and "a noun representing an entity" (notebook PC), it is effective to observe whether "the adjective" modifies "the noun". Of course, each other combination of words than "adjective showing a state of an entity" and "a noun representing the entity" has their typical dependency patterns to detect a strong semantic relation. Therefore, we prepare the typical patterns for each type of a keyword pair. In order to define the types of keyword pairs, we classify all keywords into the following 4 categories: entity, attribute, value, and phenomenon. Our system has a dictionary database which has information on the category of each word. Currently, the types of keyword pairs are defined as the combinations of these 4 categories. There are 10 possible combinations. We prepare dependency patterns for the four major combinations: "entity + entity", "entity + phenomenon", "entity + attribute" and "entity + value".

- "Entity + Entity"
  - entity noun + adnominal particle + entity noun
    Example) Osaka no hoteru (a hotel in Osaka)
  - entity noun + case-marking particle + relative clause + entity noun
    Example) Osaka ni aru hoteru (a hotel in Osaka)
  - entity noun + entity noun
    Example) Osaka hoteru (Osaka hotel)
  - entity noun + one arbitrary noun + entity noun
    Example) Osaka no shinise hoteru (a traditional hotel in Osaka)

- "Entity + Phenomenon"
  - entity noun + adnominal particle + phenomenon noun
    Example) Linux no insuto-ru (installation of Linux)
  - entity noun + phenomenon noun
    Example) Linux insuto-ru (Linux installation)
  - entity noun + case-marking particle + phenomenon noun
    Example) Linux wo insuto-ru suru (installing Linux)
- "Entity + Attribute"
  - entity noun + adnominal particle + attribute noun
    Example) Kawase no re-to (exchange rate)
  - entity noun + attribute noun
    Example) Kawase re-to (exchange rate)
- "Entity + Value"
  - entity noun + adnominal particle + value noun, or adjective
    Example) Ao no bara (a blue rose)
  - entity noun + value noun, or adjective
    Example) Ao bara (a blue rose)

In this list, it is forbidden to change the order of keywords, except for "entity + entity".

Our system decides that the keywords have a strong semantic relation if they have the prepared pattern in the sentence.

2.2 Our method for Judgment using Dependency Structures of Sentences

In this section, we show the procedure for judgment using dependency structures of sentences. In this paper, we call a sentence which contains both of two keywords a "*key-sentence*".

(1) First of all, the system removes all tags from the HTML source of the target page.

(2) *Key-sentences* are extracted from the Web page.

(3) Parenthesized phrases in the *key-sentences* are processed by the following strategies:
  - See the type of the parenthesis. Some symbols such as "< >", "≪ ≫", "[]" tend to be used only for emphasizing, so remove such symbols from the *key-sentence*.
  - If both of the keywords aren't in the parenthesized phrases, remove the phrase because the phrase tends to have no influence on the dependency structure between the keywords.
  - If both of the keywords are in the parenthesized phrases, regard the phrase as an individual *key-sentence*.
  - and so on.

(4) Over long (including conjunctive particles) *key-sentences* are divided by each conjunctive particle.

(5) A natural-language parser analyses the *key-sentence*. Before this step, the major causes of failure by our natural-language parser removed in steps (3) and (4) have. The parser used here is a Japanese parser developed by CSK CORP, and it has dictionary database (=ontology) in order to solve polysemy problem. In this connection, the top-level categories of the ontology are shown in Figure 3.

(6) The system compares dependency tree prepared for each type of input keyword pair with the parsed tree of *key-sentence*. Our system decides that the keywords have a strong

semantic relation if the matching ends successfully.

3 Judgment Using Dependency Structures in Tables

Usual HTML documents include many tables. There are relatively strong semantic relations among cells in the same row or column in the tables. Furthermore, each row and column tends to have a unique meaning, which can be seen in the first cell of each row and column [7]. Thus, table structures include much useful information for information retrieval or information extraction. However, most of conventional Web search engines don't use the information effectively [8]. Our meta-search engine analyses structures of tables in order to measure the strength of semantic relations between keywords in the tables.

We have enumerated the major patterns of positions where input keywords appear in a table, and found the following 16 patterns (Figure 1) which show the keywords have a strong semantic relation. We have also found that each pattern shows where information requested by users tends to exist. In the following patterns, "X" and "Y" means input keywords. "Headline of a table" means the first row which contains only one cell, in the table. "Headline of a row (or column)" means the first cell of the row (or column). "Parent table" means a table including another table recursively. "Child table" means the included table. "Headline table" means a table which has only the role as a headline of another table which is laid nearby it. We call the latter table "body table".

⟨*I*⟩ Both keywords are in a table.(In these cases, information requested by users tends to be contained in the table.)

- (a) X is in the headline of a row in a table + Y is in the headline of a column in the table
- (b) Both of X and Y are in the headline of the table and one of them modifies the other.
- (c) X is in the headline of a table + Y is in the table.
- (d) Both of X and Y are the in the same cell and one of them modifies the other.
- (e) X is in a parent table and works as the headline of its child table + Y is in the child table.
- (f) X is in a headline table + Y is one of the headlines in the body table.

⟨*II*⟩ Both keywords are in a table.(Requested information tends to be contained in the linked page.)

- (g) X and Y satisfy the condition (d) and each word has a link to another page.
- (h) X is in a headline table + Y is in the body table and it has a link.
- (i) X is in the headline of a table + Y is in the table and it has a link.

⟨*III*⟩ One keyword is in the title of a page and the other is in a table.(Requested information tends to be contained in the table)

- (j) X is in the title of a page + Y is in the headline of a table in the page
- (k) X is in the title of a page + Y is in the headline of a row in a table in the page
- (l) X is in the title of a page + Y is in the headline of a column in a table in the page
- (m) X is in the title of a page + Y is in a headline table in the page
- (n) X is in the title of a page + Y appears in a table in the page frequently.

⟨*IV*⟩ One keyword is in the title of a page and the other is in a table.(Requested information tends to be contained in the linked page).

- (o) X is in the title of a page + Y is in a table in the page and it has a link.

(p) X is in the title of the page + Y is in a table in the page, and there is a word having a link in the row in which Y exists.

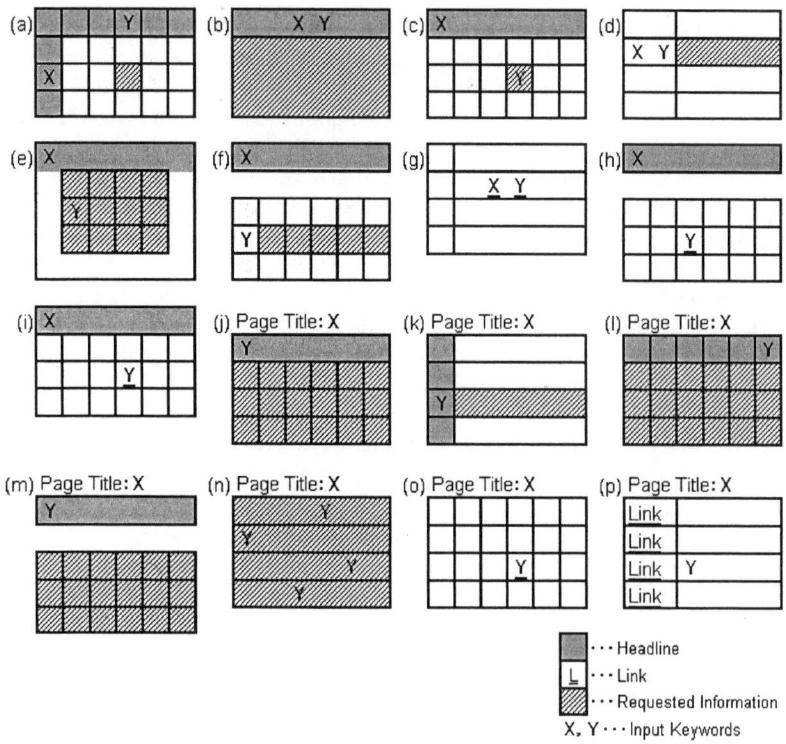

Figure 1: Type of Tables

Basically, our system has a table pattern searcher which observes tables to find the 16 patterns in order to find pages containing the requested information. However, there is a problem in that so many table tags appearing in current Web pages are not used for representing tables, but only for adjusting page layouts.

The latter tags don't represent tables semantically, so we must not apply the above mentioned method to such tags.

Therefore, we developed a judging program that classifies table tags into "semantically representing a table"(type A) and "used only for adjusting layouts"(type B). We incorporated it as pretreatment of the table pattern searcher.

As a judgment strategy, the following factors are adopted as independent variables.

- whether periods exist in the table
- whether texts except tags exist in the table
- whether specifications of a color exist in the table

- whether specifications of the width of a line exist in the table
- the number of cells
- the number of cells containing links
- the number of cells containing characters

Furthermore, we have gathered teaching data (subordinate variables) by seeing displayed pages and judging the type of table tags in the pages. Then we can create the optimal decision tree by C4.5 [9]. The decision tree was improved so that the type A table tags might not be missed as much as possible.

The result of an open test (target: 657 tables in 60 pages) is seen in Table 2. About 90% of the judgment is valid, and we think it can be put into practical use.

Table 2: Accuracy of Judgment of Type A or Type B

		Judgment by our program	
		Type A	Type B
Judgment by human	Type A	23.6%	5.8%
	Difficult to judge	0.3%	0.8%
	Type B	4.7%	64.8%

4 System Configuration

Our system is a meta-search engine using existing search engines.

First, a user inputs keywords as retrieval conditions from a Web browser, and their categories are judged using the dictionary database. Simultaneously, the keywords are given to the existing search engine and Web pages are output as the result of the retrieval. The Web pages are measured for their relevance by the unit for the sentence judgment and the table pattern searcher. The unit for the sentence judgment decides a candidate set of dependency patterns by using the categories of the keywords. Then it checks whether the *key-sentence* has one of the candidate patterns or not. The table pattern searcher classifies every table in the page into type A or type B. Moreover, it checks whether the tables come under (a) to (p) or not. Then, the ranking output by the existing search engine is resorted according to the result of the judgment by our system. That is to say, pages having semantic relations (matched patterns) should be ranked higher than the other pages. Currently, we are under investigation the following two ways of scoring strategies (cf. section 5).

(1) The system chooses the pages which have semantic relations found by either the table pattern searcher or the unit for the sentence judgment. The chosen pages are ranked higher than the others.

(2) Each of the table pattern searcher and the unit for the sentence judgment individually gives a certain score to the pages in which it finds semantic relations. The system resorts pages by the total of the scores, that is, the page regarded as a relevant by both of judgments is ranked higher than the others.

In the future, scoring strategies need to be argued in details. The resorted ranking is given to the user.

Currently, we use Google [10] as the existing search engine. As the natural language parser, we use a Japanese parser. Other units are implemented by Ruby.

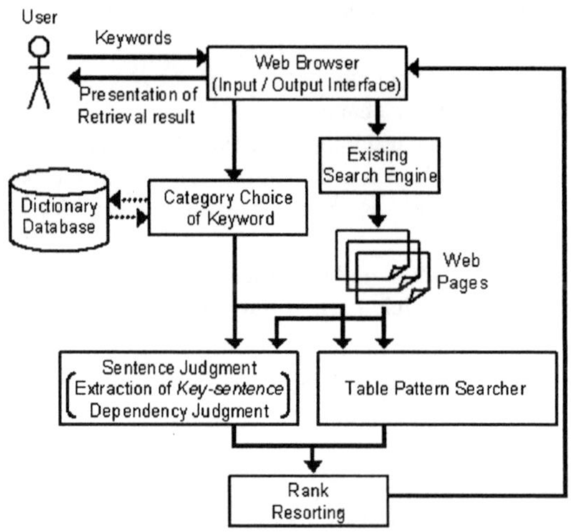

Figure 2: The System Configuration

5 Experimental Evaluation

5.1 Methods for Evaluation of Information Retrieval and Web Retrieval Systems

The traditional indices for evaluating whether retrieval is successful are precision and recall. Precision and recall are the evaluation method suitable for the traditional boolean retrieval in which the retrieved data is not graded by the relevance. However, Web search engines have to decide ranking of the retrieved pages because they are usually too many. Therefore, in order to evaluate Web retrieval, it is necessary to improve evaluation methods. In the retrieval experiment centering on TREC(Text REtrieval Conference) [11], several methods such as precision(λ), recall(λ), R-precision, average precision, etc. are currently used for evaluation of information retrieval including decision of page ranking [1]. Moreover, a user of Web retrieval tends to refer only to the results of higher ranks. Therefore, higher precision (= fewer irrelevant pages) is generally more worthy than higher recall (= fewer missed pages).

5.2 Collection of Data for Evaluation

In order to perform experimental evaluation of our system, a test collection consisting of the following data is required.

(A) a set of documents for retrieval (database)
(B) a set of retrieval queries
(C) information on whether each document is relevant to each query

Especially, it is quite costly to construct the data set of (C), because the relevance of each page must be judged by a human source, one by one. So, we have to consider how to collect a data set of adequate quality for credible evaluation by a lower cost source as far as possible. In order to satisfy this requirement, we design the set of retrieval queries (= (B)) carefully in order that the set contains not so many queries, however, the queries cover almost all of the typical patterns of keywords for Web retrieval. As a method to enumerate the typical patterns of keywords, we focused on the most typical purposes for which people use a Web search engine. The purposes can be classified into the following two types.

One is the case when the user wants to take an action, such as a purchase, make a reservation, download in Web space, so they look for the suitable pages for the action. The other is the case when the user wants to look for the pages which have information in which they are interested.

Roughly speaking, the world can be regarded as being composed of "entities" and "phenomena". Therefore, information retrieved by users must relate to entities and/or phenomena. More concretely, information concerning entities can be classified into three types: connotative characteristics of the entity, instances of the entity, and attribute values of the entity. Information concerning phenomena can be classified in the same way as entities. Then, the above mentioned purposes of Web retrieval can be subdivided as follows:

(i) finding pages in which the user can receive services he/she wants
(ii) finding pages on connotative characteristics of an entity
(iii) finding pages on attribute values of an entity
(iv) finding pages on instances of an entity
(v) finding pages on connotative characteristics of a phenomenon
(vi) finding pages on attribute values of a phenomenon
(vii) finding pages on instances of a phenomenon

We think that we can get a good quality of query set for our experimental evaluation, by gathering queries for achieving each of these purposes. We designed a questionnaire in which several situations corresponding to the purposes (i) to (vii) are set out (ex. "I want to go to Universal Studios in Japan", or "I want to buy a notebook PC"). The questionnaire asks subjects what they will input to a Web search engine as a query in each of the situations. Moreover, it asks them for which kinds of Web pages they intended to search with the query. To this date, we have gathered 1947 queries from 200 subjects who are students in Shizuoka University or Graduate School. In the gathered queries, single keyword is used in 419 cases (21.5%), double keywords are 942 (48.4%), and triple or more are 586 (30.1%).

We classified the keywords contained in the queries into 10 categories seen in Figure 3. Table 3 shows the typically used patterns of keyword combination, classified on the basis of the categories. As seen in Table 3, the most typical 12 patterns cover 75% or more of all gathered two-keyword queries. We consider that we can get a relatively practical query set by selecting queries from each of the 12 patterns in accordance with the rate of each pattern. So we selected 50 two-keyword queries from the gathered queries by the method. Numbers of selected queries for each pattern are seen in Table 4. As a result, we acquired data (B) of the test collection. Next, we input the queries to Google to get 100 retrieved pages for each query. These pages become data (A). Then we judged the relevance of each page in data (A) to construct data (C).

5.3 Result of the Evaluation

As mentioned above, the data set (A) contains 5000 pages. However, it contains binary files such as PDF. We remove binary files from the data set because such files cannot be analysed by our current natural language parser and table pattern searcher. So, the data set (A) actually used for our experimental evaluation becomes to contain 4863 pages. We evaluated the pre-

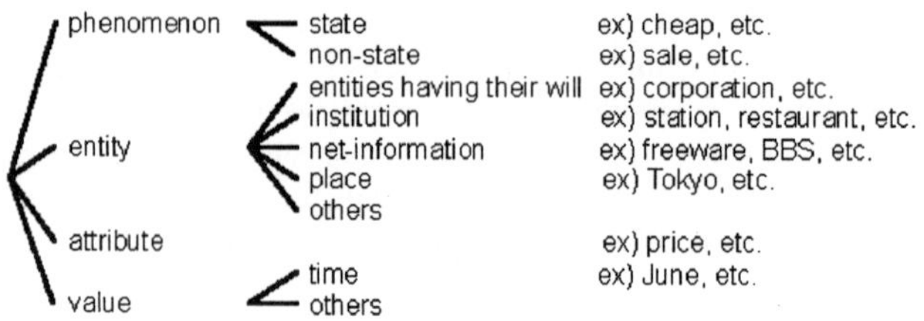

Figure 3: Categories of Retrieval Keyword

Table 3: Patterns of Keyword Combination

a	phenomenon: non-state	entity: others	153	16.5%
b	attribute	entity: others	94	10.1%
c	entity: others	entity: others	66	7.1%
d	entity: institution	entity: others	60	6.5%
e	entity: place	entity: institution	58	6.2%
f	entity: net-information	entity: others	54	5.8%
g	entity: place	attribute	53	5.7%
h	entity: entities having their will	entity: place	45	4.8%
i	phenomenon: non-state	entity: institution	38	4.1%
j	phenomenon: state	entity: others	36	3.9%
k	entity: place	entity: net-information	34	3.7%
l	entity: institution	entity: institution	34	3.7%
The sum total of higher rank 12 pattern			725	78.0%

cision (= the rate of the relevant pages) in the 4863 pages by using the data set (C). The result is 46.9%.

Next, we choose the pages which are judged as relevant by our "unit for the sentence judgment", "table pattern searcher", or their combination method. Then we evaluated the precision and recall of the chosen pages. The results can be seen in Table 5. In Table 5, T-Judgment means the judgment using table patterns, and S-Judgment means one using sentence analysis. S+T-Judgment means the following procedure:

(1) Judge a page by T-judgment. If the page doesn't contain the table pattern (a) - (p) mentioned in 3, do (2). Otherwise, judge the page as relevant ("O" in Table 5).

(2) Judge the page by S-judgment.

The pages not containing table cannot be judged by T-judgment. There are 901 such pages in the data set (A) (" △ " in Table 5). So the rate of pages containing tables is $(4863 - 901)/4863 = 81.5\%$ in the data set (A). And the rate of pages in which our system can find

Table 4: Numbers of selected queries for each keyword pattern

a	b	c	d	e	f	g	h	i	j	k	l	total
10	6	5	4	4	4	4	3	3	3	2	2	50

the table pattern (a) - (p) is $(1068 + 495)/(4863 - 901) = 39.4\%$. Similarly, the pages not containing *key-sentences* cannot be judged by S-judgment. There are 3156 such pages in (A) (" △ " in Table 5, too). So the rate of pages containing *key-sentences* is $(4863-3156)/4863 = 35.1\%$ in (A). And the rate of pages in which our system can find the typical dependency patterns between keywords is $(715 + 273)/(4863 - 3156) = 57.9\%$.

Furthermore, we rearranged the result of S+T-Judgment in Table 5 by each keyword pattern seen in Table 4. The results are shown in Table 6. We found that the precision/recall of each pattern is considerably uneven. This indicates that the precision/recall can be improved much more by developing new strategies for each keyword pattern.

Table 5: Total result for all pages

		Program's Judgment	
		O	X
User's Judgment	O	HIT	MISS
	X	FA	CR

(Unit; page)

	HIT	FA	CR	MISS	△	Precision	Recall
T-Judgment	1068	495	1364	1035	901	68.3%	50.8%
S-Judgment	715	273	294	425	3156	72.4%	62.7%
S + T-Judgment	1393	672	1910	888	—	67.5%	61.1%

△ — The number of pages which has not been included in CR-MISS since the page has neither *key-sentences* nor tables.
Precision — HIT/(HIT+FA). Recall — HIT/(HIT+MISS).

Table 6: Total result for every keyword type

	Precision	Recall
a	54.3%	45.4%
b	53.9%	49.9%
c	64.8%	68.5%
d	64.7%	51.1%
e	73.4%	69.5%
f	62.2%	42.4%
g	49.1%	61.4%
h	55.1%	77.3%
i	83.9%	46.4%
j	62.6%	73.6%
k	79.2%	52.1%
l	79.0%	78.0%

Then, we scored each page in the data set (A) by the following strategy and resorted the ranking which Google outputs. The scoring strategy is:

- If a page is judged relevant by T-Judgment, give +100 points.
- If a page is judged relevant by S-Judgment, give +100 points.
- Give (100- original (Google's) ranking) points to every page.

We compared the rankings output by Google with ones rearranged by our meta-search engine. The comparison is done for 1000 pages, composed of the best 20 pages for each of 50 queries in the data set (B). As a result, the precision of Google is 0.588, while our system 0.684. That is to say, the best 20 pages by our system contain about 2 more relevant pages than Google. The detailed result is seen in Table 7, which shows average precision, R-Precision, precision of the best 5/10/15/20/30/100 pages. Figure 4 shows the recall-precision graph [1]. They are the evaluation measures used by NTCIR-Web [12]. They are computed using trec_eval [13] which is a standard tool used by TREC [11] etc.

In addition, we compared the performance of Google with that of our system by MAP (mean average precision: average of average precision per query) [1] which is used in order to investigate performance differences between the two techniques. The advantage of our system is 0.0568. Generally saying, 0.05 or more advantage of MAP shows a significant difference of the two techniques. We also compared MAP of the two systems using the t-test. The different was significant, at the 1% level (P=0.00000008).

Table 7: Comparison of Google with our system

	Google	Our system
The number of retrieval subjects	50	50
A: Retrieved:(HIT+FA)	1000	1000
B: Relevant: (HIT+MISS)	2281	2281
C: Rel_ret: (HIT)	588	684
Precision(C/A):	0.5880	0.6840
Recall(C/B):	0.2578	0.2999
F measure:	0.3584	0.4384
Average precision (MAP):	0.1968	0.2536
Precision: At 5 docs:	0.6200	0.7600
Precision: At 10 docs:	0.5780	0.7460
Precision: At 15 docs:	0.5880	0.7107
Precision: At 20 docs:	0.5880	0.6840
Precision: At 30 docs:	0.3920	0.4560
Precision: At 100 docs:	0.1176	0.1368
R-Precision:	0.2768	0.3088

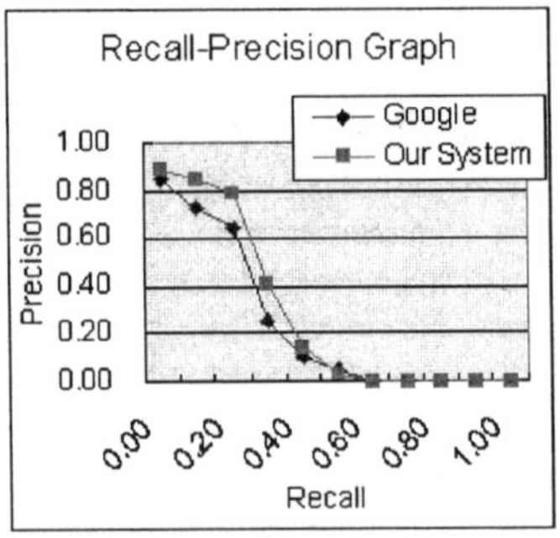

Figure 4: Recall - Precision Graph

6 Conclusion

We discussed how to measure the strength of semantic relations between keywords input as retrieval conditions. Keywords for WWW search must have a certain semantic relation and the relevant Web pages must contain the keywords which have the corresponding relation. Thus measuring the strength of semantic relations is useful to improve precision of search engines. The semantic relations are represented in some structures such as a sentence, a table, and so on. In this paper, we concentrated on sentence structure and table structure and discussed how to measure the strength of semantic relations represented by them. We have constructed a meta-search engine by using the measuring method. We have also evaluated the system and found that our system can raise the precision of the original search engine by about 10%.

Through our experiments, we inspected the pages which contain semantically related keywords but are irrelevant and we get some idea to detect such irrelevant pages. For example, a case that a sentence contains the related keywords but they are in a subordinate clause or a case that the font size of the sentence is smaller than the other sentences in the page, the clause or the sentence containing the keywords may make less contribution to the theme of the page. In such cases it may be effective that the system decrease a score of the pages. We are going to implement our ideas and evaluate the effect.

In this paper we dealt with only sentence structure and table structure, but semantic relations between keywords may be represented with the other structures. We are now paying attention to a structure of itemized document and a structure of a title and its contents. We will consider how to detect semantic relations represented with such structures and extend our meta-search engine.

In order to make a meta-search engine more convenient, strategies which do not concern semantic relations between keywords are also useful. For example, irrelevant pages tend to have typical categories of www pages such as a mailing list, a private diary, an advertisement of books, syllabus of lectures, and so on. In order to improve our meta-search engine, it may be effective to remove the pages belonging to such categories. So we are trying to design a filter which accepts users' request on unnecessary categories and removes the pages belonging to the categories. We will integrate the filter into our meta-search engine to improve our engine.

Although we have constructed and evaluated a prototype of search engine for Japanese, we think that the methods are applicable also to search engine for the other languages, such as English, because of the following reasons. The structure of table does not depend on each language. Sentences in every language can be parsed into dependency tree, and the theory that "if keywords have a close relation in dependency tree of some sentence, the page containing the sentence tends to be a relevant page" is also independent of languages.

References

[1] Kazuaki Kishida, Makoto Iwayama, Koji Eguchi: Methodology and Pragmatics of Retrieval Experiments at NTCIR Workshop, 2002.

[2] http://agents.www.media.mit.edu/groups/agents/publications/newt-thesis/subsection2_6_2_1.html

[3] Robert M.Losee: TEXT RETRIEVAL AND FILTERING Analytic Models of Performance, Kluwer Academic Publishers, 1998.

[4] Kazuhiro Kazama, Masanori Harada: Advanced Web Search Engine Technologies, Journal of Japanese Society for Artificial Intelligence, Vol.16, No.4, pp.503-508, 2001.

[5] Tsunenori Mine, Hiroki Fujitani, Makoto Amamiya: A Japanese Information Retrieval Method Using Syntactic and Statistical Information, NLPRS2001, pp.429-434, 2001.

[6] Kenji Tateishi, Naoyuki Ohba, Tsunenori Mine, Makoto Amamiya: Japanese Information Retrieval System using Syntactic Dependency Relations on the Web, Transactions of Information Processing Society of Japan, Vol.98, No.59(DD-13), pp.47-54, 1998.

[7] Minoru Yoshida, Kentaro Torisawa, Jun'ichi Tsujii: A Method for Information Extraction from Tables, The Sixth Annual Meeting of The Association for Natural Language Processing, pp.252-255, 2000.

[8] Yoshihiro Iwaguchi, Minsoo Jung, Masami Shishibori, Jun-ichi Aoe: An Indexing Method for Table Structures of HTML Format, IPSJ SIGNotes Fundamental InfologyCFI 61-22, pp.159-166, 2001.

[9] http://www.cse.unsw.edu.au/~quinlan/

[10] http://www.google.co.jp/

[11] http://trec.nist.gov/

[12] http://research.nii.ac.jp/ntcir/

[13] ftp://ftp.cs.cornell.edu/pub/smart/trec_eval.v3beta.shar

Information Modelling and Knowledge Bases XVI
Y. Kiyoki et al. (Eds.)
IOS Press, 2005

An Active Multidatabase System Architecture with an Aspect-Oriented Media Modelling Framework

Shuichi KURABAYASHI† and Yasushi KIYOKI‡
† *Graduate School of Media and Governance, Keio University*
‡ *Faculty of Environmental Information, Keio University*
5322 Endoh, Fujisawa, Kanagawa, 252-0816, Japan
{kurabaya, kiyoki}@sfc.keio.ac.jp

Abstract. In this paper, we propose a meta-level active multidatabase system architecture that provides a fundamental framework for retrieving, modeling and integrating various kinds of multimedia data. The aim of our research is to evolve the current passive multidatabase infrastructure into the active infrastructure, that allows the automatic and meaningful delivery of information that matches user requirements. The system provides active database functionalities and integrated media modelling framework over heterogeneous multimedia information resources. Our system architecture consists of three layers that comprises an aspect-oriented media modelling framework: 1) *bottom-layer: a data access layer* for accessing and retrieving heterogeneous information resources, 2) *intermediate-layer: an aspect extraction layer* for creating the metadata for meta-level computation and 3) *top-layer: a relationship evaluation layer* for dynamically computing correlations among the extracted features. By the 3-tier architecture, users can incorporate a wide variety of legacy information systems and data integration technologies into the active multidatabase environment without modification of the system. Our system applies the active database functionalities with multimedia retrieval methods to heterogeneous information resources to access and integrate them in a dynamic way. In this paper, we clarify feasibility and effectiveness of our system architecture by showing several experimental results.

1 Introduction

Today, we have opportunities to access various multimedia information resources. Those multimedia data is stored in passive information systems, such as web servers and relational database systems. The passive information systems are triggered in response to requests given by users. In order to retrieve the up-to-date multimedia information in the passive systems, users must manually and periodically check the state of data. Active database systems realize new searching/modifying operations in response to changes of contents in databases [6, 9]. By introducing the active database functionalities into the distributed multimedia information resources, in particular, users can receive various up-to-date information automatically. The active systems offer higher capability of information delivery than the passive systems.

Conventional active database systems can push up-to-date information in a specific information resource to users. However, users must manually compare, examine and integrate different information resources to obtain objective information, because it is difficult for the conventional systems to support such comparison and examination automatically among different information resources. When those information resources grow larger in scale and become more semantically complex, it is difficult to automatically check and react to changes of contents in them. In the multidatabase environment, *active behavior* is the essential missing piece that allows the automatic and meaningful delivery of information that matches user requirements.

The aim of our research is to realize the active multidatabase system that integrates autonomous and heterogeneous multimedia information resources automatically. In the multidatabase environment, various databases are newly interconnected, and contents of those databases change dynamically. Newly interconnected multimedia data could have new meanings, and the change of data contents may affect meanings of them. In such active and dynamic environments, it is difficult to statically determine meanings of data and relationships among data. To realize the automatic and meaningful delivery of multimedia information that matches requirements from users, it is important that the system provides evaluation mechanism for dynamic relationships between multimedia data and requirements from users. To realize the above new objectives, we propose **a meta-level active multidatabase system architecture that provides a fundamental framework for retrieving, modelling and integrating various kinds of multimedia data**. The meta-level system is designed in an abstract and higher level layer of local information systems, and it is constructed independently of local information systems[17]. We have proposed a meta-level multidatabase system that integrates heterogeneous databases by dynamically applying functions that compute correlation between data values of heterogeneous databases [8]. We have already proposed methods for spatial and temporal database computations [8]. The fundamental framework for computing semantic relationships with context recognition has been proposed in [12, 17]. We have also proposed the basic meta-level active multidatabase system architecture, called ADMIS, that manages and executes active rules [10, 11]. We have designed active rules for multimedia environments, called *meta-active rules*. ADMIS realizes both active database functionalities and database integration mechanisms in the meta-level system.

In this paper, we present new extensions to the meta-level active multidatabase system for actively integrating heterogeneous **multimedia** information resources. We extend the mechanism of condition evaluation to support evaluating dynamic relationships among heterogeneous multimedia data. As the extension, we design the active multidatabase system architecture with *an aspect-oriented media modelling framework*. The aspect-oriented media modelling framework provides fundamental functionalities to define meta-level schema structures, called *aspect-view*, that specifies methods to determine meanings of media data and relationships among media data. This framework is seamlessly integrated in the active rule system. The aspect-view consists of two concepts: *aspects* of multimedia data and *relationships* among multimedia data. **The aspect determines a feature set of multimedia data from the specific viewpoint. The relationship determines computation method for correlations among features extracted from the specific aspect.** In the aspect-oriented media modelling framework, the meta-active rule specifies conditions among the data retrieved via those aspect-views. It does not directly deal with legacy multimedia data. By this approach, we realize “media independence” of the meta-active rules. Media independent active rules can concentrate on meanings of media data, and those rules are not affected by physical aspects of media data, such as file format of media data, location of media data, metadata extraction methods for media data and access methods for multimedia information systems.

Those extension makes it possible to specify various situations of multimedia information resources, such as similarity of impressions among music data, in the condition evaluation. Our approach has two features as follows:

Feature 1: Our active multidatabase system makes it possible to deal with heterogeneous multimedia information resources as a single active database, because the system provides the active database functionalities in the meta-level system, that evaluates relationships among multimedia data, over those multimedia information resources. The active functionalities in the meta-level system are independent of legacy information systems. Even if those multimedia information resources are manipulated in passive systems, such as file systems or web search engines, the system actively integrates those information resources in the framework of active database, because the system provides the capability of accessing and integrating passive systems with the processes of condition evaluation or action execution.

Feature 2: The proposed architecture consists of independent three layers that comprise specialized functionalities for the meta-level information integration: 1) *a data access layer*, 2) *an aspect extraction layer* and 3) *a relationship evaluation layer*. The system architecture encapsulates behavior of each functionalities and its parameters in individual layers. The system accesses and retrieves heterogeneous multimedia information resources by the data access layer. In the aspect extraction layer, the system creates the metadata from the retrieved data for meta-level computation that the system can compute relationships among those data. The system evaluates implicit relationships among the metadata by applying correlation computation functions in the relationship evaluation layer. The system incorporates a wide variety of legacy information systems and data integration technologies into the meta-level active multidatabase environment, without modification of other layers. We also present optimization methods for the system architecture.

Our research issue, the active multidatabase system realizes automatic provision of information and cooperation of autonomous and ubiquitous multimedia information systems. Related researches can be summarized as follows: 1) passive multimedia search systems and 2) active multidatabase systems. MediaLand[4] is the passive cross-media search system that realizes the multi-paradigm multimedia querying that seamlessly integrates existing search approaches. The paper have also proposed uniform data model that statically describes all kinds of media objects and their relationships. Several active database systems for heterogeneous databases have been studied. For example, the paper [3] presented active database functionalities in the CORBA services, and this system realizes unified utilization for active database functionalities in the distributed and heterogeneous database environment. This system also includes the active database framework in information systems. The paper [5] presented a method to use heterogeneous legacy OODBs as active databases by implementing middlewares which interpret active rules. It defines data types and operations for spatial databases in the CORBA framework. This method also provides spatial database functionalities to the active databases. Those active multidatabase researches interconnects traditional relational databases as local databases.

2 Design and Concept of System Architecture

We propose the active multidatabase system with the aspect-oriented media modelling framework. The system actively integrates multimedia information resources by dynamically evaluating implicit relationships among media data in the active rule system. The aim of our research is to evolve the current passive multidatabase infrastructure into the active infrastructure, that allows the automatic and meaningful delivery of information that matches user requirements.

Toward the above new objective, we identify fundamental functions for realizing the active multimedia-multidatabase environment as follows: 1) detecting/notifying events among heterogeneous information systems and 2) evaluating conditions among heterogeneous multimedia information resources. In our active multidatabase system, it is not necessary to deal with any updating conflicts among active rules, because our active multidatabase system does not support any updating operations. Our active rules for multidatabase environment do not include updating operations and do not cause any updating conflicts among themselves. We have designed the system for applying it to the active retrieval operations on multiple and heterogeneous information resources. The fundamental operations in our system are designed for retrieving. In the paper[11], we have proposed the basic active multidatabase system architecture, and we have designed the meta-active rule language for describing events and conditions among heterogeneous information resources. This basic architecture includes the uniform event model and the event detection mechanism. Therefore, we concentrate on the methodology for evaluating conditions among heterogeneous multimedia information resources in this paper.

We present new extensions to the meta-active rule for actively integrating heterogeneous **multimedia** information resources. We extend the mechanism of condition evaluation to support the aspect-oriented media modelling. By those extension, users can specify various situations of multimedia information resources, such as similarity of impressions among music data, in the condition evaluation.

2.1 Modelling Multimedia Data in the Active Multidatabase Environment

The aspect-oriented media modelling framework provides primitive functions to define a meta-level schema structure, called *the aspect-view*, that specifies meanings of media data and relationships among media data. This framework is seamlessly integrated in the active rule system. The aspect-view deals with two issues: *aspects* of multimedia data and *relationships* among multimedia data.

The aspect determines a feature set of multimedia data from the specific viewpoint. Media data consists of various aspects. Each aspects has its own feature. Users extract features from various aspects of media data to retrieve objective data. In order to realize significant multimedia information retrieval, it is important to match the aspect required by users and the aspect covered by media retrieval techniques. For example, to retrieve music data, users may retrieve music by instructions used in music, or by impressions of music. In those cases, instructions used in music and impression words are aspects of music. For realizing a meaningful media-retrieval mechanism that is aware of various meanings of multimedia data, selecting aspects of media data is necessary to evaluate relationships among multimedia data.

As shown in Figure 1, multimedia data consists of several kinds of single media. Each single media has various aspects, and every aspect has its own meaning. Therefore, different aspects have different features. To dynamically evaluate relationships among multimedia data, a framework for dynamic feature extraction from various aspects is essentially needed.

We can not compare the specific feature with the other feature extracted from the different aspect, since the feature of aspect is extracted according to its own semantic criteria. Rela-

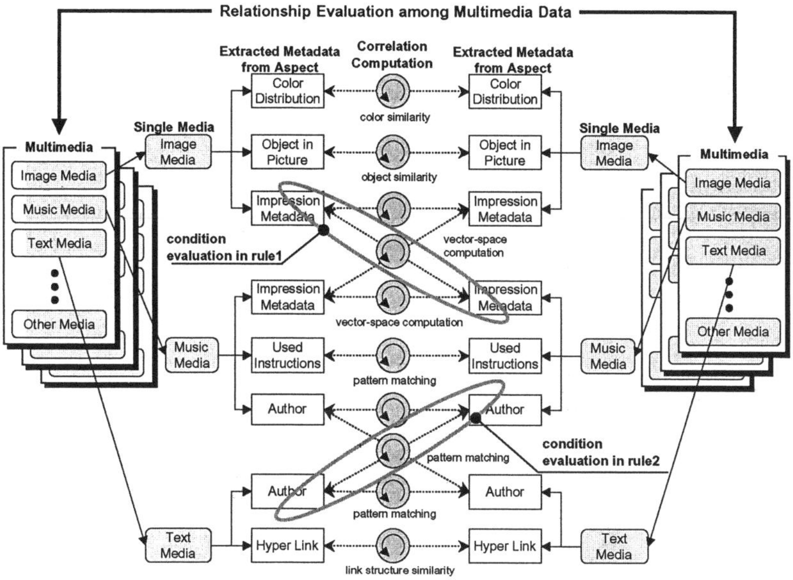

Figure 1: Overview of the Aspect-Oriented Media Modelling Framework

tionships among those multimedia data are dependent on the selected aspects. As media data has a separate meaning for every aspect, extremely wide variety of relationships exist among multimedia data. For realizing a meaningful media-retrieval mechanism that is aware of various meanings of multimedia data, selecting aspects of media data is necessary to evaluate relationships among multimedia data.

The relationship determines computation method for correlations among features extracted from the specific aspect. The essence of our approach is to describe relationships by functions. The functions computes correlations among multimedia data. No static descriptions about relationships are provided. Relationships among those multimedia data are dependent on the selected aspects. In the active multidatabase environment, where a wide variety of multimedia information resources are dynamically added/deleted, it is difficult to statically describe relationships among those information resources. Describing relationships with functions makes it possible to dynamically compute relationships among the multimedia data.

2.2 The Meta-Active Rule with The Aspect-View

The aspect-view defines the feature set of the specific aspect and relationship computation methods. The overview of the aspect-view is shown in Figure 2. The aspect-view is independent of every local information resource, because it is not for storing media data but for evaluating relationships among media data in the meta-level system. By selecting aspects of the multimedia data, the system determines the corresponding aspect-view. The system

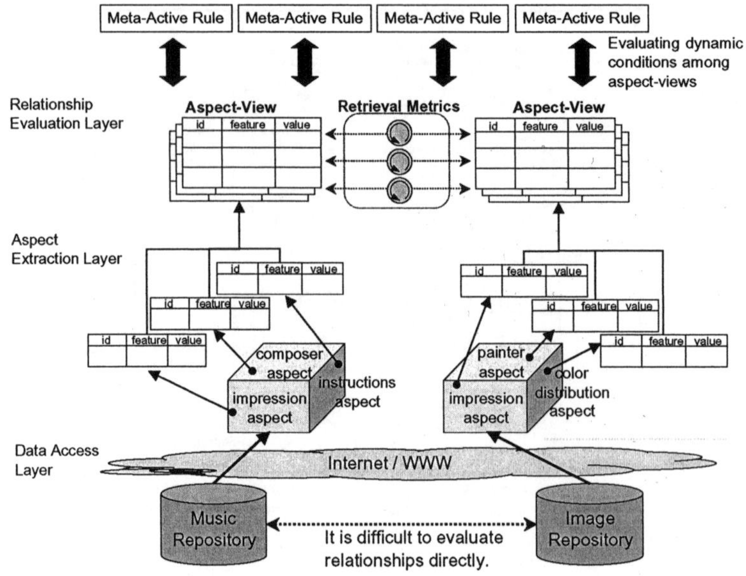

Figure 2: Overview of Aspect-View

extract the metadata from the aspect. Relationships between aspect-views are described by relationship computation functions. Primitive functions for defining the aspect-view are described in Section 2.5.

The meta-active rule is defined on the meta-level schema. In the aspect-oriented media modelling framework, the meta-active rule specifies conditions among the data retrieved via those aspect-views. The meta-active rule does not directly evaluate conditions among legacy multimedia data, but indirectly evaluates conditions among aspect-views. To detect the meaningful conditions in the multimedia data, multimedia data should be described in the metadata for meta-level computation. In our approach, those heterogeneous multimedia data are mapped into the aspect-view, and indirectly integrated in the meta-level system. By this approach, we realize "media independence" of the meta-active rules. The feature of media independence is very important in the active multidatabase environment, because types of media are increasing and new multimedia data is always added.

In conventional active database systems, which evaluate relationships among simple textual data by pattern-matching techniques, the change of pattern in the data means the change of relationships among other data. Those systems detect situational changes in data contents by detecting the change of pattern in the data. In multimedia database systems, which evaluate relationships among multimedia data by various media-retrieval techniques, the change of pattern in the data does not directly mean the change of relationships among the multimedia data. To detect the meaningful condition in the multimedia data, the active rule must have capabilities to deal with aspects and relationships dynamically.

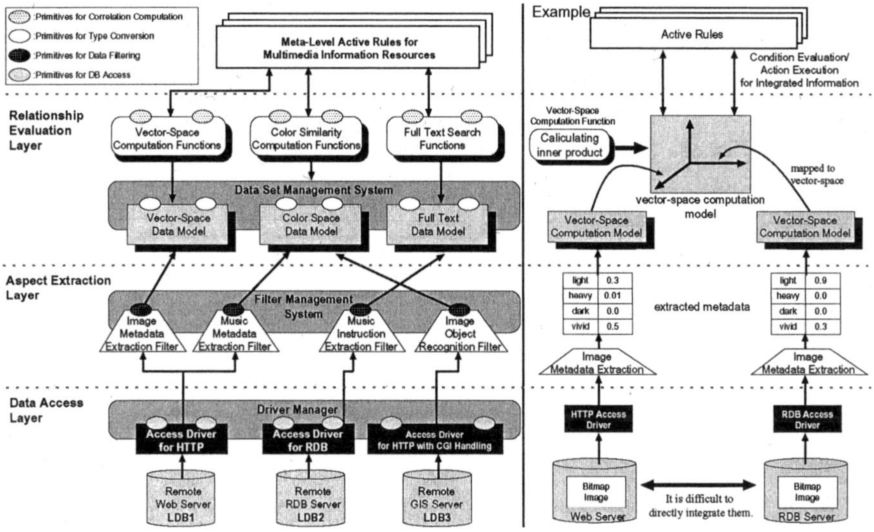

Figure 3: System Architecture

2.3 3-tier System Architecture

In the active multidatabase system, we identify fundamental functions for evaluating relationships among multimedia data as follows: 1) accessing multimedia data stored in heterogeneous information systems, 2) extracting features from the specific aspect, 3) evaluating relationships among multimedia data according to the extracted features. We present a 3-tier architecture of the active multidatabase system. The system architecture makes it possible to combine above functions according to situational changes in legacy multimedia databases. As shown in Figure 3, the proposed architecture consists of three layers: 1) *a data access layer*, 2) *an aspect extraction layer* and 3) *a relationship evaluation layer*. Each layer corresponds to the specific function of the system. Those layers are highly independent of each another, and they are specialized to their own functionalities. The system architecture encapsulates behavior of each functionalities and its parameters in individual layers. By this architecture design, our system realizes incorporation among a wide variety of legacy information systems and data integration technologies into the active multidatabase system, with minimum modifications for the system. Unlike our approach, conventional approaches, such as wrapper/mediator approaches, that use those functionalities in some ad-hoc ways, it is difficult to dynamically process the active rule that requires combination of integration functions for multimedia data.

2.3.1 The Data Access Layer

The data access layer is *the world of data retrieval.* The main functionality of this layer is transparently accessing information resources and retrieving data from them. By providing this layer in the bottom level, other layers can transparently access all the local information systems. The system can connect the newly implemented information systems, by introducing appropriate access drivers into the layer.

2.3.2 The Aspect Extraction Layer

The aspect extraction layer is *the world of data modelling*. To realize aspect-oriented media retrieval, this layer dynamically creates the metadata according to the aspect specified by a user. Various information resources are stored in forms which are not suitable for information integration, such as HTML files, JPEG image format and WAV sound format. In these data forms, it is difficult for the system to dynamically compute implicit relationships among the data. To integrate those information resources, the system provides a method to create *the metadata for meta-level computation*, that the system can compute relationships among them. The layer provides a framework for creating the metadata of media data. The metadata creation methods and their parameters are encapsulated in the layer. By this encapsulation, other layers are independent of methods for data modelling. The system can recognize new aspects of media data, by introducing new feature extraction filters into the layer. For example, to connect an image data into the active multidatabase system, the data filter which generates metadata based on color distribution of the image and the data filter which generates metadata based on objects in the image are applicable. We have implemented several data filters as shown in Table 1.

Table 1: Implemented Filters

Data Source	Metadata	Details
Raster image	Vector representation	this data filter extracts metadata from raster images according to the color information of images[13].
MIDI music	Vector representation	this data filter creates metadata from MIDI files according to human's impression. This data filter is based on the automatic metadata generation method for music, which is proposed in [2].
Google	Table	The data filter can be utilized to incorporate existing search engines.

2.3.3 The Relationship Evaluation Layer

The relationship evaluation layer is *the world of computation*. This layer has two important functions: 1) computing implicit correlation among the data and 2) editing data set. To evaluate correlations among media data, the system provides *correlation computation functions* in this layer. Users can specify relationships among those heterogeneous information resources by using various functions and parameters for them. By this approach, we do not need to introduce any static and explicit relationship specifications, such as relationship-relations. The data set management system provides data set manipulation functionalities. In the proposed system architecture, that manages and integrates data in the meta-level system, it is necessary for the meta-level system to have a data set management facility. The system provides the data set management system as a separate module, in order to change the data set management system according to requirements of applications.

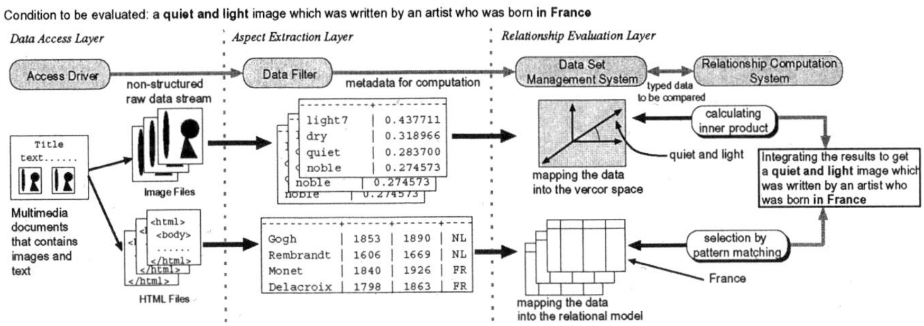

Figure 4: Data Flow of the System

2.4 *Example of the Aspect-Oriented Relationship Evaluation*

Figure 4 shows the data flow of our system architecture that evaluates conditions among multimedia documents. We will consider an example that evaluates conditions heterogeneous multimedia documents, which contain multiple images and text. Firstly, the system retrieves multimedia data from remote information systems using access drivers. To deal with the impression aspect of images, the aspect extraction layer generates the metadata of images using the specific filter. Each metadata represents impression words of an image. In the relationship evaluation layer, the generated vector metadata are mapped into the vector space model, such as shown in the paper [16]. To deal with biographical aspects of multimedia data, another data filter extracts text data from multimedia documents, and transforms the extracted text as tables. The tables are mapped into the relational model in the relationship evaluation layer. The relationship evaluation layer combines two searching techniques: calculating inner product in the vector space model and the selection operator in the relational model. By the combination of various relationship computation systems, the system realizes condition evaluation among multimedia information.

2.5 *Design of Primitive Functions*

As a basic operation set of the proposed multidatabase system, we define a primitive functions in the active multidatabase system. The primitive functions are independent of legacy information systems implementation. We have designed the primitive functions for the active rule processing[11]. Therefore, we concentrate the primitive functions for defining the aspect-view that dynamically evaluates conditions among heterogeneous multimedia information resources.

2.5.1 Primitive Function for Data Retrieval

A primitive function **MDP1** is the operation supported by the local information resources. MDP1 retrieves data from specified information resources, and returns retrieved data as the data stream. It is defined as:

MDP1: $f_{access}(R_a, A_i(V)) \rightarrow U^{out}$

The parameter R is information resources specified by an identifier a. In order to connect to the information resource R_a, the function $f_{access}()$ uses an access driver A_i, where i is a identifier of the access driver. The parameter V specifies options for the access driver. The result of the function $f_{access}()$ is the data stream U^{out}. The data stream is raw representation of data in local information systems. The function $f_{access}()$ does not support converting the format of data stream. The format of data stream is dependent on the local information system.

2.5.2 Primitive Function for Feature Extraction

A primitive function **MDP2** extracts the feature from the specified aspect of the data stream. The data stream is retrieved by MDP1. MDP2 is defined as:

$$\textbf{MDP2: } f_{extract}(U^{in}, M_x(P)) \rightarrow V_1^{out}$$
$$V_1^{out} = \{S_{[1,1]} \cdots S_{[d,m]}\}$$

The parameter U^{in} is the data stream provided by MDP1. The parameter $M_x()$ is the specific feature extraction method and P is its parameter. To transform the data stream U^{in} into metadata, the function $f_{extract}()$ applies the parameterized function $M_x(P)$ to the data stream U^{in}. Users can apply different feature extraction methods into the same U^{in}. The result of $f_{extract}()$ is the data set $\{S_{[1,1]} \cdots S_{[d,m]}\}$, which represents the feature of the data stream. S is a primitive datum in the system. The parameter d represents maximum size of attribute of the feature. The parameter m represents maximum size of tuples of the feature.

2.5.3 Primitive Function for Binding Data

A primitive function **MDP3** binds an anonymous data set to a specific name. MDP3 is defined as:

$$\textbf{MDP3: } f_{bind}(V_1^{in}, N_e) \rightarrow V_2^{out}$$
$$V_2^{out} = \{N_e, \{S_{[1,1]} \cdots S_{[d,m]}\}\}$$

The parameter V_1^{in} is the anonymous data set returned by the function $f_{extract}()$. The parameter N_e is a name for the data set. The function $f_{bind}()$ returns the data set with a unique and valid name in the system.

2.5.4 Primitive Function for Naming Field

A primitive function **MDP4** assigns a name to each field of the data set. MDP4 is defined as:

$$\textbf{MDP4: } f_{field}(V_2^{in}, \{F_{[1]} \cdots F_{[d]}\}) \rightarrow V_3^{out}$$
$$V_3^{out} = \{N_e, \{F_{[1]} \cdots F_{[d]}\}, \{S_{[1,1]} \cdots S_{[d,m]}\}\}$$

The parameter V_2^{in} is the data set returned by the function $f_{bind}()$. The parameter $F_{[d]}$ is a name for the field d. The result of the function $f_{field}()$ is a data set, whose fields have valid names.

2.5.5 Primitive Function for Assigning Type Information

A primitive function **MDP5** assigns type information to each field of data set by applying type conversion functions. MDP5 is defined as:

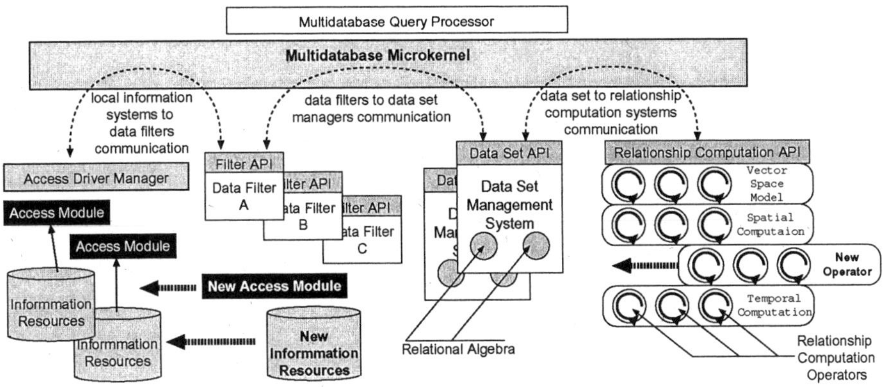

Figure 5: Overview of the System Implementation

MDP5: $f_{type}(V_3^{in}, \{T_{[1,j,C_k(W)]} \cdots T_{[d,j,C_k(W)]}\}) \rightarrow D^{out}$
$D^{out} = \{N_e, \{F_{[1]} \cdots F_{[d]}\}, \{T_{[1,j,C_k(W)]} \cdots T_{[d,j,C_k(W)]}\}, \{S_{[1,1]} \cdots S_{[d,m]}\}\}$

The function $f_{type}()$ accept the parameter $T_{[d,j,C_k(W)]}$ as type information. The parameter j is type ID. To determine the type information, the function $f_{type}()$ applies the type conversion function $C_k()$ and its parameter W into the raw data set $S_{[d,m]}$ of V_3^{in}.

2.5.6 Primitive Function for Integrating Multimedia Data

A primitive function **MDP6** integrates multimedia data by evaluating relationships among them. MDP6 is defined as:

MDP6: $f_{apply}(\{D_1^{in} \cdots D_n^{in}\}, G_e(P), R_k(N)) \rightarrow D^{out}$

The function $f_{apply}()$ accept the parameter $\{D_1^{in} \cdots D_n^{in}\}$ as data to be integrated. The parameter G_e specifies the function of data set management systems, and P is its parameters. The parameter $G_e(P)$ determines the method of data editing. The parameter R_k specifies the function of correlation computation systems, and N is its parameters. By specifying $R_k(P)$, relationships among each value in the data set $\{D_1^{in} \cdots D_n^{in}\}$ can be detected. The function $f_{apply}()$ returns a data set D^{out} that is generated by the data set manipulation.

3 System Implementation and Performance Tuning

For realizing the system to integrate various multimedia data on the WWW, the extensibility of the system is very important. On the WWW environment, information systems for multimedia data and types of multimedia data increase, and users' requirements for integrating multimedia data drastically change. One of the important issue to be solved is to realize a **fast** and **extensible** system. In this section, we present an implementation methodology for improving performance of the system.

As shown in Figure 5, our meta-level system provides fundamental functionalities proposed in Section 2 as modules for the kernel. The microkernel of the implemented system

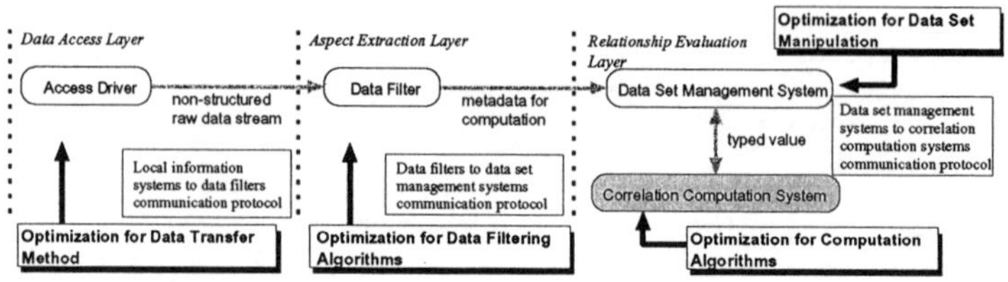

Figure 6: Optimization Points in the System

manages four types of modules: 1) access drivers, 2) data filters, 3) data set management systems and 4) relationship computation systems. The microkernel architecture increases backward compatibility with existing systems, and forward compatibility with systems to be developed in the future. By this system design, we incorporate a wide variety of legacy information systems and data integration technologies into the meta-level multidatabase system. The microkernel architecture simplifies each module. Users can improve the overall performance of the system by separately optimizing the individual module, because each module has the highly independent functionality.

3.1 System Performance Optimization

As shown in Figure 6, our system has four *hot-spots* which are key points for effective system optimization. Users can detect bottlenecks of the system, by analyzing the consumed time in each hot-spot to perform a multidatabase query. Each hot-spot can be independently optimized according to its own circumstances, without affecting other layers. It is important and effective to optimize each layer according to its individual characteristics.

The data transfer method is the hot-spot which causes I/O communication overhead. When the bottleneck of the system is the amount of data transfer, it is necessary to introduce a optimization mechanism for the data transfer method, such as a data cache mechanism and a data compression mechanism. The aim of optimization for data transfer is to decrease the communication overhead between the access driver and the local information system.

The data filter is the hot-spot which requires the CPU resource for data processing, such as metadata extraction from images. In the viewpoint of the system implementation, media types of the data stream that are to be filtered is one of the important factor for optimizing the data filter. Optimization to be applied is dependent on types of the data stream. The metadata extraction filter for image files requires fast image processing, and the metadata extraction filter for music files requires fast wave sound processing.

The data set management system is the hot-spot in which data set manipulations, such as the join operation, are performed. As the system can incorporate multiple data set management system which are based on different data models, optimization for the data set management system is dependent of its data model. To resolve bottlenecks in this hot-spot, it is effective to introduce existing optimization methods for legacy database technologies, such as fast join algorithms and main-memory database technologies. As the prototype system implementation, we have used a main-memory database approach for optimizing data set manipulation methods. We have extended PostgreSQL database system as the main-memory database system, and have plugged it into the meta-level system. By this implementation, the system can utilize the complete set of relational algebra, and the data integration process in our system completely works in main-memory without disk access.

As relationship computation is a CPU centric process, the relationship computation system is the hot-spot where complicated computations for relationship evaluation between data are performed. Application specific optimization approaches are effective for optimizing relationship computation in the relationship evaluation layer.

4 Evaluation

In this section, we show several experiments of results with regard to improvement of the system performance. The aim of experiments is to verify two characteristics of the system: 1) optimization for a layer do not affect the other layers and 2) by resolving the bottleneck of the individual layer independently, users can improve the overall performance of the system.

4.1 *Experimental Environment*

We have performed four experiments in the viewpoint of system performance. We examine the execution time in the individual layer, when the system processes four types of queries: 1) *a data retrieval centric query*, where processing time is mostly consumed in the data access layer, 2) *an aspect extraction centric query*, where processing time is mostly consumed in the aspect extraction layer, 3) *a relationship evaluation centric query*, where processing time is mostly consumed in the relationship evaluation layer and 4) *a query that requires overall performance of the system*, where processing time is consumed in the all layers. The aim of the study is to verify that separated optimization for the individual layer improves the total processing performance of the system, without processing overhead in other layers. We have implemented the system on a PC/AT compatible machine which specification is shown in Table 2.

Table 2: System Specification

Category	Instantiation
CPU	Intel Mobile Pentium4 1.6GHz
Main Memory	512MB
Hard Disk	IDE 40GB (5,400rpm, UltraATA100, 16MB Cache)
NIC	100Base-TX
OS	Linux Kernel version 2.4.21
File System	ext3

4.2 *Experiment-1: The Data Retrieval Centric Query*

In the experiment-1, we evaluate optimization methods for data retrieval centric queries. We submit five queries that integrate three large text files, in which an operating system's log information is written. One text file contains various log information, such as messages by system services and applications. Each query uses different data filters that extract the application specific information from log files. One filter extracts the kernel's log information, and another filter extracts the HTTP daemon's log information. It is necessary to apply different

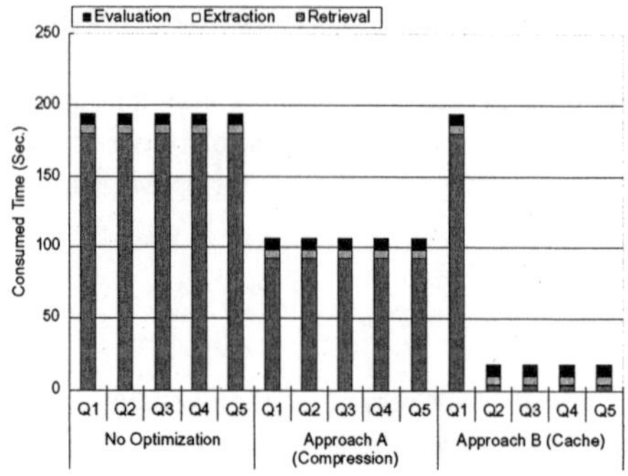

Figure 7: Effect of Optimization for the Data Access Layer (Experimental Result 1)

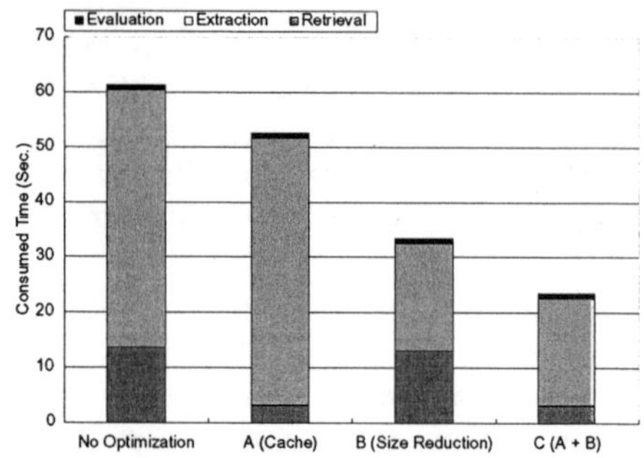

Figure 8: Effect of Optimization for the Aspect Extraction Layer (Experimental Result 2)

data filters, because a schema of each log information is different according to applications. The size of each text file is 41M bytes. The total size of those text files is 123M bytes. In this experiment, 82% of the processing time is consumed for the data retrieval.

We have implemented two optimization methods in the data access layer. One is a data cache mechanism, that decreases communication overheads between the access driver and local information systems by keeping the retrieved data. The other is a data compression mechanism, that compresses the communication between local information system and the data access layer by looseless data compression algorithm, such as the deflate algorithm, to reduce the amount of the data transfer. In order to verify that the optimization method can be applied without affecting other layers, we compare the default system implementation and those optimization methods.

Figure 7 shows the result of the experiment-1. In this graph, "no optimization approach" shows the default query processing time required in the system. The consumed time for processing each query is shown as Q1 to Q5. The approach-A, which uses the data compression mechanism, has decreased 42% of the processing time in all the queries. The approach-B, which uses the cache mechanism, has decreased 82% of the processing time after Q2 in the data access layer. As the transfered data required by Q1 is cached, other queries need not to transfer the data again. The optimization in the data access layer has not increased the consumed time in the aspect extraction layer and the relationship evaluation layer. Because those queries require the same data, the approach-B is more effective than the approach-A. When queries require different data, the approach-A will be more effective than the approach-B.

4.3 Experiment-2: The Aspect Extraction Centric Query

In the experiment-2, we submit 30 queries for evaluating conditions among thirty image files. Each query computes a correlation between one image and other 29 JPEG image files by using metadata extraction filter for images. The metadata extraction filter generates vector metadata that represents impression features of the image. The system computes correlation by calculating inner product between image files. The size of each image file is 640×520 pixels. The average file size is 47K bytes. The total size of those image files is 1.45M bytes. In this experiment, 77% of the processing time is consumed for the aspect extraction, because they require complicated computation, such as the metadata extraction from images.

We have implemented two optimization methods. One is the data cache mechanism in the data access layer. The other is a image size reduction mechanism in the aspect extraction layer, that reduces several pixels in image files. By the reduction, the image size is reduced to 81 × 64 pixels. The average file size of the reduced image is 2K byes. In order to verify that each optimization method can be applied without affecting other layers, we compare three optimization approaches: 1) *Approach-A* that uses the data cache mechanism in the data access layer, 2) *Approach-B* that uses the image size reduction mechanism in the aspect extraction layer, and *Approach-C* that uses both the data cache mechanism and the image size reduction mechanism.

Figure 8 shows the result of experiment-2. In this graph, "no optimization approach" shows the default query processing time required in the system. The approach-A, which uses the optimization method in the data access layer, has decreased 77% of the processing time in the data access layer. This approach has not increase the processing time of other layers. The approach-B, which uses the optimization method in the aspect extraction layer, has decreased 58% of the processing time in the aspect extraction layer. This approach has not increase the processing time of other layers. This result indicates that each optimization method reduces the processing overhead in each layer without affecting other layers. The approach-C, which optimizes both the data access layer and the aspect extraction layer, has decreased the total processing overhead than the approach-A and the approach-B. This result shows that optimization methods in the different layers do not affect each other, and that those optimization methods can be combined.

4.4 Experiment-3: Relationship Evaluation Centric Query

In the experiment-3, we evaluate optimization methods for relationship evaluation centric queries. We submit a query which integrates two databases, DB-A and DB-B, that store geographical data. The query is submitted to test if the interior of one polygon in the DB-A intersects the interior of polygon in DB-B. Each database has 10000 tuples, and each tuple consists of a complex polygon data object and its OID. The average size of the polygon data is 896 bytes. To test intersection of polygons, we have implemented two functions: *function-A "intersect_polygon()"* that strictly calculates intersection area between two polygons, and *function-B "intersect_boundingbox()"* that calculates intersection areas between bounding boxes of two polygons. The function-A introduces complicated matrix computation to obtain a precise solutions, but the function-B always uses two rectangles, which are bounding boxes of polygons, to quickly obtain solutions. When an application does not require strict solutions, the function-A is effective to gain performance of the query processing. We have compared performance of the query processing in two cases: the case of using the function-A and the case of using the function-B. By this experiment, we verify that users can independently choose a optimization method for the relationship evaluation layer and the optimization process does not affect other layers.

Figure 9 shows the result of the experiment-3. In this graph, "no optimization approach" shows the processing time required in the system which uses the function-A. The optimized system, which uses the function-B, has decreased the overhead of processing in the relationship evaluation layer, as computation by the function-B is much more simple than the computation of the function-A. We have verified that the optimization in the relationship evaluation layer did not affect the execution time in the data access layer and the aspect extraction layer.

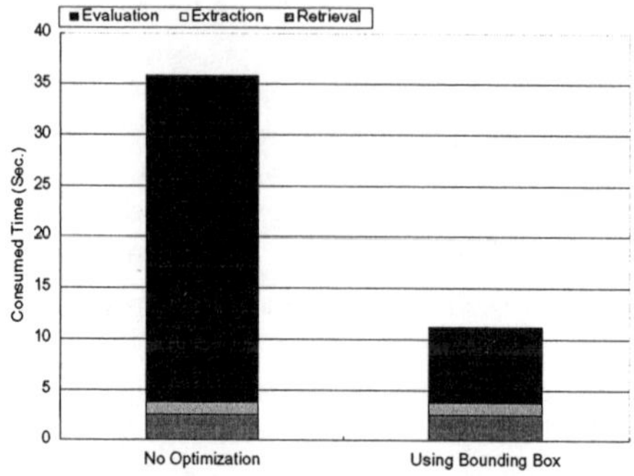

Figure 9: Effect of Optimization for the Relationship Evaluation Layer (Experimental Result 3)

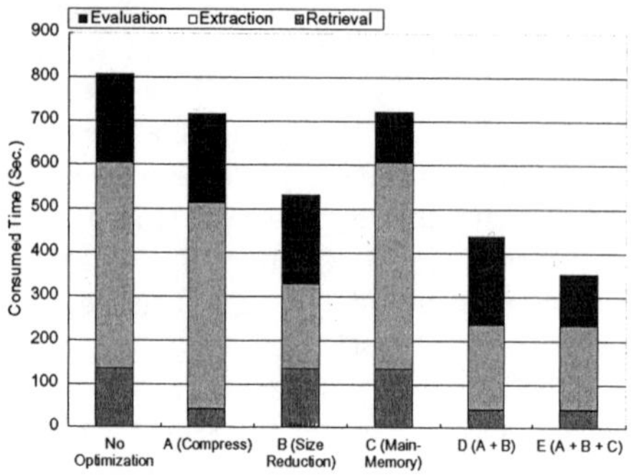

Figure 10: Effect of Optimization for All Layers (Experimental Result 4)

4.5 *Experiment-4: Query that Requires Overall Performance*

By evaluating several optimization approaches for the query that utilizes overall layers, we verify that optimization for one layer does not affect other layers and that the overall performance of the system is improved by separately resolving the bottleneck of each layer. In the experiment-4, we submit a query that integrates two large image databases. Each local database includes 300 JPEG image files, and the size of each image file is 640 × 520 pixels. The average file size is 50K bytes. The system maps those images using metadata extraction filter described for images.

The query is submitted to join those databases with the condition that correlations between two images, which calculates inner product of two vectors, are over the specified threshold. This query accesses a large amount of images in the data access layer, and uses the complicated data filter in the aspect extraction layer, and applies the join operation, that is one of the heaviest operation in relational algebra with dynamic computation of inner product in the relationship evaluation layer. We have implemented three optimization methods: 1) The data compression mechanism, which is the same as described in Section 4.2, in the data access layer. 2) The image size reduction mechanism, which is the same as described in Section 4.3, in the aspect extraction layer. 3) The main-memory database technology, which is described in Section 3.1, in the relationship evaluation layer. In order to show that each optimization method can be applied without affecting other layers, we compare five optimization approaches as shown in Table 3. By comparing those approaches in the viewpoint of system performance, we verify that optimization for one layer does not affect other layers, and that the overall performance of the system is improved by separately resolving the bottleneck of each layer.

Figure 10 shows the result of the experiment-4. In this graph, "no optimization approach" shows default query processing time required in the system. The approach-A, which uses the optimization method in the data access layer, has decreased the overhead of the processing in the data access layer. This approach does not affect the processing time of other layers. The approach-B, which uses the optimization method in the aspect extraction layer, has decreased the overhead of the processing in the aspect extraction layer. This approach does not affect the processing time of other layers. The approach-C, which uses the optimization method in the relationship evaluation layer, has decreased the overhead of the processing in the relationship

Table 3: Optimization Approaches in the Experiment-4

Approach	The Data Access Layer	The Aspect Extraction Layer	The Relationship Evaluation Layer
Approach-A	Optimized (data compression)		
Approach-B		Optimized (size reduction)	
Approach-C			Optimized (main-memory db)
Approach-D	Optimized	Optimized	
Approach-E	Optimized	Optimized	Optimized

evaluation layer. This approach does not affect the processing in other layers. Those results indicate that each optimization method contributes to decrease the overhead of the processing in each layer without affecting other layers. The approach-D, which optimizes both the data access layer and the aspect extraction layer, has decreased the overhead of the total processing in comparing with the approach-A and the approach-B. The approach-E, which combines three optimization methods, has decreased the overhead of the total processing time in comparing with the other approaches. This result indicates that those optimization methods in the different layers do not affect each other, and the system could independently use them.

In the result of the experiment-1, 2 and 3, we have shown that our system effectively performs the query which has the bottleneck in the specific layer, by independently optimizing individual layers. In the experiment-4, we have shown that the system effectively performs the query which requires operations in the three layers. Those experimental results show that optimization in each layer does not affect the performance of other layers. Those results clarified the feasibility and effectiveness of our system architecture.

5 Conclusion

The aim of our research is to evolve the current passive multidatabase infrastructure into the active infrastructure, that allows the automatic and meaningful delivery of information that matches user requirements. In this paper, we propose the meta-level active multidatabase system architecture that provides a fundamental framework for retrieving, modeling and integrating various kinds of multimedia data. The system provides active database functionalities over heterogeneous multimedia information resources. The essential feature of the system is that the system architecture is designed with aspect-oriented media modelling framework, that is seamlessly integrated in the active rule system. The aspect-oriented media modelling framework deals with two important issues: 1) dealing with "aspects" of media data and 2) evaluating "relationships" among multimedia data. We have implemented the system architecture and clarified feasibility and effectiveness of the system through the experimental study from the viewpoint of the system performance. As the future work, we will develop actual applications related to mobile computing and P2P network environments.

Acknowledgments

I wish to express my gratitude to Ms. Asako Ijichi (Graduate School of Media and Governance, Keio University) for valuable discussions and helpful criticisms on this study.

References

[1] A. Bouguettaya, B. Benatallah and A. Elgermid, "An Overview of Multidatabase Systems: Past and Present", In A. Elgermid, M. Rukinkiewiez, and A. Sheth (Ed.), *Management of Heterogeneous and Autonomous Database Systems*, Morgan Kaufmann, 1998, 1–27.

[2] A. Ijichi, Y. Kiyoki, "An Automatic Metadata Generation Method for Music Retrieval-by-Impression Dealing with Impression-Transition",*Proc. the 7th IASTED International Conference on Internet and Multimedia Systems and Applications*, USA, August, 2003.

[3] A. Koschel, R. Kramer, G. V. Bultzingsloewen, T. Bleibel, P. Krumlinde, S. Schmuck and C. Weinand, "Configurable Active Functionality for CORBA", *11th ECOOP'97 Workshop #7: CORBA: Implementation, Use, and Evaluation.* Jyvaskyla, Finland, June 10th, 1997.

[4] Ji-Rong Wen, Qing Li, Wei-Ying Ma and Hong-Jiang Zhang, "A multi-paradigm querying approach for a generic multimedia database management system", *ACM SIGMOD Record*, Volume 32, Issue 1, March 2003, pp.26–34.

[5] K. C. Kim, S. B. Yoo, K. W. Ko and S. K. Cha, "Active System for Heterogeneous ODBMS Using Mobile Rule Codes", *ADBIS-DASFAA 2000*, Prague, Czech Republic, September 5-8, 2000, 93–106.

[6] K. R. Dittrich, S. Gatziu and A. Geppert, "The Active Database Management System Manifesto: A Rulebase of ADBMS Features", *Proc. 2nd International Workshop on Rules in Databases (RIDS)*, Athens, September 1995, 101–115. Springer-Verlag LNCS-985.

[7] M. W. Bright, A. R. Hurson and S. Pakzad, "A Taxonomy and Current Issues in Multidatabase Systems", *Computer*, Vol. 25, No. 3, 1992, 50–60.

[8] N. Ishibashi, H. Hosokawa and Y. Kiyoki, "A Spatial and Temporal Data Integration Method for Heterogeneous Database Environments", *Proceedings of the IASTED International Conference on APPLIED INFORMATICS Symposium 2. Networks, Parallel and Distributed Processing, and Applications*, 2001, 323–330.

[9] N. W. Paton, and O. Díaz, "Active Database System", *ACM Computing Surveys*, Vol. 31, No. 1, 1999, 63–103.

[10] S. Kurabayashi, N. Ishibashi and Y. Kiyoki, "A Multidatabase System Architecture for Integrating Heterogeneous Databases with Meta-Level Active Rule Primitives", *Proceedings of the 20th IASTED International Conference on Applied Informatics*, 2002, 378–387.

[11] S. Kurabayashi and Y. Kiyoki, "A Meta-Level Active Multidatabase System Architecture for Heterogeneous Information Resources", Information Modelling and Knowledge Bases (IOS Press), Vol. 15, June 2003.

[12] T. Kitagawa and Y. Kiyoki, "The mathematical model of meaning and its application to multidatabase systems", *Proc. 3rd IEEE Int. Workshop on Research Issues on Data Engineering: Interoperability in Multidatabase Systems*, 1993, 130–135.

[13] T. Kotani, Y. Kiyoki, T. Kitagawa, "An automatic extraction method for impression words' from images, and its application to the mathematical model of meaning", *Proc. the 9th IEICE Data Engineering Workshop*, 1998, (in Japanese).

[14] Y. Kiyoki, A. Miyagawa and T. Kitagawa, "A multiple view mechanism with semantic learning for multidatabase environments," Information Modelling and Knowledge Bases (IOS Press), Vol. IX, May, 1997.

[15] Y. Kiyoki and T. Kitagawa, "Application of a Semantic Associative Search Method to Multidatabases for Environmental Information," Information Modelling and Knowledge Bases (IOS Press), Vol. XI, May, 1999.

[16] Y. Kiyoki, T. Kitagawa and T. Hayama, "A metadatabase system for semantic image search by a mathematical model of meaning", *ACM SIGMOD Record*, Vol.23, No.4, pp.34–41, Dec. 1994.

[17] Y. Kiyoki, T. Kitagawa and Y. Hitomi, "A fundamental framework for realizing semantic interoperability in a multidatabase environment", *Journal of Integrated Computer-Aided Engineering*, Vol. 2, No. 1(Special Issue on Multidatabase and Interoperable Systems), John Wiley & Sons, Jan. 1995, 3–20.

Information Modelling and Knowledge Bases XVI
Y. Kiyoki et al. (Eds.)
IOS Press, 2005

Web Applications' Reliability Problem Revisited

Kenji Wakabayashi, Tetsuya Suzuki, and Takehiro Tokuda
{waka, tetsuya, tokuda}@tt.cs.titech.ac.jp
Dept. of Computer Science,
Tokyo Institute of Technology,
Meguro, Tokyo 152-8552, Japan

Abstract. The dubious submission problem is a problem of Web applications in which users are not sure whether the submission of a program execution was actually established or not. In daily life, this means we are not sure about the establishment of the payment of membership fees or the contract of the travel insurance. We present a solution to the dubious submission problem to make Web applications more reliable.

1 Introduction

Web applications are widely used for many critical purposes such as reservation, shopping and banking in daily life. The purpose of this paper is to point out that the unreliable nature of current Web applications based on http (and also https). We present a problem called dubious submission problem to show that users are not sure whether the submission of a program execution was established or not. Then we give a solution to the dubious submission problem based on a new approach called acknowledged http. In daily life, the dubious submission means the double payment of the membership fees, even if the user believes that the payment was done once normally.

Related but slightly different approaches to reliable Web applications are Applet approach proposed by M. C. Little et al. [10], Simple HTTP extension approach proposed in RFC 2310 [11], and HTTP payload approach used in HTTPR proposed by IBM [12].

The organization of the rest of this paper is as follows. In Section 2 we explain the dubious submission problem of current Web applications based on http. In section 3 we explain the current http mechanism and give the reason why the dubious submission problem may take place. In section 4 we propose a new framework of the acknowledged http. In section 5 we give a solution to the dubious submission problem. In section 6 we give concluding remarks.

2 Dubious Submission Problem

We present a problem called dubious submission problem. First let us describe a normal submission of Web applications based on http.

[Normal submission case]
A user tried to pay the membership fee by clicking the submit button of the Web page. Then the user received the normal resulting page.

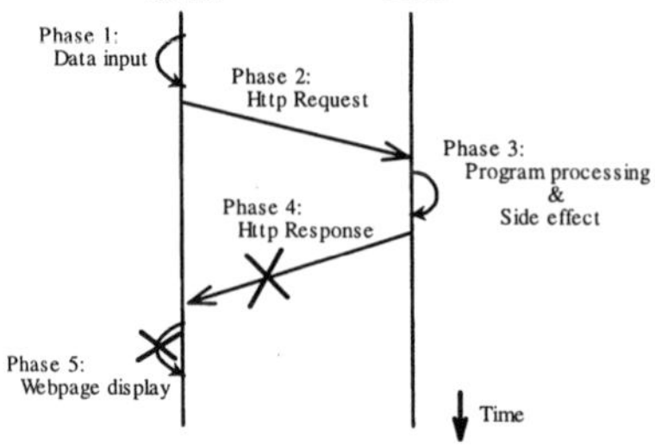

Fig. 1(a) Dubious Submission Problem

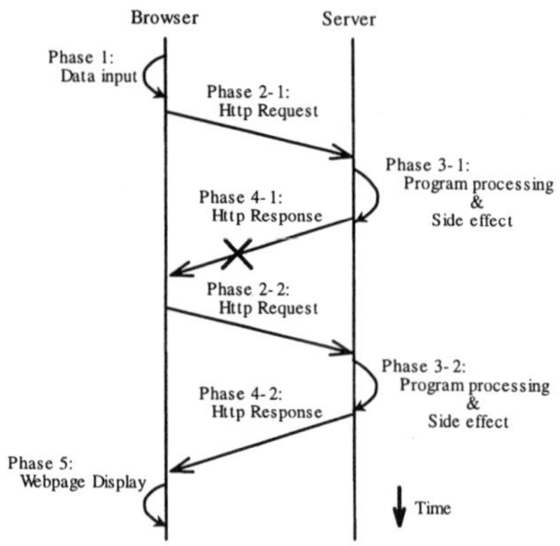

Fig. 1(b) Dubious Submission Problem

In this case the user may safely think that the membership fee payment was done normally. In the following cases, we are not sure the payment was done or not.

[Dubious submission problem Case 1] (submit and cancel)
A user tried to pay the membership fee by clicking the submit button of the Web page. After waiting for some amount of time, the user became uncertain and clicked the cancel button of the browser. The user received no resulting page (Fig.1(a)).

In this case 1, the user might think that the payment was canceled safely. However there exist following two cases

1) The membership fee payment was canceled safely.

2) The membership fee payment was done once despite that the user did not receive the normal resulting page.

[Dubious submission problem Case 2] (submit and submit)
A user tried to pay the membership fee by clicking the submit button of the Web page. After waiting for some amount of time, the user became uncertain and clicked the submit button of the Web page again. Then the user received the normal resulting page (Fig.1(b)).

In this case 2 the user might think that the payment was done once safely. However there exist following two cases

1) The membership fee payment was done once safely.

2) The membership fee payment was done twice despite that the user received the resulting page once.

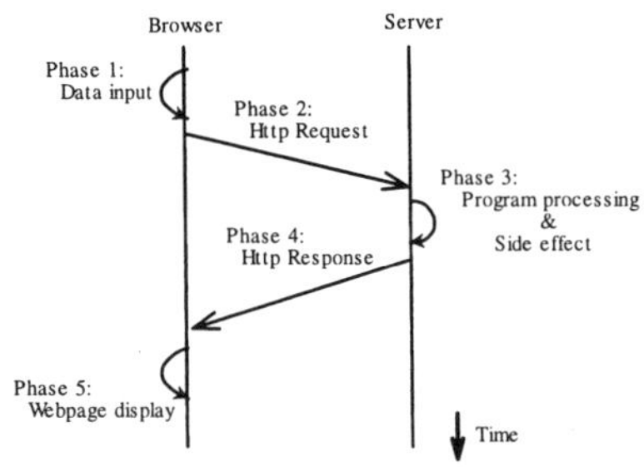

Fig. 2 Execution of one Web Application

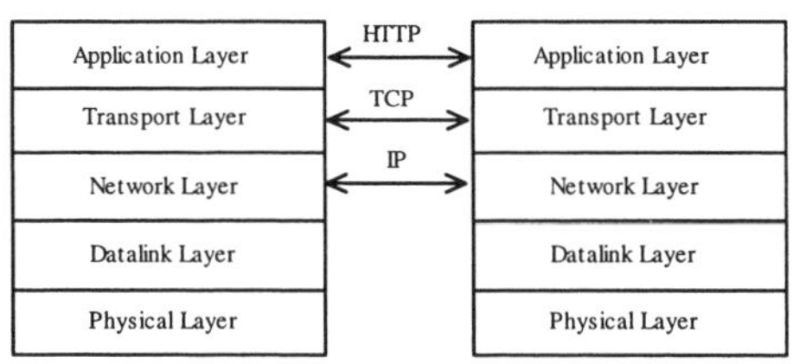

Fig. 3 Ordinary Layered Approach to Web Applications (http)

3 Ordinary Layered Approach

We explain the current approach to Web applications based on http and give the reason why the dubious submission might take place.

The ordinary mechanism of Web application program execution is as follows.

[Ordinary HTTP]

1) Data input is done on the client side.

2) HTTP request is sent from the client side to the server side.

3) Program processing is done for permanent storage on the server side.

4) HTTP response is sent from the server side to the client side.

5) Display of the response on the client side.

The process of Web application execution is shown in Fig.2. The layers of programs supporting Web applications are shown in Fig.3.

The dubious submission is caused by the existence of disconnection operation of the TCP connection between the client side and the server side. There are two types of disconnection operations.

1) Spontaneous disconnection This disconnection is made by the user by clicking the submission button of a Web page or clicking the cancel button of the browser or showing some other page in the browser during the TCP connection.

2) Time-out disconnection This disconnection is made by application programs, such as browsers and Web servers, or TCP programs, because they do not receive the reply within the expected time and they perform time-out operations..

Now we can fully explain what happened to cases 1 and 2 of the dubious submission as follows.

[Case 1 (submit and cancel)]

1) Data input is done on the client side.

2) HTTP request is sent from the client side to the server side.

3) Program processing is started for permanent storage on the server side. Spontaneous disconnection takes place by clicking the cancel button.

4) HTTP response cannot be sent, because there exists no connection.

5) Display of the response cannot take place on the client side. because there exists no connection.

[Case 2 (submit and submit)]

1) Data input is done on the client side.

2) HTTP request is sent from the client side to the server side.

3) Program processing is started for permanent storage on the server side. Spontaneous disconnection takes place by clicking the submit button again. No further steps are performed for the first connection.

4) HTTP request is sent again from the client side to the server side.

5) Program processing is done for permanent storage on the server side.

6) HTTP response is sent from the server side to the client side.

7) Display of the response is done on the client side.

4 Acknowledged layered Approach

We show our acknowledged layered approach to Web applications. First we need to state our goal as follows.

[Our Goal]
Our executions are either (Side-effect is established on the server side and normal resulting

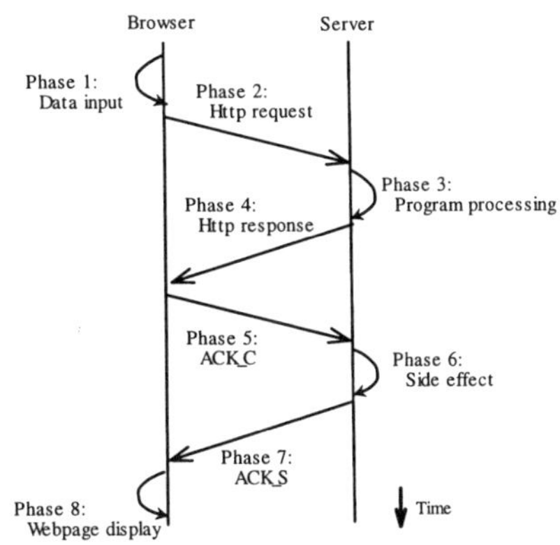

Fig. 5 Acknowledged layered Approach to Web Applications

page is displayed on the client side) or (side-effect is not established on the server side and normal resulting page is not displayed on the client side).

Here is our acknowledged HTTP to achieve this goal.

[Acknowledged HTTP]

1) Data input is done on the client side.

2) HTTP request is sent from the client side to the server side.

3) Program processing is done for non-permanent storage on the server side.

4) HTTP response is sent from the server side to the client side.

5) ACK_C is sent from the client side to the server side.

6) Side-effect is established by copying the results from non-permanent storage to permanent storage.

7) ACK_S is sent from the server side to the client side.

8) Display of the response is done on the client side.

5 A Solution to Dubious Submission Problem

Now we present a solution to the dubious submission problem.

[Solution to Dubious Submission Problem]

1) Data input is done on the client side.

2) HTTP request is sent from the client side to the server side.

3) Program processing is done for non-permanent storage on the server side.

4) If the TCP connection is already disconnected, then no further step is performed. If not, HTTP response is sent from the server side to the client side.

5) ACK_C is sent from the client side to the server side. If ACK_C is not received within the expected time, then HTTP response from the server side to the client side is sent again.

6) Side-effect is established by copying the result from non-permanent storage to permanent storage.

7) ACK_S is sent from the server side to the client side. If ACK_S is not received within the expected time, then ACK_C from the client side to the server side is sent again.

8) Display of the response is done on the client side.

Fig.7 shows a part of the source code of acknowledged http handling ACK_C from the client side and the disconnection detection. Now we can show what will happen to case 1 and case 2 of the dubious submission problem respectively in Fig.6(a) and Fig.6(b), if we use our solution. No payment takes place in case 1 and exactly one payment takes place in case 2.

6 Conclusion

We have presented our acknowledged layered approach to web applications. By slightly modifying the current processes of http, browsers, and Web servers, we can give a solution to the dubious submission problem. This will be the great improvement of reliability of Web applications.

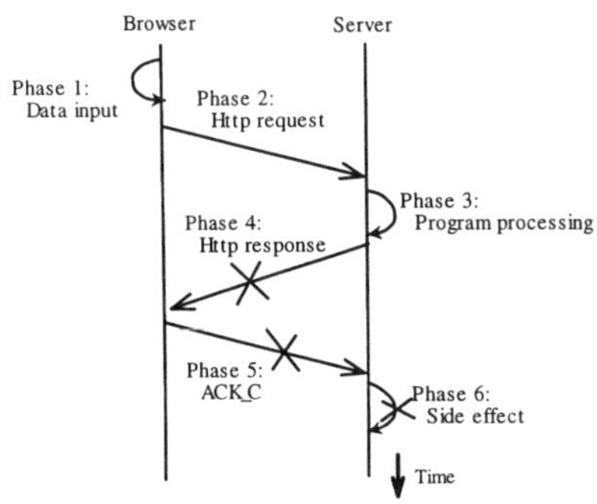

Fig. 6(a) A Solution to the dubious submission Problem

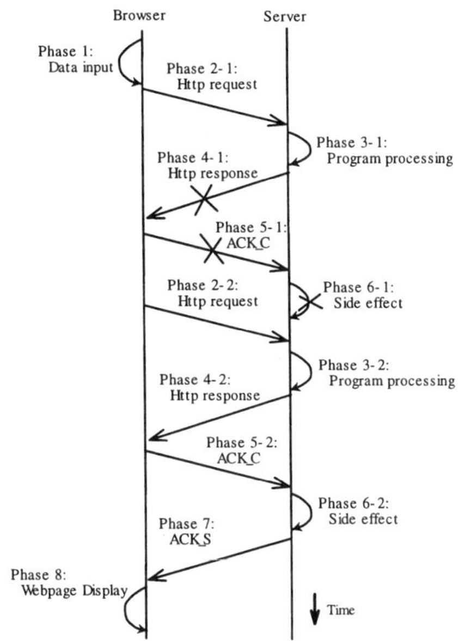

Fig. 6(b) A Solution to the dubious submission Problem

```
byte[] buff = new byte[BUFF_SIZE];
int length;
try{
    length = readAck(buff); //ACK_C Reading
    if(length == -1){
        //Transaction Rollback
        programInstance.doRollback(requestManager);
        return;
    }
}
catch(SocketException se){
    //Transaction Rollback
    programInstance.doRollback(requestManager);
    return;
}

//Transaction Commit
programInstance.doCommit(reqeustManager);
```

Fig. 7 A part of the source code of the solution

References

[1] HTTP - Hypertext Transfer Protocol, http://www.w3.org/Protocols/ .

[2] R. Fielding, et al., HTTP/1.1, http://www.ietf.org/rfc/rfc2616.txt, 1999.

[3] TRANSMISSION CONTROL PROTOCOL, http://www.ietf.org/rfc/rfc793.txt , 1981.

[4] W. Richard Stevens, TCP/IP Illustrated, Volume 1: The Protocols, Addison-Wesley, 1994, ISBN 0-201-63346-9.

[5] W. Richard Stevens, TCP/IP Illustrated, Volume 3: TCP for Transactions, HTTP, NNTP, and the UNIX Domain Protocols, Addison-Wesley, 1996, ISBN 0-201-63495-3.

[6] W. Richard Stevens, UNIX Network Programming, Volume 1, Second Edition: Networking APIs: Sockets and XTI, Prentice Hall, 1998, ISBN 0-13-490012-X.

[7] Elliotte Rusty Harold, Java Network Programming, 2nd Edition, O'Reilly, 2000, ISBN 1-565-92870-9.

[8] Andrew Tanenbaum, Computer Networks, 3rd Edition, Prentice Hall, 1996, ISBN 0-13-349945-6.

[9] Philip A. Bernstein and Eric Newcomer, Principles of Transaction Processing, Morgan Kaufmann Publishers, 1997, ISBN 1-55860-415-4.

[10] M.C. Little, S.K. Shrivastava, S.J. Caughey and D.B. Ingham, Constructing Reliable Web Applications using Atomic Actions, Computer Networks and ISDN Systems, Vol. 29, 8-13, 1997.

[11] RFC 2310, The Safe Response Header Field, http://www.faqs.org/rfcs/rfc2310.html.

[12] IBM, HTTPR Specification, Draft Proposal, http://www-106.ibm.com/developerworks/webservices/library/ws-phtt/httprspecV2.pdf.

Agile Process Modeling in a Process Engineering Environment

Lachana INCHAIWONG and A. S. M. SAJEEV
University of New England, Armidale. Australia
[lachana | sajeev]@turing.une.edu.au

Abstract. Software process engineering environments (SPEEs) are expected to be flexible enough to cope with changes occurring in modern software development and provide more automated support to software process activities. ProceeD (Process engineering and Design) is an object-oriented environment to support modeling and design of processes, in particular, agile processes. ProceeD, using graphical notations and a metaphor of factory production, represents software process activities in different levels of abstraction, thus, making it easy to design software processes. It facilitates semi-automation of process design by providing a library of templates and adaptable process elements.

1. Introduction

Considerable research has gone into automated support for software design and development. CASE (Computer Aided Software Engineering) tools supporting various design methods including modern ones such as UML (Unified Modeling Language) provide software engineers with support for designing and analyzing software systems. In comparison, tool support for design, analysis and management of software *processes* is much less. This is in spite of the fact that a well defined software process is critical in developing reliable software within budget and on time[1, 2].

Traditionally, the Waterfall model [2-4] was the common process used in large projects. However, it did not take into consideration the wide variation in the nature of software projects including, size, duration, manpower and stability of requirements.

Agile methodologies are the new frontier in software processes [5, 6]. Many software projects turned to agile software development as traditional development methodologies failed to satisfy their requirements. Most of the agile methodologies involve minimally defined processes, frequent tests and integration, anticipation of changes and less documentation. Having fewer phases and minimally defined processes do not mean that the task of defining software processes in agile software development environments can be overlooked.

The availability of different process methodologies help software designers choose and adapt software processes according to organizational and project specific needs. In order to do this efficiently, they need tool support that will automate routine tasks of process selection and adaptation and help them design processes and analyze them.

In this paper, we describe a Process Engineering Environment called ProceeD (Process Engineering and Design) that enables developers to create, manage and reuse software process models, in particular, in an agile environment. ProceeD also facilitates analysis of software processes, however, in this paper we concentrate more on the design and reuse facilities based on a modeling language called *Order* [7]. Since Agile Processes require customer involvement in the software development process, a key characteristic of ProceeD is a visual process modeling language that uses a familiar factory metaphor.

Section 2 of the paper outlines the role of Process Centered Software Engineering Environments in process improvement. Section 3 describes the architecture of ProceeD. Section 4 outlines support for automation in process modeling. Section 5 describes the associated process specification language. How process design can be automated in ProceeD is described in Section 6, and Section 7 concludes the paper.

2. CMM and PSEE in Process Improvement

Process improvement is a key element of the Capability Maturity Model (CMM) in assessing the maturity of a software organization. Process-centered Software Engineering Environments (PSEEs) can help organizations improve software processes[1, 2]. In this paper, we focus on PSEEs that are built to support software process engineering tasks. We refer to this kind of PSEEs as Software Process Engineering Environments (SPEEs).

E^3 [8], for instance, provides a visual environment for object oriented process modeling. Its design aims at model understanding, model structure and reuse. SPADE [9] is another SPEE which uses SLANG, an extension of Petri nets, as a process modeling language to create, enact and evolve process models. Funsoft [10] nets are used to describe software processes in a PSEE called MELMAC, which provides animation, simulation and analysis of process models. SPADE and Funsoft nets use mathematical models to represent software processes whereas E^3 uses simpler graphical notations. Rigorous notations are suitable for software process analysis tasks. However, they are less intuitive to use, and require considerable training. ProceeD, on the other hand, uses a factory metaphor to make process design simpler and intuitive.

Spearmint [11] is a modeling language and a graphical tool developed to support the descriptive modeling of large, complex, real-world processes. It also supports the generation of web-based Electronic Process Guides for process guidance. Compared to ProceeD, Spearmint does not have a Process Modeling Language as such, however, it is possible to model processes using product flow graphs.

MILOS [12] is a process-support environment aimed to assist Agile production processes, however, the emphasis is on providing collaboration and coordination technology in a distributed environment. Similarly, Endeavors [13] is an open, distributed, extensible process execution environment that provides integration of tools, customization and reuse, dynamic changes and distribution. The emphasis once again is on distributed development.

SPEEs may help improve software processes by providing functionalities corresponding to the Key Process Areas (KPAs) in CMM [2]. SPEEs can provide support to most KPAs required in higher levels of organizational maturity. The differences in characteristics of both organizations and projects mean we need different kinds of software processes. Whereas waterfall model is suitable for large scale projects where requirements seldom change, more often software companies work on products that require flexibility and adaptability in software processes; agile processes provide one way to meet this need. Kelther, Monecke and Schild [14] identify the important aspects in developing software tools to support challenges in Agile software development, i.e., communication, coordination, cooperation, planning, development, documentation, configuration management and process automation.

Most PSEEs support process modeling in the large. In contrast, agile processes support the use of smaller artifacts and should be provided with tools that are simple to use and are flexible [15].

3. The Architecture of ProceeD

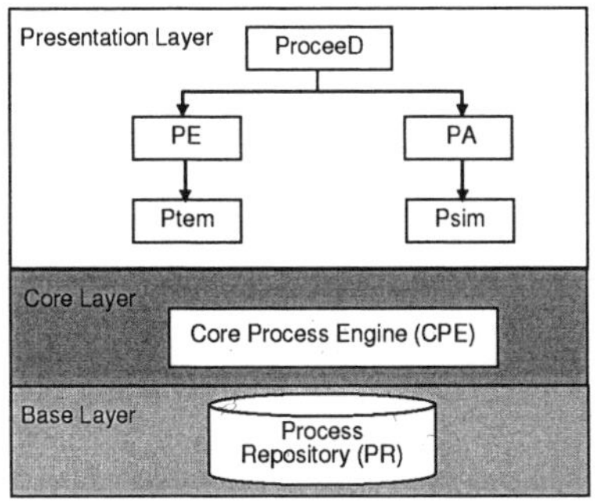

Fig. 1. Main components in ProceeD

Figure 1 illustrates the architecture of ProceeD which consists of a Process Editor (PE) module and a Process Analyzer module (PA) at the top layer. Process Editor provides templates, thus, making it easier for developers to customize software process models. Process Simulation Manager helps developers to verify consistency and check behavior of their designs. Process repository is the central database designed to keep all process artifacts that can be accessed for modification or adaptation in later stages, thus, facilitating process evolution. At the core layer, Core Process Engine (CPE) deals with requests coming from presentation layer, and control accesses to Process Repository (PR) at base layer.

4. Modeling Support

ProceeD provides semi-automated support for process engineers at three levels: *process modeling, analysis* and *management.*

4.1 Process Modeling

One of the advantages of modern processes such as agile methodologies is the flexibility they provide. Processes can be adapted to meet organization needs as well as project needs. Thus, an organization may choose a process method (e.g. Extreme Programming) as the organization level process, but may make changes to it to fit particular projects. For instance, if the organization is embarking on a project that is safety critical, the emphasis on verification and validation activities needs to be much higher.

ProceeD provides a process specification language called *Order* to specify processes and a library to store process specs. The environment allows retrieval of a process specification from the library for reuse and adaptation.

4.2 Process Analysis

While modeling or adapting a process, an engineer needs to make sure that the specification meets the process requirements [1, 2, 16]. The Analysis tool can automatically generate process flow diagrams indicating any bottlenecks. It can also help answer various what-if questions on the design.

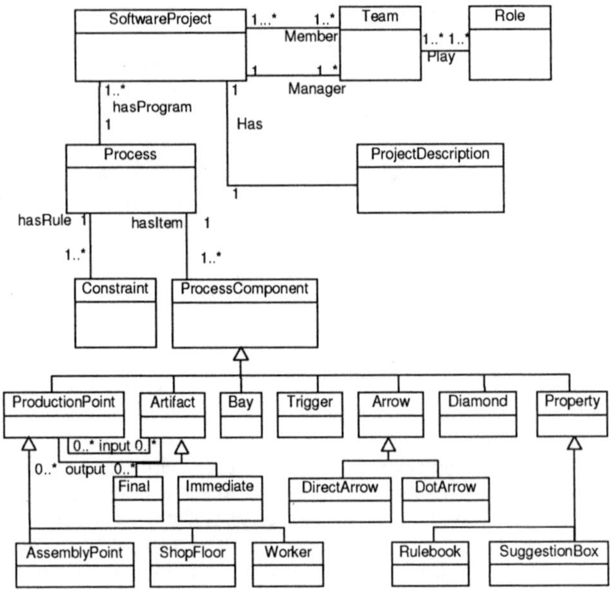

Fig. 2. UML model of components used in Order

4.3 Process Management

Once a process is designed, analyzed and put into operation, engineers need to track the progress of the project. The Process Management tool of ProceeD allows process-data (e.g. dates, completed phases) to be entered and the process to be managed. The management tool can generate reports on the current status of the project, including any warning signs, as well as, steps to be taken for the preparation of the coming activities.

In this paper, we concentrate on ProceeD's support for processing modeling, and therefore do not discuss analysis and management tools further.

5. Order Specification Language

The presentation layer of Figure 1 provides support for process modeling using a graphical language called *Order* [7]. It uses the metaphor of production line in a factory. The metaphor is that a software project is a factory that produces software and associated artifacts as products. Several kinds of production points in *Order* represent software development activities that result in products—intermediate or final. Each production point has a *rulebook* for specifying rules and constraints that the point should obey, and a *suggestion box* to provide guidelines; as the name implies, suggestions are not mandatory on a production point. Each artifact used in a software factory has a property page listing its attributes. *Order* provides three kinds of process control—sequential (arrow), selection (diamond) and fire (trigger). In this section, we describe *Order* using UML and its associated Object Constraint Language (OCL) [2, 17]. Figure 2 shows the UML model of *Order*.

Associated with a software project in *Order* are a project description and a process. A software project consists of teams that can be specialized into different teams depending on the process chosen. For example: extreme programming team (xpTeam), scrum team (scrumTeam) etc. Each process contains constraints specific to the process and several process components defining the process (See Figure 2).

The rule books and suggestion boxes contain constraints and rules expressed in natural or formal language. For formal specification, *Order* uses OCL [17]. OCL is the formal language associated with UML. For example, in OCL, we can specify that a ProductionPoint can start production when at least one of its inputs is ready as follows.

```
Context ProductionPoint::start():integer
pre: self.inputSet -> (exists i : i.ready)
```

[Read this as: a precondition of starting a production point is that there exists an input i among the set of inputs for the Production Point, such that i.ready is true.]

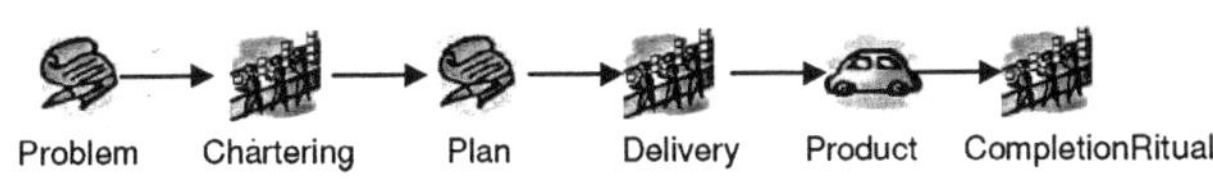

Fig. 3. High- level view of Crystal Clear process

6. Support for Process Modeling

ProceeD provides a visual environment and graphical notations to simplify process modeling for process engineers and non-technical persons. ProceeD also provides predefined templates of process models. Thus, designers can create new process models in less time. It applies the object oriented paradigm to allow process designers to treat process models in an efficient way. Process Modeling happens in three ways – decomposition, specialization and aggregation. It's also possible to use a combination of the three.

6.1 Process Decomposition

Process decomposition is when a process is defined at the highest level first, and then decomposed to include further details. For example, Figure 3 shows the Crystal Clear [5] process at the highest level. Icons used in Order come from its factory metaphor.

Crystal Clear is a part of the Agile alliance which includes Extreme Programming, Scrum, DSDM, ASD, etc [6]. It is the lightest methodology in the Crystal family designed for small teams. The methodology encourages all projects to have three properties – Frequent delivery, Close communication and Reflective Improvement. A project in Crystal Clear has three parts – a chartering activity, a series of one or more delivery cycles and a completion ritual. The model shown in Figure 3 can be decomposed to provide further details as shown in Figure 4. The *Chatering* phase in Crystal Clear is a single cycle process (indicated by the rectangle) with an input bay, I1 and an output bay, I2. I1 is where the *Problem* (collection of software requirements) comes into the *Chatering* phase and I2 is where the resulting *decision* - continue the project with plans or cancel the project - goes out of the *Chatering* phase. We can further decompose each activity, to include further details. The decomposition continues until all aspects of the process are captured in the model. Each process element is stored separately in the ProceeD library.

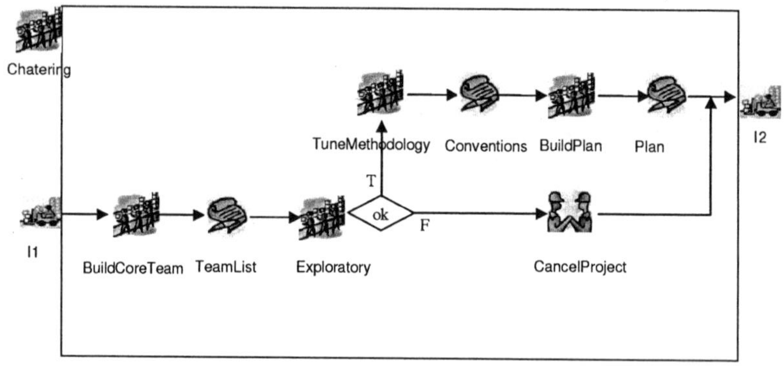

Fig. 4. Crystal Clear decomposition: Chatering process

6.2 Processing Modeling by Specialization

Given a process element, specialization allows the creation of another element which has some properties common with the original element. This provides an easy path to adapt existing process elements to new situations. The Object-Oriented Method uses the principle of inheritance to model specialization.

In ProceeD, we can use inheritance to create new process elements, say P', from existing ones, say P. We expect process engineers to follow the accepted principles of specialization in using inheritance.

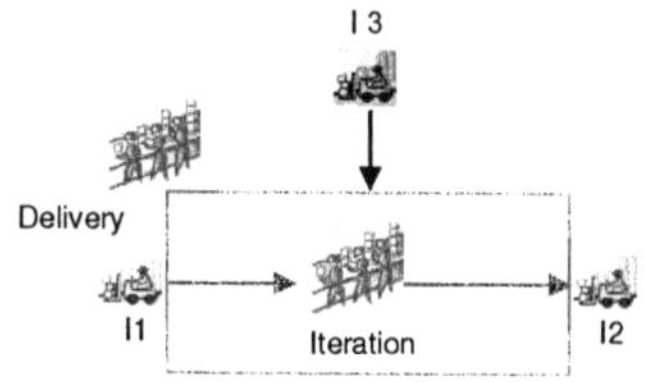

Fig. 5. Process specialization

As an example, in Figure 3, we have a process element modeling the *Delivery* phase of Crystal Clear. If we want to use Crystal Clear in a component-based software engineering environment, the process may need to be extended to include another input bay, say I3, through which the production point can access a component library. Instead of having to design a new production point from scratch, we can specialize the *Delivery* process as shown in Figure 5. The steps to do that are:

* Retrieve the *Delivery* production point from library
* Add a new bay to the production point (Figure 5)
* Specify constraints and guidelines in rules and suggestion boxes
* Store the specialization as a new process element in the library

6.3 Processing Modeling by Aggregation

Aggregation is the reverse of decomposition. In the case of aggregation, a process engineer designs process elements P1, P2, P3 and use them in the modeling of a process P0. Aggregation is useful especially in cases where a process reuses elements from previously modeled processes. The modeling steps are:

* Search library for process elements required
* Retrieve the required process elements
* Design process elements that are not available in the library
* Interconnect them using connection symbols of Order

As an example, there are several similarities between different agile methodologies. If an organization wants to move from one agile process (say, Crystal Clear [5]) to another (say, XP [18]), several of the process elements, such as unit tests can be reused instead of having to design all elements from scratch.

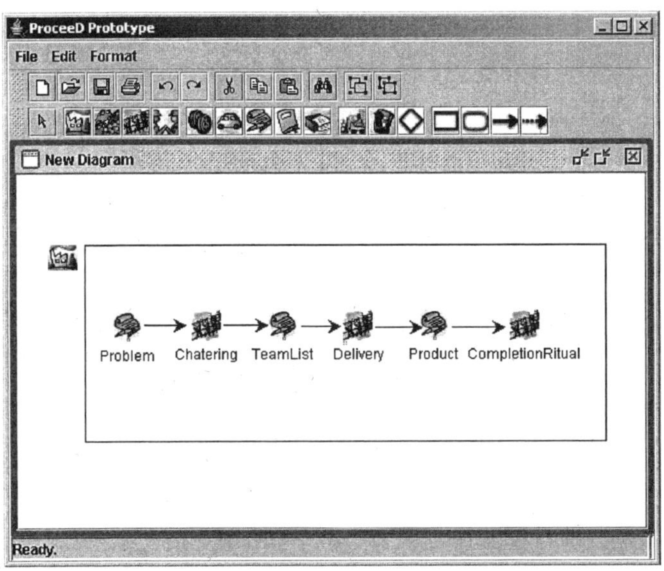

Fig. 6. Process Modeling Interface

7. Conclusions

Automated support for process design has not captured as much attention as support for other aspects of software engineering such as software design. However, process improvement is an important element of the maturity of an organization as indicated by the Capability Maturity Model. Besides, flexible processes such as Agile methodologies are becoming more common. Process engineering environments for such methodologies should provide simpler methods to understand and design processes since customer involvement in software development is a key characteristic of such processes. SPEEs should also provide easy mechanisms to change and design new processes based on existing ones.

In this paper, we described a process engineering environment called ProceeD which achieves these aims by (a) providing a visual modeling environment (b) using a commonly understood metaphor of factory production for processing modeling and (c) using principles of object-orientation for process composition. A prototype of ProceeD is currently being developed. Figure 6 shows the interface for process modeling.

References

1. Ellmer, E. *Improving Software Process.* in *Proceedings of Software Engineering Environments Conference.* 1995.
2. Finkelstein, A., J. Kramer, and B. Nuseibeh, eds. *Software Process Modeling and Technology.* Advanced Software Development Series, Research Studies Press/John Wiley & Sons. 1994. 279-292.
3. Boehm, B.W., *Software Engineering Economics.* Prentice-Hall: Englewood Cliffs, NJ. 1981.
4. Royce, W.W., *Managing the Development of Large Software Systems: Concepts and Techniques.* Proceeding 1970 WESCON Technical Papers, Western Electronic Show and Convention, 1970. 14.
5. Cockburn, A., *Agile Software Development.* The Agile Software Development Series, ed. A. Cockburn and J. Highsmith. 2002: Addison-Wesley.
6. Highsmith, J. and A. Cockburn, *Agile Software Development: The Business of Innovation,* in *IEEE Computer.* 2001. p. 120-122.
7. Sajeev, A.S.M. and L. Inchaiwong. *Step-wise Decomposition of Software Processes.* in *Proceeding 6th IASTED International Conference on Software Engineerng and Applications.* 2002.
8. Baldi, M., et al., *E3: Object Oriented Software Process Model Design in Software Process Modeling Technology.* Advanced Software Development Series, Research Studies Press/John Wiley & Sons, 1994: p. 279-292.
9. Bandinelli, S., et al., *SPADE: An Environment for Software Process Analysis, Design and Enactment in Software Process Modeling and Technology.* Advanced Software Development Series, Research Studies Press/John Wiley & Sons, 1994: p. 223-247.
10. Deiters, W. and V. Gruhn, *The FUNSOFT net approach to software process management.* Journal of Software Engineering and Knowledge Engineering, 1994. 4(2): p. 229-256.
11. Louise, S., et al. *An Evaluation of the Spearmint Approach to Software Process Modelling.* in *Software Process Technology (EWSPT 2001).* 2001.
12. Bowen, S. and F. Maurer, *Process Support and Knowledge Management for Virtual Teams Doing Agile Software Development.* 2002.
13. Bolcer, G. and R. Taylor. *Endeavors: a process system integration infrastructure.* in *Proceedings of the 4th International Conference on the Software Process.* 1996: IEEE Computer Society Press.
14. Kelter, U., M. Monecke, and M. Schild, *Do We Need 'Agile' Software Development Tools in Objects, Components, Architectures, Services, and Applications for a Networked World.* Springer Lecture Notes in Computer Science 2591, 2002: p. 412.
15. Hansen, K.M. *Agile Environments - Some Patterns for Agile Software Development Facilitation.* in *Proceedings of VikingPLoP:Pattern Languages of Programs 2002.* 2002. Denmark.
16. Deiters, W. and V. Gruhn, *Software Process Model Analysis based on FUNSOFT Nets.* Mathematical Modeling and Simulation, 1991. 8.
17. Object Management Group, *OMG Unified Modeling Language Specification.* 2003.
18. Beck, K., *Extreme Programming Explained: Embrace Change.* Addison Wesley. 2000.

Proactive Agents That Support Children's Exploratory Learning

Marjatta KANGASSALO*, Roope RAISAMO**,
Pentti HIETALA**, Janne JÄRVI**, Kari PELTOLA*,
Rami SAARINEN**, Eva TUOMINEN*, Arto HIPPULA**

**Department of Teacher Education, Early Childhood Education, FIN-33014 University of Tampere, Finland*
***Department of Computer and Information Sciences, TAUCHI, FIN-33014 University of Tampere, Finland*

Abstract. In this research project[1] we investigate, model and implement intelligent proactive agents to support the concept learning process of visually impaired children and children with normal eyesight. Our approach to achieve this goal is to utilise computer-based learning and playing, which would provide a shared and networked multimodal environment. Each agent models and scaffolds a child with respect to his/her own personal traits and capabilities. This system supports collaboration in pairs and in groups that consist of visually impaired and seeing children. The system provides children with possibilities for common social interaction and activities. The proactive assistance facilitates children's building of mutual confidence and self-efficacy in learning. Collaboration in pairs and groups consisting of visually impaired and seeing children thus becomes very important and also has social significance.

1. Introduction

Computers enter the lives of our children very early. In addition to the joy of discovery and play, computers offer at their best opportunities for learning important skills. These benefits concern all children, and especially disabled children. Computer programmes have been utilised for the rehabilitation of children with mental retardation for a long time. Pedagogical computer play contributes to the development of self-efficacy in learning and conceptual

[1] The research consortium (grants 202179 and 202180) is funded by the Academy of Finland as a part of the Academy's research programme on Proactive Computing (PROACT) for the period of 2002-2005.

thinking. New computer hardware and software technologies enable new and effective ways for exploratory individual and collaborative learning. The central task of the present project is to create pedagogically significant operational opportunities to bridge the gap between children with different potentials in pre-school and primary education and to specifically focus on visually impaired children in integrated pre-school groups. This can be accomplished by utilising the technology of proactive agents to support each child at an optimal level with respect to his/her personal profile as well as by applying multimodal interface technology, especially haptic interaction.

The domain of the environment is nature and the physical environment of the children. Our design and development will make use of an earlier prototype called PICCO [1, 2], which was funded by the Academy of Finland and Tekes. The PICCO programme is a pictorial simulation that has been extensively used in day-care centres and schools. The agent architecture will also benefit from our experience gained in another project, the EduAgents collaborative agent-based learning environment [3, 4, 22]. This prototype emphasised domain-independent architecture for the use of several agents, including both teacher agents and learning companion agents. As regards to the related work in the area of pedagogical agents, we can mention the EU funded NIMIS project: Networked Interactive Media in Schools [5, 6]. This is generic intelligent agent architecture for early literacy and story creation for primary school level.

The research on self-efficacy in learning is based on our earlier research findings on self-directed learning readiness. This phenomenon has been rigorously investigated by L.M. Guglielmino [7] in her doctoral dissertation. She also created a measure called SDLRS (self-directed learning readiness scale), which has been used in our research. Our earlier results imply that the phenomenon of self-directed learning readiness can be divided into several different factors. One of them can be called the factor of self-efficacy in learning [8], which will also be used in this new study.

Experience with children with special needs has been acquired in an ongoing project in which an innovative multimodal interaction environment is being designed for visually impaired children. These children would benefit from the agent environment because of the possibility to produce information that is missing due to sensory disability.

2. Theoretical Background and Aims of the Research

The objective of our project is to research intelligent agents in the context of pre-school and elementary school children. For this purpose, we are producing a proactive and multimodal agent-based system that would support children's cognitive development both in learning and play. The system should also be useful for children with special needs either alone, or preferably, in a group with other children. Our aim is to construct the same interface for both normal and visually impaired children, which would enable equal participation in joint collaborative tasks. The specialisation will be implemented using proactive agents that personalise the use of the system.

The agent-based environment predicts and provides dedicated support starting from the child's own standpoint, needs, interests and the problem situation. The support aims to raise the child's thinking, problem-solving and advancing via different kinds of activation: questions, explanations, and other feedback (e.g., through touch and voice). The non-visual modalities would be essential in respect to the needs of visually impaired children. Various elements of excitement, activation and captivation are also utilised. The actions and interface of the agent system are constructed to support exploratory learning and development, as well as usability and accessibility for children with special needs.

2.1. Children's exploratory and collaborative learning and conceptual thinking

One part of our research project is to investigate children's conceptual learning and thinking in the social context of a proactive intelligent agent system. The theoretical and methodological foundations are laid by theories and a set of concepts derived from cognitive and social psychology and cognitive science. The research will give an important approach integrated with the earlier and ongoing research in this subject area by Kangassalo and her cooperative projects (e.g., [9, 2, 10]). The skill of thinking and analysing the complex phenomena of life is extremely valuable for man. Thinking skills, as well as individual and collaborative exploratory activities, have been raised to a significant position in today's rapidly changing technological environments and will have versatile future applications as well.

This research focuses on the development of the conceptual learning and thinking of 5- to 10-year-old children regarding natural phenomena. It examines children's reciprocal interaction as well as the potential of the learning environment in such activity environments where children can use information technology in addition to traditional materials and tools. With regard to the development of children's conceptual thinking concerning natural phenomena, our specific research interests focus on the formation processes of children's conceptual models of natural phenomena, conceptual changes and the construction of knowledge.

In the light of current views on learning, it appears to be crucial that instructional practice supports learners' active participation and engagement in knowledge construction and communal thinking [11]. This can be achieved by creating meaningful and motivating learning activities in which the learning material is presented in authentic and multiple forms. It can also be achieved through learning activities that promote the formulation of representations and explanations, as well as communal sharing and knowledge building, via different mediating tools [23].

In the present collaborative research project, the natural phenomena which will be the subject domain of the proactive intelligent agent system are such astronomical phenomena as the solar system, the Sun and Earth, day and night, and the seasons. The developing proactive intelligent agent system utilises knowledge and experiences that have been collected earlier when developing and using the pictorial computer based simulation programme-PICCO in research projects [1, 12]. When we selected the natural phenomenon for the simulation PICCO programme, it was essential that the chosen phenomenon was important and significant with regard to life. It had to awaken sufficient interest in the children and efficiently utilise the possibilities offered by the computer technology. And finally, it cannot be easily and illustratively presented in any other way. Based on the earlier research results achieved by Kangassalo's projects (e.g., [9, 2, 13]) it is possible to say that children's exploration of the phenomenon using the PICCO programme could assist children in improving the knowledge in keeping with current scientific knowledge of the phenomenon.

Exploratory and collaborative learning is especially important with visually impaired children. These children have sensory restrictions and, therefore, limited accessibility to explorations. Usually it is assumed that they are inclined to learn false or restricted conceptions of the world, which can also lead to social discrimination. Visually impaired children could therefore considerably benefit from sharing learning experiences and the knowledge itself with other children in the group. Hence the means of collaboration development and exploration are of special value for these children as the target group of this project. The proactive and intelligent agent system as well as the special hardware for haptic interaction is thought to add special value for this problem setting.

Our group has experience in research on constructing systems supporting collaboration of two different communities of learners, although in an adult context [14, 15]. We believe both our theoretical and practical insight in this area will benefit the project.

2.2. *Self-efficacy in learning*

According to our studies, the construction of self-efficacy in learning is a factor in the self-directed learning readiness phenomenon [8]. Self-efficacy in learning is a psychological dimension by nature. We have been investigating self-directed learning readiness phenomenon with a method called SDLRS [7]. It is a self-assessment scale for adults that includes 40 different items or statements. Our earlier data consists of adult and pre-adult respondents in Finnish secondary schools and universities. The data was collected in 1996.

The construction of self-efficacy in learning acts as a conceptual model for one's own learning potential. Self-efficacy in learning represents the beliefs of the learner in his/her own learning capacity trust in one's own learning skills and abilities to learn new things in new situations. Our earlier observations have showed that self-efficacy in learning provides a significant explanation for the user profiles of information technology, including the extent of use and human-computer relations such as attitudes and beliefs [8]. Thus, a deep self-efficacy in learning predicts a many-faceted and non-problematic human-computer relation. We have measured these attitudes and beliefs with CAS (Computer Attitude Scale) [16]. Our psychological explanatory models for computer use have been constructed according to earlier investigations related to this theme (e.g. [17]).

It is extremely important to explore and assess this kind of self-efficacy in learning in children, both among normal children and children with special needs. Visual restrictions make it much more difficult for impaired children to understand and learn what matters. Therefore, self-efficacy in learning is an even more decisive factor in the learning development of disabled children than it is for average children. The nature of this kind of research will be closely related to developmental psychology.

2.3. *Proactive intelligent agent system*

Research problems in constructing proactive intelligent agent systems are related both to building the artificial intelligence (AI) component and the user interface. Proactive functionality is implemented in the AI component based on creating student profiles in which information on their learning is stored. Based on this long -term rule -based user logging, the agents initiate proactive actions. They may, for example, adapt the level of teaching or the presentation of the data based on observating the child. The inferencing and learning capabilities of the agents will enable them to produce proactive behaviour, that is, to autonomously suggest interesting or important information at appropriate times.

The actions of the agents will be non-obtrusive so that they do not interfere with the learning process. Since the targeted users of the system have either just started school or are in their first years of school, the most useful way to represent agent feedback is speech synthesis. This is also very suitable for visually impaired children. Naturally, speech is the main agent presentation channel for visually impaired children. If the child so wishes, hyphenated words can also be shown and spoken/spelled for visually impaired users. Haptic presentation is used to support visual or auditive presentation. Tactile or force feedback can suggest recommended choices in the user interface. For the visually impaired, haptics is used to represent the whole interface together with auditory information.

3. Proactive Computer-Based Simulation System

As mentioned earlier the goal of the research project is to develop a computer-based simulation environment for both visually impaired children and children with normal eyesight at pre-school and primary school. Our aim is to support children's conceptual learning and thinking concerning the natural phenomena in question and to create new opportunities for collaborative learning in integrated school groups for children. The system to be constructed utilizes multimodal interface technology, with exploratory learning as the pedagogical approach and proactive agent technology as a tutoring tool. Next, the system is examined with regard to these aspects.

3.1. Structure of the Simulation

The structure of the simulation system is based on the selected natural phenomena. The simulation consists of micro worlds which a child can explore according to his or her interests. The micro worlds are the Earth, the solar system, the Earth and the Sun, atmosphere, and the interior layers of the Earth. There is also a study room where different tasks accomplished in other micro worlds will be gathered to be replayed or redone.

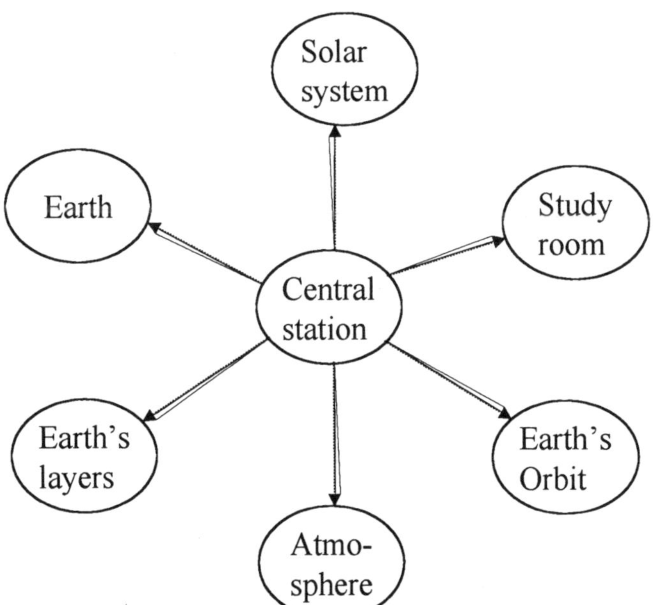

Figure 1. The structure of the simulation

Figure 1 shows the structure of the simulation. The simulation starts from the Central station. From the doors of the Central station, the user can navigate to all the micro worlds in the simulation. When the user wants to navigate from one micro world to another, he/she mainly goes through the Central station. This decreases the chance of getting lost in the simulation because the user is always only one step away from the Central station. The user can return to the Central station by pressing the corresponding Central station button.

In each micro world, the user can spend as much time as he or she wishes, and explore the phenomenon represented there. The micro worlds differ in their structure; some include more narration, and others are based more on the user's own active exploration. Mainly, however, the user's own activity is required. In some micro worlds, the user is provided with opportunities to move from one micro world to other. This is when the phenomena in different micro worlds are naturally connected to each other; for example, when the user is exploring planet Earth in the 'Solar system' -micro world, he or she is offered a possibility to travel with a space rocket to the surface of the Earth (to the 'Earth' –micro world).

3.2. Multimodal Interface Technology

The system that will be constructed utilizes multimodal interface technology. This means that in addition to visual and auditory feedback, the sense of touch is being used as the third feedback channel. For visually impaired children, haptic and auditory feedbacks are the main presentation channels. For children with normal eyesight, on the other hand, the system provides opportunities to receive and use the information obtained through multiple sense modalities.

The simulation is created with a Reachin display (http//:www.reachin.se) consisting of the Phantom (by Sensable technologies, USA; http//:www.sensable.com/) robot arm system and a stereo display with a semi-transparent mirror, Figure 2. The robot arm system produces the haptic interaction to the end point of a stylus, resembling a thick pen, which is attached to the robot arm and can be grasped by the user. The user can feel three-dimensional entities that are defined in VRML scenegraph and simultaneously see a three-dimensional stereo picture of the same object on the mirror above the working space. Furthermore, spatial sound can be added to produce complementary information. By using the stylus and looking at the 3D picture, the user experiences a multi-sensory co-centric virtual 3D environment. A solid object surface is felt through haptic interaction due to the dynamic counter force produced by the programmed calculation of forces of the robotic arm and thus felt by the user in the stylus at the exact geometric contact point in the working space. This effect can be used in some other ways to produce other kind of haptic information for the user, e.g. as a force guiding the user to an intended point in the working space.

Figure 2. Phantom device and Reachin display

3.3. Exploratory Learning as a Pedagogical Tool

The pedagogical approach of the computer-based simulation system is based on exploratory learning. A child can explore the selected phenomena in the simulation according to his/her own interests. The theoretical foundations for exploratory learning are derived from the inquiry-models that emphasize the role of questions as a starting point for the inquiry. Such a model is the interrogative model of inquiry [18, 19, 20] (see also [21]). It's a model of scientific reasoning, where scientific procedure is viewed as information seeking by questioning. Scientific procedure refers to a series of questions which divide the initial question or problem into smaller questions. The inquirer tries to derive an answer to his/her initial question by using the existing knowledge and by formulating and seeking answers to smaller questions. In the present research, applying this model means that a child's learning is viewed as an active process guided by his/her own questions and previous knowledge. Different micro worlds provide opportunities to explore and experiment subordinate questions and seek answers to questions in which the child is interested.

Children need guidance and support for their explorations for making progress and deepening their understanding. In our system, programmable proactive agents have been used to support children's questioning and explorations. Proactivity can be defined as anticipative computer-technology. It adapts according the user and situation, predicts the user's intentions and acts accordingly. The agents' operations in the system will be mainly based on auditory and haptic feedback, since the system is developed especially for visually impaired children.

3.4. Proactive Agent Technology as a Tutoring Tool

Proactive agents support children's explorations in the simulation by asking questions and encouraging the child's own questioning and formation of hypotheses as an aim to guide the child's exploration process towards scientific inquiry. Proactive agents scaffold each child with respect to his/her capabilities and exploration paths, guide the explorations from familiar everyday contexts towards the causes and explanations of the phenomena, and direct the child's attention to the objects and events, as well as to the properties and the relations of the phenomenon. The agents don't make any decisions for the child or force him or her in any particular exploration path. At any moment, the child can choose either to listen to what the agents ask or suggest, or to ignore them. The agents' support is based on the child's exploration paths and user-profiles. As the child's explorations proceed, the agents' support may decrease step by step. The agents in each micro world have different imaginary characters and different names and voices. Narration and play have been used to some extent when designing the agents' action and nature. The structure of the agent system will be seen in Figure 3.

Proactive agents scaffold a child's explorations by assisting him/her to find the phenomena that are connected to the micro world the child is currently exploring. The agents can also offer hints about things the child could explore in the current micro world.

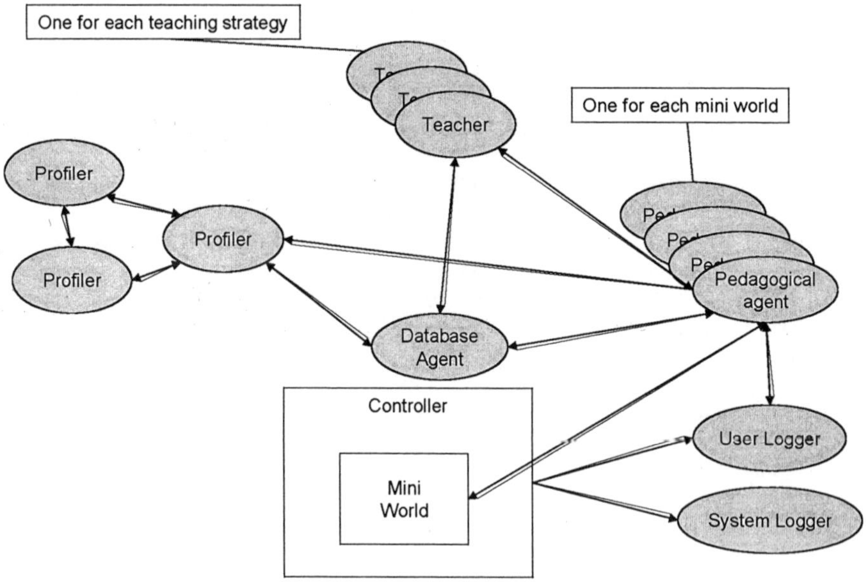

Figure 3. The structure of the agent system

3.5. An Example

The following provides a brief example of one part of the simulation system. The example concerns the exploration of the Earth and the concept of gravity. The spherical Earth can be felt three-dimensionally with the stylus. When exploring the Earth, the child can 'feel' the gravity in the stylus. The stylus in the child's hand is pulled slightly towards the Earth. When exploring the Earth's surface, the child can hear the names of the continents and people's voices. The sounds of sea can be heard above the oceans. The agent may, for example, direct the child's attention to gravity by saying (when the child is moving from space to the earth's surface): "Did you notice that when you came through the atmosphere you were pulled to the earth's surface? It's the same reason why we people stay on the ground: the earth's gravity pulls us as well as other objects." The agent may also challenge the child's thinking with different arguments like "What happens to different objects when you throw them in the air or drop them?" As a visual feedback, it is possible to view the spherical Earth, Figure 4.

Figure 4. A three-dimensional stereo picture of the spherical Earth on the mirror above the working space

Also, when exploring the Earth, the agent activates the child to find his/her home country, Finland, as well as to locate the north and south poles. In addition, the agent also directs the child's attention to the rotation of the Earth around its axis.

4. Pilot Tests and Research Experiments

Innovative research in human-computer interaction involves a combination of constructive and empirical research. This is particularly important when the users of the computer environment are children. In addition to the observation of children, we shall involve the children, both normally seeing and visually impaired, into the design of our environment from the very beginning.

The development and testing of the proactive intelligent agent system started at the beginning of January 2003 and will be continuing throughout the project. In the first phase, we have tested a manuscript and the usability of the system with the assistance of 5-9-year-old children, both visually impaired children and children with normal eyesight. The tests and research experiments of the next pilot version will be carried out both in the laboratory situations and normal school situations.

The research data will be collected in day care centres, primary schools and in the special laboratory conditions constructed for this research. The research data consists of observations and interviews of both children and their teachers. The research will be integrated as a natural part of the children's normal everyday life and action. The tasks in the computer system and the interviews with the children contain action similar to children's normal preschool and school situations.

5. Concluding Remarks

The final goal of the research project is to create pedagogically significant operational possibilities to bridge the gap between children with different potentials in pre-school and primary education and especially to focus on visually impaired children in integrated school groups. We envision the following main results of the project:

1. A tutoring system for children applying proactive agents.
2. Knowledge on how proactive agents can be used in tutoring systems.
3. Knowledge on children's exploratory and collaborative learning and thinking in novel technological learning environments representing natural science.
4. Knowledge on children's developmental process of self-efficacy in learning in use of information technology

The results can be applied in many different fields, not just in the tutoring system developed as a research prototype in this project. We envision a classroom where normal and visually impaired children play and learn together, using the proactive interfaces best suited for each child's needs. Their collaboration will be strengthened by a shared computer environment where they carry out investigations of their own choice and preference with the guidance and activation by proactive software agents. This common environment would also enable them to create, tell and share personal stories and constructions in and outside the environment. We believe that this kind of equal collaboration will be of utmost importance for all children. Our project will contribute to turning this vision into reality.

Acknowledgements
We express our great thanks the children and their parents for participating in our pilot experiments.

References

[1] Kangassalo, M. (1992) The pictorial computer-based simulation in natural sciences for children's use. In Ohsuga, S., Kangassalo, H., Jaakkola, H., Hori, K. and Yonezaki, N. (Eds.) Information Modelling and Knowledge Bases III: Foundations, Theory and Applications. Amsterdam: IOS Press, 511-524.

[2] Kangassalo, M. (1997) The formation of children's conceptual models concerning a particular natural phenomenon using PICCO, a pictorial computer simulation. Acta Universitatis Tamperensis 559. University of Tampere. 188 pp.

[3] Hietala, P.& Niemirepo, T. (1995) A framework for building agent-based learning environments. In Proceedings AI-ED 95: Artificial intelligence and education, Washington, DC, August 1995, AACE, 578.

[4] Hietala, P. & Niemirepo, T. (1998) The competence of learning companion agents. International Journal of Artificial Intelligence in Education, vol.9, no. 3-4, 178-192.

[5] NIMIS (2002) Nimis Project Pages. Available on the WWW:http://collide.informatik.uni-duisburg.de/Projects/nimis/ (referenced 4.8.2002)

[6] Hoppe U. et al. (2000) Supporting collaborative activities in computer integrated classrooms - the NIMIS Approach. In Proc. of 6th International Workshop on Groupware, CRIWG 2000. Madeira, Portugal, 18 - 20 October 2000, IEEE CS Press.

[7] Guglielmino, L.M. (1977) Development of the self-directed learning readiness scale. Dissertation. University of Georgia.

[8] Tuomivaara, S. (2000) Vapaa-ajan ja työn tietokonesuhteet ja käyttöhalukkuusmallit (Human-computer relations and psychological models for computer use at work and leisure). Akateeminen väitöskirja (Doctoral dissertation). Psykologian laitos. Tampereen yliopisto. Tampere: Juvenes Print.

[9] Kangassalo, M. (1994) Children's independent exploration of a natural phenomenon by using pictorial computer-based simulation. Journal of Computing in Childhood Education 5 (3/4), 285-297.

[10] Kangassalo, M. and Kumpulainen, K. (2003) The dynamics of children's science learning and thinking in a social context of a multimedia environment. In Hannu Jaakkola, Hannu Kangassalo, Eiji Kawaguchi and Bernhard Thalheim (Eds.) Information Modelling and Knowledge Bases XIV. Amsterdam: IOS Press, 188-197.

[11] Goldman, S. R. (1997) Learning from text: Reflection on the past and suggestions for the future. Discourse Processes, 23, 357-398.

[12] Kangassalo, M. (1999) PICCO Day and Night, Seasons – the Earth and the Sun. Cd-rom. ISBN 951-98035-0-5. Piccos Ohjelmat Oy.

[13] Kangassalo, M. (2001) Explorative learning in PICCO -environment. In Hannu Jaakkola, Hannu Kangassalo and Eiji Kawaguchi (Eds.) Information Modelling and Knowledge Bases XII. Amsterdam: IOS Press, 259-261.

[14] Hietala, P. (2001) Enthusiasm meets experience: collaboration of two communities through computer conferencing. In Montgomerie, G., Viteli, J. (Eds.), ED-MEDIA 2001: Proceedings of the 13th World Conference On Educational Multimedia, Hypermedia & Telecommunications, 25th - 30th June, Tampere,Finland, 745-750.

[15] Koivunen. K. (2002) Knowing who we are - supporting companion awareness in discussion forums. ED-MEDIA 2002: the 14th World Conference On Educational Multimedia, Hypermedia &Telecommunications, 24th - 29th June, 2002, Denver, Colorado.

[16] Loyd, B.H. & Gressard, C. (1984) Reliability and factorial validity of computer attitude scales.

Educational and Psychological Measurement 44:501-505.

[17] Davis, F.D. (1993) User acceptance of information technology: system characteristics, user perceptions and behavioral impacts. International Journal of man-machine studies, 38, 475-487.

[18] Hintikka, J. (1985) True and false logic of scientific discovery. In Hintikka, J. (Ed.) Logic of discovery and logic of discourse. New York: Plenum.

[19] Hintikka, J. (1988) What is the logic of experimental inquiry? Synthese, 74, 173-190

[20] Sintonen, M. (1999) Why questions, and why just why-questions? Synthese 120, 125-135.

[21] Hakkarainen, K. & Sintonen, M. (2002) The interrogative model of inquiry and computer-supported collaborative learning. Science & Education 11, 25-43.

[22] Hietala, P.& Niemirepo, T. (1998) Multiple artificial teachers: how do learners cope with a multi-agent learning environment? Workshop Proceedings on Current Trends and Applications of Artificial Intelligence in Education, Mexico City, Mexico, 16-20 March, 1998. ITESM, 33-40.

[23] Vosniadou, S. (1994) From cognitive theory to educational technology. In E. DeCorte & H. Mandl, (Eds.) Technology-based learning environments. Psychological and educational foundations, pp. 11-17. NATO ASI Series F: Computer and Science Systems, Vol. 137. New York: Springer.

Information Modelling and Knowledge Bases XVI
Y. Kiyoki et al. (Eds.)
IOS Press, 2005

A Prototype Implementation of Adaptive Metadata Generation to Digital Images

Hideyasu Sasaki & Yasushi Kiyoki
Keio University, 5322 Endo Fujisawa, Kanagawa, 252-8520 Japan
hsasaki@alumni.uchicago.edu kiyoki@mdbl.sfc.keio.ac.jp

Abstract. In this paper, we present an adaptive metadata generation method by exemplary textual labeling to digital images as a prototype system for content-based metadata indexing. Instead of costly human-created metadata, our method ranks sample images by distance computation on their structural similarity to query images, and automatically generates *metadata* as textual labels that represent geometric structural properties of the most similar sample images to the query images. In the case of ranking conflict, the proposed method generates metadata that are adaptive to selected domains and identify query images within at most two geometric types of visual objects. The objective labeling to sample images provides performance evaluation with objective measure on appropriate labeling. Our experimental results assure that the proposed prototype system has generated appropriate metadata to digital images of geometric figures with 80% preciseness in overall.

1 Introduction

In this paper, we present an adaptive metadata generation method by exemplary textual labeling to digital images as a prototype system for content-based metadata indexing. Instead of costly human-created metadata, our method ranks sample images on the basis of similarity to query images, and automatically generates metadata as textual labels that represent geometric structural properties of the most similar sample images to the query images, as outlined in Fig. 1 [1]. In the case of ranking conflict, the proposed method generates metadata that are adaptive to selected domains and identify query images within at most two geometric types of visual objects. Here, we simply define metadata as data about other data and objects [2] in the human cognizable form such as textual labels as well as the HTML tags, *etc.*

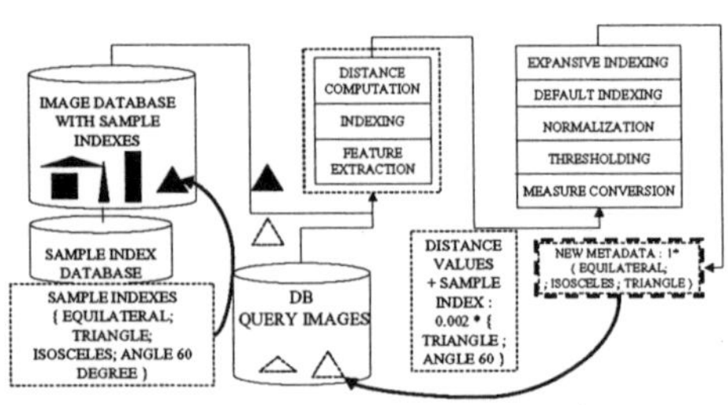

Figure 1: Architecture of the adaptive metadata generation.

1.1 Motivation

Images contain valuable information within themselves for content-based multimedia analysis, though image processing of visual features does not generate any textual labels as metadata that allow efficient and intuitive keyword-based retrieval. Those in the multimedia research field demand content-based metadata indexing that remedies the following four issues; Index automatically-generated metadata to digital images; Generate metadata of human cognizable information, *i.e.*, textual labels; Include external information that is not available from image processing of data or objects; Also, accept external knowledge of the expertise in certain domains in metadata generation, *e.g.*, brain disease images for expert physicians.

Digital images *per se* do not contain any textual labels for metadata generation as different from documents that contain direct sources for labeling, *i.e.*, textual indexing. Image processing approach as content analysis automatically extracts visual features out of images, and dynamically generates binary indexes, without any human cognizable meanings, to images. Image processing focuses on internal information stored in images, though external knowledge is indispensable to metadata generation that refers to categorization of metadata. We need a solution for content-based metadata indexing that is based on image processing techniques as content analysis and that is also accessible to external knowledge on structural properties of visual objects for textual labeling.

1.2 Background

Metadata indexing is classified in two main approaches regarding indexing techniques: keyword-based indexing and content-based indexing. Those individual indexing approaches correspond to two types of analysis object: textual information and visual features, respectively [3]. Table 1 gives an overview on taxonomy of metadata indexing in this research field.

Keyword-based indexing was developed as an automatic solution to document indexing. Human-created indexes are costly in selecting and indexing proper labels [4]. Documents *per se* contain textual labels that are easy to automatically extract [5, 6]. Keyword-based indexing is classified in two subclass approaches: textual closed caption [7] and associated text incorporation [8]. This classification is regarding associative/secondary indexing techniques, and corresponds to different attributes on primary analysis object: textual information in documents and textual information associated with images, respectively. Typical studies on textual closed caption include automatic content annotation and indexing to documents based

Table 1: Taxonomy of Metadata Indexing Approaches

Approach	Primary Analysis Object	Primary Indexing Technique	Associative Indexing Technique	Index Format	Application Domains
Textual Closed Caption	Textual Information in Documents	Keyword-based Indexing	Human Language Technology, Ontology, *etc.*	Textual Labels	General Domains
Associated Text Incorporation	Textual Information associated with Images	Keyword-based Indexing	Term Categorization	Textual Labels	General Domains
Image Processing	Visual Features in Images	Content-based Indexing	Visual Feature Extraction	Binary Indexes	General Domains
Categorization Modeling for Content Analysis	Visual Features in Images	Content-based Indexing	Model-based Categorization	Textual Labels	Model-Specified Types of Specific Domains
Adaptive Metadata Generation (**Our Approach**)	Visual Features in Images	Content-based Indexing	Exemplary Textual Labeling based on Structural Distance Computation	Textual Labels	Various Types of Specific Domains

on human language technology (HLT) [9], multilingual cataloging of document databases by vocabulary indexes [10], unfamiliar metadata searching by entry vocabulary modules [11], HTML tags based WWW document organization [12], and content table creation by HTML tags [13]. Typical studies on associated text incorporation include HTML source code analysis by reference information, *e.g.*, tags or metadata to images [14], term linkage with images in PDF documents like maps and descriptions [15], categorical term analysis for metadata indexing to images [16], semi-automatic textual labeling to images for classification by using lexical databases [17], and similar approaches are found in audio indexing [18] and speech indexing [19]. Keyword-based indexing realizes summarization of documents and image/document commingling objects with textual labels. The latter approach is, nonetheless, not a direct solution for indexing images. Images *per se* contain visual information that is different from textual but peripheral information associated with those images.

Content-based indexing was introduced to the multimedia research field as a solution for direct indexing to images with image processing techniques as content analysis that are often employed in content-based image retrieval (CBIR) systems [20, 21]. In CBIR systems, image processing techniques are used to automatically generate binary indexes to images by feature extraction and classify those generated indexes by computation of structural distance on visual features in general domains. Content-based indexing is classified in several subclass approaches: image processing, categorization modeling for content analysis, *etc.* This classification is regarding associative/secondary indexing techniques, and corresponds to individual types of application domains: general domains, model-specified types of specific domains, *etc.* Typical studies on image processing include multispectral image indexing [22] in pattern analysis with a number of applications of CBIR: QBIC [23, 24], WebSEEK [25], MARS [26], *etc.* Similar content-based approaches are also found in speech analysis [27]. Image processing automatically generates binary but not textual indexes to images. More advanced solutions come from combining keyword-based indexing and content-based indexing [28]. The categorization modeling approach realizes textual labeling to images. Its typical studies are found in the field of medical image indexing and retrieval. For example, semantic, *i.e.*, knowledge modeling for textual labeling to brain tumor images is employed in a brain disease image retrieval system [29]. A number of studies have employed such modeling as methods that discover direct linkages between the results of content analysis of visual objects and the prospective metadata that often describe sophisticated categories. For example, a small black bulb of extracted features is associated with fatal stroke in neoplastic region. However, that "direct-linkage" involves not only its costly implementation but also the limitation of its applications to model-specified types of specific and narrower domains that have hierarchical categories on image classification, *e.g.*, a metadata categorization in a field of brain tumor images. In addition, the direct linkage is not intuitive in human recognition mechanisms, in which geometric or simple objects contained in images are perceived and associated with their basic structural properties. For example, an image of a tower aside a bridge contains a small horizontal rectangle and a tall vertical rectangle. Those properties are integrated and associated with another image stored in knowledge stock, which represents more sophisticated concepts such as Big Ben. It is reasonable and supportive to put an intermediate layer of metadata indexing based on structural properties of extracted features between a layer of content-analysis results of visual objects and another layer of sophisticated metadata association.

To remedy such issues and limitations, a desirable solution is the combination of content-based indexing as primary indexing approach and textual labeling as secondary indexing technique that is adaptive to various types of specific domains. In Section 2, we formulate our adaptive metadata generation method. In Section 3, we provide our method with mathematical formulation. In Section 4, we present a prototype system that is based on the proposed formulation by employing general image processing techniques. In Section 5, we describe several experiments on performance evaluation of our prototype system. In Sections 6 and 7, we discuss the experimental results with conclusion and our future work.

2 Adaptive Metadata Generation

We formulate the proposed adaptive metadata generation method with following seven processes: two preprocesses for collection of images & labels, and distance computation, and five main processes for indexing, as described in Fig. 1. The proposed method allows cost-efficient system implementation by employing two general functions in image processing techniques: feature extraction out of images and structural distance computation of extracted visual features. Our method automatically indexes query images in the form of the combinations of textual labels and numerical indicators as *metadata* by using results of distance computation on structural similarity between query images and sample images indexed with sample textual labels. Sample textual labels represent structural properties of sample images. Numerical indicators represent the proportion of structural properties that query images contain in comparison with sample images by using the results of distance computation. Also, when distance computation evaluates a single query image as similar to several different types of sample images, the proposed method generates new adaptive metadata that narrow down the scope of identification of the query image within at most two types of structurally similar images. The proposed method with well-defined sets of sample images and textual labels is adaptive to various types of specific domains without any anterior modeling on metadata categorization. The purpose to focus on geometric structural properties in metadata indexing is that automatically-generated metadata can be used as the objects of text-based analysis, which reflects external and expert wisdom that are accessible in various types of specified domains, and generates new metadata as represent more sophisticated concept.

Process – 1. Collection of Images & Labels

The purpose of this process to collect images and labels is to specify a target domain with a set of sample images for metadata generation to visual objects. This process as manual preprocessing involves four steps. The first step selects well-defined sample images as represent typical types of structural properties that are found in the specified domain, *e.g.*, typical shapes of architectures like water pools and baseball parks in urban surroundings. The second step selects a set of well-defined textual labels as examples that describe structural properties on selected sample images, *e.g.*, { Triangle, 70° ; Isosceles ; . . . } to a top roof of a certain type of a baseball park. The third step stores sample images into an image database, sample textual labels, which correspond to the stored sample images, into another label database, and query images requested for metadata generation to the other image database, respectively. Its final step combines sample textual labels with sample images.

Example; A sample image of right angle triangle has a set of sample textual labels: { Triangle ; Angle 90° ; Isosceles }. Another sample image of obtuse isosceles triangle has a set of sample textual labels: { Obtuse ; Isosceles ; Triangle ; Angle 150° }.

Process – 2. Distance Computation

The purpose of this process is to compare query images with sample images indexed with sample textual labels. This process as automatic preprocessing applies visual feature extraction and distance computation functions available in image processing for content analysis to query images and sample images indexed with sample textual labels. This process involves three steps: Store sample images and query images into a CBIR system; Fire distance computation functions in the CBIR system; Get results of structural distance computation from the

Table 2: Sample Data Processing in Default Indexing (*e.g.* Query: Right Angle Triangle)

SAMPLE IMAGES	SAMPLE TEXTUAL LABELS	DISTANCE VALUES	A/F NORMALIZATION
RIGHT ANGLE TRIANGLE	TRIANGLE ; ANGLE 90° ; ISOSCELES	0.00	0.50
OBTUSE ISOSCELES TRIANGLE	TRIANGLE ; OBTUSE ; ANGLE 150° ; ISOSCELES	0.00	0.50
RHOMBUS	RHOMBUS ; QUADRILATERAL ; QUADRANGLE ; POLYGON	0.000160	0.00
ELLIPSE	ELLIPSE	0.00256	0.00
ACUTE ISOSCELES TRIANGLE	TRIANGLE ; ACUTE ; ANGLE 30° ; ISOSCELES	0.244	0.00
TRAPEZOID	TRAPEZOID ; QUADRANGLE	0.256	0.00
CIRCLE	CIRCLE	0.310	0.00
EQUILATERAL TRIANGLE	TRIANGLE ; EQUILATERAL ; ANGLE 60° ; ISOSCELES	0.877	0.00
RECTANGLE	RECTANGLE ; QUADRANGLE ; POLYGON	0.974	0.00
SQUARE	SQUARE ; RECTANGLE ; QUADRANGLE ; POLYGON ; QUADRILATERAL	3.603	0.00

CBIR system. In this process, a CBIR system is selected from several CBIR systems. That selected system provides query images with sets of values that represent structural distance of query images from sample images. Those returned values represent certain proportion of structural properties that query images contain against what sample images represent in the form of sample textual labels.

Example; As described in Table 2, a query image of right angle triangle has the same number of structural distance value with "Right Angle Triangle" and "Obtuse Isosceles Triangle" as 0.00, which means the selected CBIR system evaluates those images as the least distant images to each other, though that result demands improvement in indexing. Meanwhile, the larger numbers correspond to the more distant images.

Process – 3. Measure Conversion

The purpose of this process is to convert the results of distance computation into more intuitive values that represent structural similarity between query and sample images. The values on structural distance take 0 to the least distant and most similar images, while the larger numbers correspond to the more distant and less similar images. This process converts the current unintuitive measure on distance computation to another intuitive measure on similarity which evaluates the least distant and most similar images as 1 and the largest distant images as 0 . After measure conversion, structural distance values of 0.00 are converted to 1.00 when query images coincide with proper sample images.

Process – 4. Thresholding

The purpose of this process of thresholding is to cut off "noise" information after measure conversion and emphasize the values that represent meaningful structural properties of query images as numerical indicators to sample textual labels. A number of values after the measure conversion correspond to insignificant structural properties while those values take lower size than a certain threshold. This process narrows down the range of the values after measure conversion by cutting off the values of lower size than a certain threshold, *e.g.*, 0.95, 0.40, *etc.*, and assimilates the values after measure conversion over the certain threshold to 1. In addition, when some values are assimilated to 1, this process evaluates all the other values as "noise" and converts them to 0.

Process – 5. Normalization

The purpose of this process is to normalize the results after thresholding as they constitute numerical indicators of proportional structural properties that sample textural labels represent.

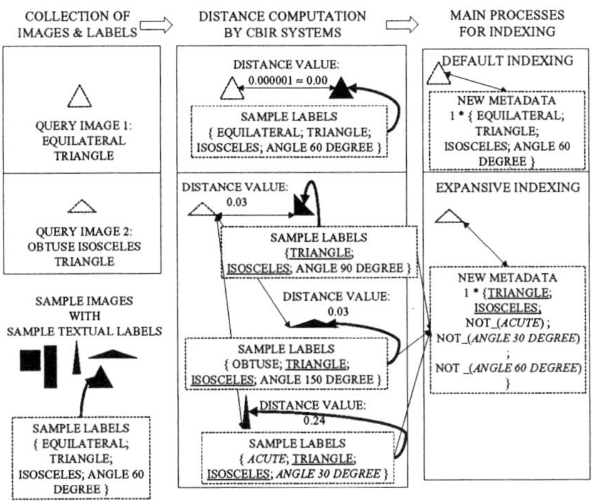

Figure 2: Exemplary default and expansive indexing.

Process – 6. Default Indexing

The purpose of this process is to index query images with the combinations of textual labels and numerical indicators as *metadata*. This process combines normalized values as numerical indicators of proportional structural properties with sample textual labels that represent structural properties, and then constitutes newly generated metadata while those metadata appropriately represent structural properties of query images.

Meanwhile, default indexing process generates conflicting textual labels that may represent several different types of sample images. A single query image corresponds to different structural types of sample labels, when several normalized values take the same largest size. In this case, default indexing process simply discards conflicting textual labels from metadata candidates, and does not generate any appropriate metadata (*See* Table 2 and Fig. 2).

Process – 7. Expansive Indexing

The purpose of this process is to generate new metadata that narrow down the scope of identification of a query image within at most two types of structurally similar images, when that distance computation evaluates a single query image as similar to several different types of sample images.

This process of expansive indexing involves two steps. The first step extracts "common labels" within conflicting textual labels of all sample images that default indexing inappropriately evaluates structurally "similar" to query images. Common labels constitute a major class that includes a subclass of the discussed sample images and other subclasses of sample images that include the common labels but are not evaluated as similar to the query images. The second step extracts "distinctive labels" from sample textual labels in the latter other subclasses of sample images that include the common labels but are not evaluated as similar to the query images. In the major class, distinctive labels classify the query images into the former subclass of sample images, and differentiate the query images from the latter other subclasses of sample images.

Example; As outlined in Fig. 2, a query image of obtuse isosceles triangle is evaluated as "similar" to two sample images: right angle triangle and obtuse isosceles triangle. In the instant case, the two sample images have 0.00 as structural distance values, 1.00 as the values after measure conversion, 0.50 as normalized values to the single query image, respectively.

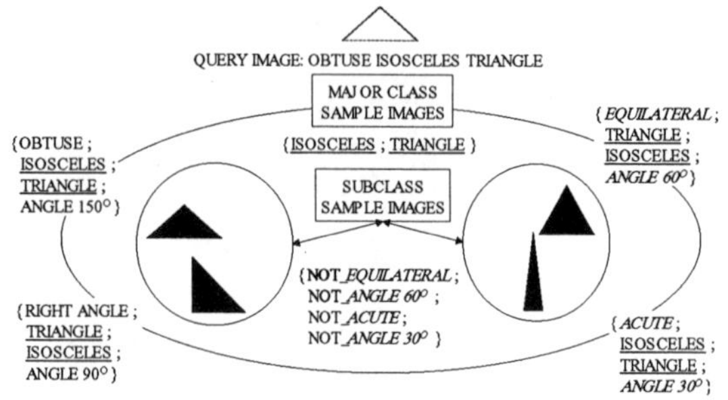

Figure 3: Exemplary major class and subclass categorization.

First, this process extracts common labels from textual labels tentatively generated in default indexing. Those common labels represent greatest common measure of generated labels that correspond to the same largest size of normalized values to the single query image of obtuse isosceles triangle as "TRIANGLE" and "ISOSCELES". Other sample images, which have normalized values lower than the largest normalized values, contain those common labels. Sample images of equilateral triangle and acute isosceles triangle contain those two common labels, here, "TRIANGLE" and "ISOSCELES" within their sample textual labels. Those two textual labels constitute a group of ISOSCELES TRIANGLE that represents a major class including right angle triangle and obtuse isosceles triangle with equilateral triangle and acute isosceles triangle. Second, this process finds distinctive factors in addition to common labels, as outlined in Fig. 3. Those two sample images with the lower normalized values also have other sample textual labels which are different from the common labels: "ACUTE", "EQUILATERAL", " ANGLE 30°", and "ANGLE 60°". Those different sample textual labels work as distinctive factors that constitute a group to represent a subclass within the major class. Distinctive factors are added with negative sign, *e.g.*, "NOT" to constitute distinctive labels against common labels: { NOT ANGLE 30° ; NOT EQUILATERAL ; NOT ACUTE ; NOT ANGLE 60° }. In expansive indexing, a combination of common labels and distinctive labels constitutes a set of new metadata that represents a subclass of right angle and obtuse isosceles triangles, which is different from the other subclass of acute isosceles and equilateral triangles within the major class of isosceles triangles.

3 Mathematical Formulation

In this section, we mathematically formulate the proposed adaptive metadata generation method.

Process – 1. Collection of Images & Labels

The process to collect images & labels consists of following four steps of collections of query images, sample images and sample textual labels, and combinations of sample textual labels with sample images.

Let define related sets of images and labels:

Definition 1. (a set of query images) As i is a continuous natural number, then $\{\mathcal{I}_i\}$ is defined as a set of query images:

$$\{\mathcal{I}_i\} := \{y | y = i, i \in \mathcal{N}\}$$

Definition 2. (a set of sample images) As j is a continuous natural number, then $\{\hat{\mathcal{I}}_j\}$ is defined as a set of sample images:

$$\{\hat{\mathcal{I}}_j\} := \{x | x = j, j \in \mathcal{N}\}$$

Definition 3. (a set of sample textual labels) As j and k are continuous natural numbers, respectively, then $\{\hat{\mathcal{W}}_{j_k}\}$ is defined as a set of sample textual labels:

$$\{\hat{\mathcal{W}}_{j_k}\} := \{z | z = \prod_{j \in \mathcal{N}} \{j \times \{\prod_{k \in \mathcal{N}} k\}\}, j \in \mathcal{N}, k \in \mathcal{N}\}$$

Here, $\prod$ represents production operation (same as below).

Let define combinations of sample textual labels with sample images:
Definition 4. (combinations of sample labels with sample images)

$$\{\hat{\mathcal{I}}_{j_{\hat{\mathcal{W}}_j}}\} := \{\prod_{j \in \mathcal{N}} \{\hat{\mathcal{W}}_j \times \hat{\mathcal{I}}_j\}\}$$

Process – 2. Distance Computation

Let define a set of structural distance values and a function of distance computation process as follows:
Definition 5. (a set of structural distance values)

$$\{\mathcal{D}_{i_j}\} := \{d | d = d_{i_j} \in \mathcal{R} : \{i_j\} = \{\prod_{i \in \mathcal{N}} \{i \times \{\prod_{j \in \mathcal{N}} j\}\}\}$$

Definition 6. (a function of distance computation between query images and sample images indexed with sample textual labels)

$$\phi : \{\prod_{i \in \mathcal{N}} \{\mathcal{I}_i \times \{\prod_{j \in \mathcal{N}} \{\hat{\mathcal{I}}_{j_{\hat{\mathcal{W}}_j}}\}\}\} \to \{\mathcal{D}_{i_j}\}$$

Here, $\to$ represents function operation.

Process – 3. Measure Conversion

Let define values after measure conversion by using the first normalization of reciprocal structural distance values, as follows:
Definition 7. (values after measure conversion)

$$\mathcal{S}_{i_j} := \begin{cases} 1 & (when \mathcal{D}_{i_j} = 0) \\ \dfrac{\frac{1}{\mathcal{D}_{i_j}}}{\sum_{j \in \mathcal{N}} \frac{1}{\mathcal{D}_{i_j}}} & \text{(otherwise)} \end{cases}$$

Here, $\sum$ represents summation operation.
Also, subject to the following computation rules:

$$\frac{1}{0} := \infty; \ \frac{\infty}{\infty} := 1; \ \frac{0}{0} := 1$$

Here, the distance computation is converted into an intuitive measure on similarity.

Process – 4. Thresholding

Let define values after thresholding the results of measure conversion with a certain selected threshold value t:

Definition 8. (values after thresholding)

$$\bar{\mathcal{S}}_{i_j} := \begin{cases} 0 & (when\ {}^{\exists}\mathcal{S}_{i_j} < t; 0 < t \leq 1) \\ 1 & (\text{otherwise}) \end{cases}$$

Here, thresholding works as noise cut-off mechanism.

Process – 5. Normalization

Let define values after normalizing the results of thresholding:

Definition 9. (normalized values)

$$||\bar{\mathcal{S}}_{i_j}|| := \frac{\bar{\mathcal{S}}_{i_j}}{\sum_{j \in \mathcal{N}} \bar{\mathcal{S}}_{i_j}}$$

Here, all the values are normalized to represent numerical indicators of proportional structural properties.

Process – 6. Default Indexing

Let define combinations of normalized values with sample textual labels in default indexing:

Definition 10. (combinations of normalized values with sample textual labels in default indexing)

$$\{\mathcal{L}_{i_{j_k}}\} := \{ \prod_{i \in \mathcal{N}} \{ \prod_{j \in \mathcal{N}} \{||\bar{\mathcal{S}}_{i_j}||\} \times \{ \prod_{j \in \mathcal{N}} \{ \prod_{k \in \mathcal{N}} \{\hat{\mathcal{W}}_{j_k}\}\}\}\}$$

Let define new metadata generated in default indexing:

Definition 11. (new metadata generated in default indexing)

$$\mathcal{M}_i := \begin{cases} \mathcal{L}_{i_{j_k}} \text{ for } {}^{\exists}j, \text{ such that } \bar{\mathcal{S}}_{i_j} = 1 & (\text{when} \sum_{j \in \mathcal{N}} \{\bar{\mathcal{S}}_{i_j}\} = 1) \\ 0 & (\text{otherwise}) \end{cases} \quad (\text{for } {}^{\exists}i)$$

Process – 7. Expansive Indexing

Let define common labels generated in expansive indexing:

Definition 12. (common labels)

$$\mathcal{L}_i^c := \begin{cases} \cap_j \{\mathcal{L}_{i_{j_k}}\} \text{ for } {}^{\forall}j, \text{such that } \bar{\mathcal{S}}_{i_j} = 1 & (\text{when} \sum_{j \in \mathcal{N}} \{\bar{\mathcal{S}}_{i_j}\} \geq 2) \\ 0 & (\text{otherwise}) \end{cases} \quad (\text{for } {}^{\exists}i)$$

Here, ∩ represents cap product operation.

Let define distinctive labels in expansive indexing:

Definition 13. (distinctive labels)

$$\mathcal{L}_i^d := \begin{cases} \overline{\{\{\cup_j \{\mathcal{L}_{i_{j_k}}\}\} \setminus \{\mathcal{L}_i^c\}\}} & \text{for } {}^{\forall}j, \text{such that} \{\{\mathcal{L}_{i_{j_k}}\} \cap \{\mathcal{L}_i^c\}\} \neq \emptyset \\ & (\text{when} \{\mathcal{L}_i^c\} \neq \emptyset) \\ 0 & (\text{otherwise}) \end{cases} \quad (\text{for } {}^{\exists}i)$$

Here ∪ represents cup product operation. The bar over a set represents not complementary set but special operation as each element in the set has an index, "NOT", as negative sign.

Let define a set of newly generated metadata in expansive indexing by combining common labels and distinctive labels with metadata generated in default indexing:

Definition 14. (a set of new metadata generated in expansive indexing)

$$\{\hat{\mathcal{M}}_i\} := \{\{\mathcal{M}_i\} \cup \{\mathcal{L}_i^c\} \cup \{\mathcal{L}_i^d\}\}$$

4 System Implementation

We have implemented our method based on a typical CBIR system that is QBIC enclosed in the DB2 package. [23, 24]. Although the selected system offers feature extraction functions on not only texture but also color histogram of visual objects, we have focused just on texture in our prototype implementation. The entire system is implemented in a local host machine (CPU 600MHz; Memory 128 MB). The entire algorithm is programmed in Java and PHP.

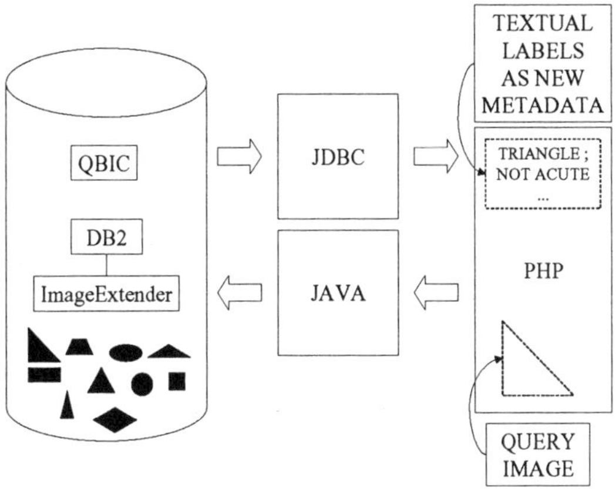

Figure 4: System implementation and data processing.

The entire data processing is outlined in Fig. 4. The images and labels are transferred from sample and query image databases, and a sample label database to QBIC by booting Apache in the local host machine. In DB2, images are transferred via JDBC to QBIC. After cataloging the images stored in QBIC, each path-linkage locates the images and QBIC defines their data structure. QBIC computes and returns structural distance values between query images and sample images. Those returned values are normalized and combined with sample textual labels to generate new metadata by our algorithms programmed in PHP. The results are delivered as outputs in HTML format.

5 Experiments

The overall experimental results show that the implemented system has generated appropriate metadata in the form of textual labels to 80% of the total 900 query images (360 original images were repeatedly used for a series of several experiments.) in 9,866 milliseconds per image in gross average: 8,840 for indexing and 1,026 for distance computation.
As experimental objects, we have prepared ten types of basic geometric figures for generating a set of sample images, *e.g.*, triangle, rectangle, ellipse, *etc.*, as described in the Appendix A. Query images are automatically generated by translating or affine-transporting those ten types of basic geometric figures, as displayed in Fig. 5.

As a prototype implementation, the bottom line for its performance evaluation is whether the implemented system automatically generates metadata to requested query images as textual labels of "objective appropriateness" without any models on categorization in a specific domain and within reasonable elapsed time. Actual images of scenery pictures, *e.g.*, an image of the rising Sun over an ocean, are divided into several parts of geometric figures: ellipse,

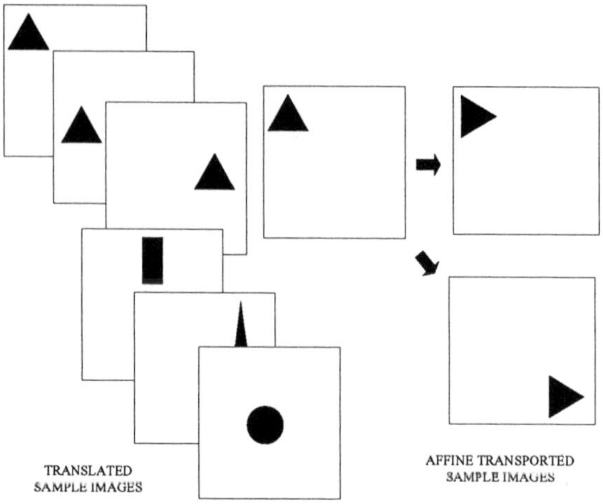

Figure 5: Query images generated by translation and affine transportation of sample images.

trapezoid, *etc.* That image could be labeled as "Sun", "Ocean", otherwise, "Horizon" so that a single metadata selection and description to the image among several candidates of sophisticated metadata inevitably brings subjective factors into performance evaluation [30]. The discussed geometric figures provide sample images with labels of clear definition on individual structural elements of actual images. This objective labeling to sample images provides performance evaluation with objective measure on appropriate labeling.

5.1 Experiment 1: Performance Evaluation

We have measured how many new metadata generated by default indexing follow mathematical definitions of query images in two cases of thresholding, 0.95 and 0.40, respectively. We have also measured improved performance by introduction of expansive indexing process with common and distinctive labels. Here, indexing performance is defined as a ratio of query images, which are indexed with appropriate metadata as textual labels. Those metadata should follow mathematical definitions of geometric figures represented in query images. This experiment has been processed to translated and affine-transported query images in the form of bitmap and jpeg formats, respectively. The results are described in the following Table 3 and Fig.'s 6, 7, 8 and 9 .

Table 3: Indexing Performance Evaluation

Image Format (Default/Expansive Indexing; Threshold)	Metadata Generation to Translated Images		Metadata Generation to Affine-transported Images	
	# Query Images	Ratio	# Query Images	Ratio
BMP (Default; Threshold 0.95/0.40)	90	0.80	90	0.80
BMP (Expansive; Threshold 0.95/0.40)	90	1.00	90	1.00
JPG (Default; Threshold 0.95)	90	0.60	90	0.20
JPG (Default; Threshold 0.40)	90	0.60	90	0.40
JPG (Expansive; Threshold 0.40)	90	0.80	90	0.40

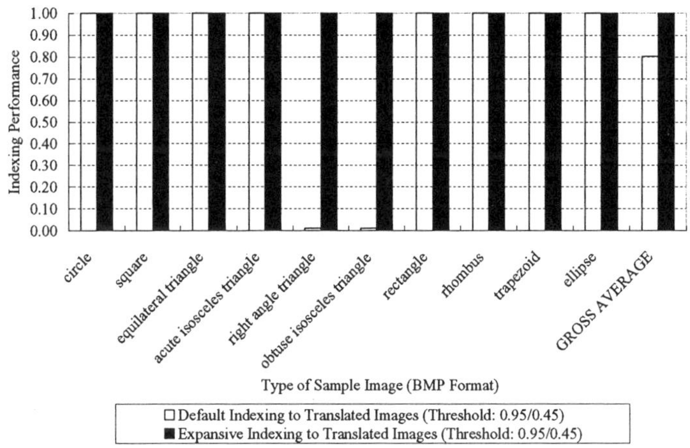

Figure 6: Performance evaluation of indexing to translated query images in bitmap format.

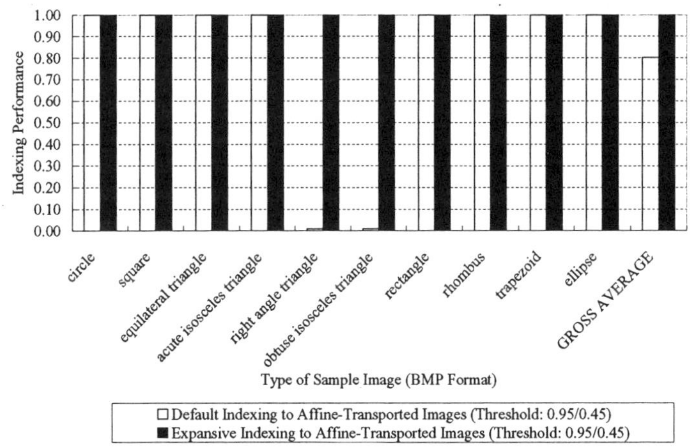

Figure 7: Performance evaluation of indexing to affine-transported query images in bitmap format.

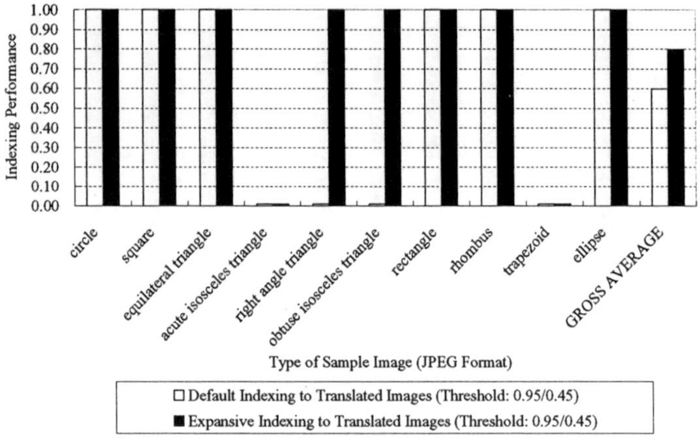

Figure 8: Performance evaluation of indexing to translated query images in jpeg format.

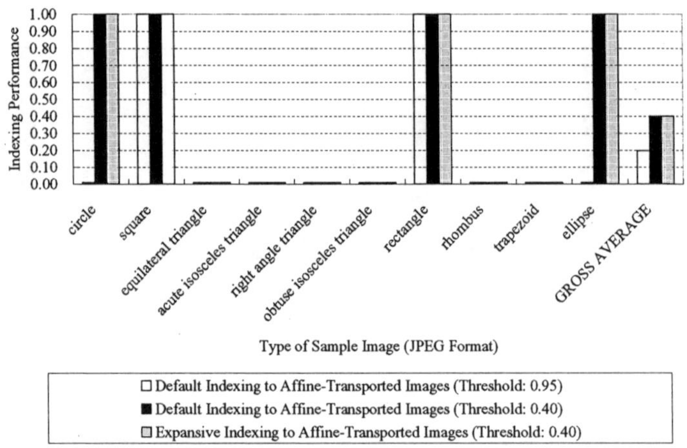

Figure 9: Performance evaluation of indexing to affine-transported query images in jpeg format.

5.2 Experiment 2: Thresholding Adjustment

Thresholding process (*See* § 2.4 & § 3.4) often cuts off important information in textual labeling with excessively high threshold values. Adjusting threshold values from 0.95 to 0.40 has accepted more informative proportion of structural properties that sample images represent. Even after thresholding adjustment, our experiment has shown that the inclusion of more common and distinctive labels has not realized any negative impact to default indexing. Its typical improvement was found in the case of default indexing to affine-transported images in jpeg format, especially on two cases of query images: circle and ellipse. Table 4 describes a typical case that lowered threshold values have improved performance by 20% on default indexing to affine-transported query images of "circle" in jpeg format. In overall, thresholding adjustment has improved default indexing performance by 5% from 60% to 65%.

Table 4: Performance Improvement by Thresholding Adjustment (Query Image: Circle)

Type of Figure for Distance Computation	Structural Distance Value	Value a/f Measure Conversion	Normalized Value (A/F Thresholding: 0.95)	Normalized Value (A/F Thresholding: 0.40)
Circle	0.011032048	0.416665	0	1
Acute Isosceles Triangle	0.02085747	0.220385	0	0
Trapezoid	0.02085747	0.220385	0	0
Equilateral Triangle	0.08342987	0.055096	0	0
Rectangle	0.11652601	0.039448	0	0
Ellipse	0.36474708	0.012602	0	0
Right Angle Triangle	0.41387418	0.011106	0	0
Obtuse Isosceles Triangle	0.41387418	0.011106	0	0
Rhombus	0.44834933	0.010252	0	0
Square	1.55569112	0.002955	0	0

5.3 *Experiment 3: Expansive Indexing*

Expansive indexing process (*See* § 2.7 & § 3.7) has extracted common and distinctive labels for translated and affine-transported query images in bitmap format, and improved indexing performance by 20% from default indexing, respectively. Similar improvement was found in the case of affine-transported query images in jpeg format. Expansive indexing has improved overall performance by 15% from 65% to 80% after thresholding adjustment. Table 5 describes which geometric figure has been drastically improved.

Table 5: Performance Improvement by Expansive Indexing

Case of Image	Type of Figure (Query Image)	A/F Default Indexing	A/F Expansive Indexing
BMP/Translated	Right Angle Triangle	Non-Zero Values a/f Thresholding but No Metadata	Generate Right Labels
BMP/Translated	Obtuse Isosceles Triangle	Non-Zero Values a/f Thresholding but No Metadata	Generate Right Labels
BMP/Affine-Transported	Right Angle Triangle	Non-Zero Values a/f Thresholding but No Metadata	Generate Right Labels
BMP/Affine-Transported	Obtuse Isosceles Triangle	Non-Zero Values a/f Thresholding but No Metadata	Generate Right Labels
JPG/Translated	Rectangle	Non-Zero Values a/f Thresholding but No Metadata	Generate Right Labels
JPG/Translated	Obtuse Isosceles Triangle	Non-Zero Values a/f Thresholding but No Metadata	Generate Right Labels

5.4 *Experiment 4: Elapsed Time*

We have measured elapsed time of default and expansive indexing, respectively. Here, elapsed time is defined as difference between a starter time and its closing time of indexing process in millisecond. This experiment has been processed to translated and affine-transported query images in bitmap and jpeg formats, respectively. The results are described in Table 6. Elapsed time of distance computation as preprocessing is described in the Appendix B, separately.

Table 6: Elapsed Time of Indexing Process

Format (Default/Expansive Indexing)	Elapsed Time of Indexing to Translated Images (Millisecond)	Elapsed Time of Indexing to Affine-transported Images (Millisecond)
BMP (Default)	613	11,748
BMP (Expansive)	6,441	5,524
JPG (Default)	8,154	15,040
JPG (Expansive)	8,156	15,040

6 Discussion

Overall indexing performance has been improved by thresholding adjustment and expansive indexing with common and distinctive labels. Artificial objects like architectures and drawings often consist of squares and rectangles with clear angles. The experimental results assure that our prototype system has feasibility as adaptive metadata generation by exemplary textual labeling to digital images of geometric figures.

Table 7: Unimproved Performance on Affine-Transported Images in JPEG Format

Type of Figure (Query Image)	A/F Default Indexing	A/F Expansive Indexing
Right Angle Triangle	8/9 Non-Zero Values a/f Thresholding; Bad Labels (Ellipse)	Not Generate Right Labels
Trapezoid	3/9 Non-Zero Values a/f Thresholding; Bad Labels (Circle)	Not Generate Right Labels
Equilateral Triangle	Non-Zero Values a/f Thresholding; Bad Labels (Rectangle)	Not Generate Right Labels
Acute Isosceles Triangle	Non-Zero Values a/f Thresholding; Bad Labels (Circle)	Not Generate Right Labels
Rhombus	Non-Zero Values a/f Thresholding; Bad Labels (Ellipse)	Not Generate Right Labels

However, indexing performance has not been improved on affine-transported query images in jpeg format even after expansive indexing. Table 7 describes which geometric figure has not found any common or distinctive labels. We consider two candidates as solutions: the introduction of a module selecting image-rotation sensitive CBIR systems for distance computation preprocessing and the inclusion of a learning module generating more sophisticated common and distinctive labels.

We have implemented the proposed method as an intuitive solution for adaptive metadata generation to digital images by exemplary textual labeling to geometric structural properties of visual objects. By sophisticated text-based analysis of metadata, our automatically-generated metadata can be used as a source to create more sophisticated conceptual metadata, which are able to reflect external and expert wisdom that are accessible in various types of specified domains. Our method is, therefore, feasible as a general approach of adaptive metadata generation by textual labeling to images with well-defined sample labels.

7 Conclusion and Future Work

In this paper, we have presented an adaptive metadata generation method by exemplary textual labeling to digital images. One of our main contributions is to automatically generate metadata in the form of textual labels to query images, instead of costly human-created metadata. The automatically-generated metadata take the form of combinations of textual labels and numerical indicators of structural similarity between query images and sample images indexed with sample textual labels. Another contribution is cost-efficient and reasonable implementation of the proposed content-based indexing method by employing general image processing mechanisms used in CBIR systems. The other one is expansive indexing that generates new adaptive metadata that narrow down the scope of identification of query images within at most two types of structurally similar images in the case of conflicting results given by distance computation functions. Based on our intuitive assumptions on human recognition, text-based analysis and categorization are applicable to our automatically-generated metadata, which can reflect external and expert knowledge in individual domains. The proposed method is applicable to various types of specific domains with well-defined sets of sample images and textual labels that are selected as represent typical structural properties of visual objects in individual target domains. Our experimental results assure that the proposed prototype system has generated appropriate metadata to digital images of geometric figures with 80% preciseness in overall. Those features and results show that the proposed method is feasible in the applications to various types of multimedia data, *e.g.*, images of artificial objects like architectures and drawings with well-defined sample textual labels.

We have several plans to extend the scope of its applications of the proposed method for adaptive metadata generation to various types of multimedia data in a number of attractive domains, ones of which are brain disease radiological and MRI images of tumors and strokes.

8 Acknowledgements

We would like to thank Kenichiro Kato for his contributions to our project of content-based metadata indexing while he had been working as an undergraduate student at our laboratory.

A Experimental Data Generation

Sample and query images have been automatically generated as experimental data by using "Image Function" available in PHP ver. 4.2.3 released in FreeBSD ver. 4.4. Those experimental data have the same size of 300×300 pixels in bitmap and jpeg formats, respectively. Each 300×300 sized datum has 3×3 grids inside as nine boxes of the size of 100×100 pixels. Geometric figures in sample and query images occupy a single box inside those nine boxes.

Ten types of geometric figures are given for generating sample and query images. Those types consist of circle, ellipse, equilateral triangle, acute isosceles triangle, right angle triangle, obtuse isosceles triangle, square, rectangle, rhombus, and trapezoid. Original geometric figures for sample images are drawn in the center box of the nine boxes. Each upper left of the center boxes is the start point for drawing each geometric figure. Those geometric figures are drawn by linking several predefined points, as given in Table 8. Query images are generated by translating and 90° right rotation with translating, *i.e.*, affine-transporting original geometric figures of sample images by using "ImageMagick" ver. $5.4.8-3$ available in PHP released in the FreeBSD.

Table 8: Linking Points for Image Data Generation

Circle	Center: $(50, 50)$; Diameter 60 pixels
Ellipse	Center: $(50, 50)$; Minor Axis 30, Major Axis 60 pixels
Equilateral Triangle	$(10, 90)--(50, 12)--(90, 90)$
Right Angled Triangle	$(10, 90)--(90, 90)--(50, 50)$
Acute Isosceles Triangle	$(30, 90)--(70, 90)--(50, 10)$
Obtuse Isosceles Triangle	$(10, 90)--(90, 90)--(90, 60)$
Square	$(10, 10)--(10, 90)--(90, 90)--(90, 10)$
Rectangle	$(30, 10)--(30, 90)--(70, 90)--(70, 10)$
Rhombus	$(10, 50)--(50, 30)--(90, 50)--(50, 70)$
Trapezoid	$(30, 30)--(70, 30)--(90, 70)--(10, 70)$

B Distance Computation Time

The proposed method employs distance computation mechanisms used in a CBIR system to compute similarity between sample and query images. We have measured elapsed time of distance computation of query images in bitmap and jpeg formats, respectively. Elapsed time of distance computation is defined as difference between a starter time and its closing time of distance computation as preprocessing in millisecond. Table 9 describes gross average elapsed time for each distance computation of query images in bitmap and jpeg formats, respectively.

Table 9: Gross Average Elapsed Time of Distance Computation

Case of Image	Gross Average (millisecond)
Computation to Translated Images (BMP)	1,066
Computation to Affine-Transported Images (BMP)	1,003
Computation to Translated Images (JPEG)	1,072
Computation to Affine-Transported Images (JPEG)	964

References

[1] H. Sasaki and Y. Kiyoki, A Prototype Implementation of Metadata Generation for Image Retrieval. In: *Proc. of the 2004 International Symposium on Applications and the Internet - Workshops (SAINT 2004 Workshops).* IEEE Computer Society Press. Tokyo, Japan, March 26–27, 2004, pp. 460 466.

[2] D. Rusch-Feja, Metadata: Standards for Retrieving WWW Documents (and Other Digitized and Non-digitized Resources). In: U. Grothkopf, H. Andernach, S. Stevens-Rayburn, and M. Gomez (eds.), *Library and Information Services in Astronomy III ASP Conference Series*, Vol. 153, 1998, pp. 157–165.

[3] A. Sheth and W. Klas, *Multimedia Data Management: Using Metadata to Integrate and Apply Digital Media*. McGraw-Hill, New York, 1998.

[4] C.C. Marshall, Making Metadata: A Study of Metadata Creation for a Mixed Physical-digital Collection. In: *Proc. of the* 3rd *ACM International Conference on Digital Libraries (ACM DL)*, Pittsburgh, Pennsylvania, June 23-26, 1998, pp. 162–171.

[5] M. Kobayashi and K. Takeda, Information Retrieval on the Web, *ACM Computing Survey* **32(2)** (2000) 144–173.

[6] F. Sebastiani, Machine Learning in Automated Text Categorization, *ACM Computing Survey* **34(1)** (2002) 1–47.

[7] M.G. Brown, J.T. Foote, G.J.F. Jones, K.S. Jones, and S.J. Young, Automatic Content-based Retrieval of Broadcast News. In: *Proc. of ACM Multimedia*, November, 1995, pp. 35-44.

[8] R.K. Srihari, Z. Zhang, and A. Rao, Intelligent Indexing and Semantic Retrieval of Multimodal Documents, *Information Retrieval* **2(2/3)** (2000) 245–275.

[9] K. Bontcheva, D. Maynard, H. Cunningham, and H. Saggion, Using Human Language Technology for Automatic Annotation and Indexing of Digital Library Content. In: M. Agosti and C. Thanos (eds.), *Research and Advanced Technology for Digital Libraries:* 6th *European Conference on Digital Libraries, ECDL 2002 Proceedings*, Lecture Notes in Computer Science, Springer-Verlag, No. 2458, Rome, Italy, September 16-18, 2002, pp. 613–625.

[10] R.R. Larson, F. Gey, and A. Chen, Harvesting Translingual Vocabulary Mappings for Multilingual Digital Libraries. In: *Proc. of the* 2nd *ACM/IEEE-CS Joint Conference on Digital libraries (JCDL) '02*, Portland, Oregon, July 13-17, 2002, pp. 185–190.

[11] F. Gey, H.M. Chen, B. Norgard, M. Buckland, Y. Kim, A. Chen, B. Lam, J. Purat, and R. Larson, Advanced Search Technologies for Unfamiliar Metadata. In: *Proc. of the* 25th *Annual International ACM SIGIR Conference on Research and Development in Information Retrieval*, Tampere, Finland, 2002, pp. 455–456.

[12] B. Doan, M. Beigbeder, J. Girardot, and P. Jaillon, Using Metadata to Improve Organization and Information Retrieval on the WWW. In: *Proc. of WebNet98*, Orlando, Florida, November, 1998.

[13] R.H. Dolin, M. Boles, R. Dolin, S. Green, S. Hanifin, B. Hochhalter, R. Inglesis, M. Ivory, D. Levy, K. Nadspal, M.S. Rae, C.J. Rucks, A. Snyder, T. Stibolt, M. Stiefel, and V. Travis, Kaiser Permanente's "Metadata-driven" National Clinical Intranet. In: V. Patel, *et al.* (eds.), *Proc. of MEDINFO 2001*, IOS Press, Amsterdam, 2001, pp. 319–323.

[14] Y. Tsymbalenko and E.V. Munson, Using HTML Metadata to Find Relevant Images on the World Wide Web. In: *Proc. of Internet Computing 2001*, Las Vegas, Nevada, June, 2001, pp. 842–848.

[15] D.A. Smith, A. Mahoney, and G. Crane, Integrating Harvesting into Digital Library Content. In: *Proc. of the 2nd ACM/IEEE-CS Joint Conference on Digital libraries (JCDL) '02*, Portland, Oregon, July 13-17, 2002, pp. 183–184.

[16] K.P. Yee, K. Swearingen, K. Li, and M. Hearst, Faceted Metadata for Image Search and Browsing. In: *Proc. of ACM CHI 2003*, Ft. Lauderdale, Florida, April 5-10, 2003.

[17] Y. Kiyoki, Y. Kitagawa, and T. Hayama, A Metadatabase System for Semantic Image Search by a Mathematical Model of Meaning, *ACM SIGMOD Record* **23(4)** (1994) 34–41.

[18] I. Magrin-Chagnolleau, Indexing Telephone Conversations by Speakers. In: *Proc. of the 2000 IEEE International Conference on Multimedia and Expo (ICME)*, NY, July 30–August 2, 2000, pp. 881–884.

[19] G. Amato, C. Gennaro, and P. Savino, Indexing and Editing Metadata for Documentary Films On Line: The ECHO Digital Library. In: D. Bearman and F. Garzotto (eds.), *Proc. of the International Cultural Heritage Informatics Meeting (ICHIM'01)*, Vol. 2, Milan, Italy, September 3-7, Archives & Museum Informatics, Pittsburgh, Pennsylvania, 2001, pp. 147–155.

[20] A.W.M. Smeulders, M. Worring, S. Santini, A. Gupta, and R. Jain, Content-based Image Retrieval at the End of the Early Years, *IEEE Trans. on Pattern Analysis and Machine Intelligence* **22(12)** (2000) 1349–1380.

[21] Y. Rui, T.S. Huang, and S.F. Chang, Image Retrieval: Current Techniques, Promising Directions and Open Issues, *Journal of Visual Communication and Image Representation* **10(4)** (1999) 39–62.

[22] J. Barros, J. French, W. Martin, P. Kelly, and J.M. White, Indexing Multispectral Images for Content Based Retrieval. *University of Virginia Technical Report*, CS-94-40, October 26, 1994.

[23] M. Flickner, H. Sawhney, W. Niblack, J. Ashley, Q. Huang, B. Dom, M. Gorkani, J.Hafner, D. Lee, D. Petkovic, D. Steele, and P. Yanker, Query by Image and Video Content: The QBIC System. *IEEE Computer* **28(9)** (1995) 23–32.

[24] J. Ashley, M. Flickner, J. Hafner, D. Lee, W. Niblack, and D. Petkovic, The Query by Image Content (QBIC) System, *ACM SIGMOD Record* **36(7)** (1995) 475.

[25] J.R. Smith and S.F. Chang, Visually Searching the Web for Content, *IEEE Multimedia* **4(3)** (1997) 12–20.

[26] M. Ortega, Y. Rui, K. Chakrabarti, S. Mehrotra, and T.S. Huang, Supporting Similarity Queries in MARS. In: *Proc. of ACM Multimedia*, 1997, pp. 403–413.

[27] L. Wilcox and J.S. Boreczky, Annotation and Segmentation for Multimedia Indexing and Retrieval. In: *Proc. of the 31st Annual Hawaii International Conference on System Sciences (HICSS)*, Vol. 2, January 6-9, 1998, pp. 259–266.

[28] P.J. Moreno, J.M. van Thong, E. Logan, and G. Jones, From Multimedia Retrieval to Knowledge Management, *IEEE Computer* **35(4)** (2002) 58–66.

[29] W.W. Chu, I.T. Ieong, and R.K. Taira, A Semantic Modeling Approach for Image Retrieval by Content, *VLDB Journal* **3** (1994) 445–477.

[30] S.S. Layne, Some Issues in the Indexing of Images, *Journal of the American Society for Information Science* **45(8)** (1994) 583–588.

Information Modelling and Knowledge Bases XVI
Y. Kiyoki et al. (Eds.)
IOS Press, 2005

Time-sensitive relationship management in technical manuals Case: Maintenance schedules

Anneli Heimbürger
Tampere University of Technology, Pori, Advanced Multimedia Center
P.O. Box 300, FIN – 28101 Pori, Finland
anneli.heimburger@tut.fi

Abstract. We introduce a method for time-sensitive relationship management which provides a systematic way of identifying, modelling and implementing time dependent relationships in application domains of knowledge management such as technical manuals. Technical manuals model structural, operational, maintenance and safety issues of technical objects in different contexts. Examples of technical objects are paper machines, base stations for mobile phones, and rock grinders. Contexts are defined by user groups, tasks, situations, environments, user devices and time. The application domain of our research is technical manuals and maintenance schedules for rock grinders. We have extended the temporal relationship category of the Relationship Navigation Analysis (RNA) by means of Allen's time interval relations for formalizing temporal conditions and describing them in natural language. Our implementation is based on the XML Linking Language (XLink) syntax. We have extended XLink's multiple links and multiple link networks with timerule attributes.

1. Introduction

1.1 Background

Knowledge management is a challenge for today's business, and it affects the position of enterprises in international competition. Formal and informal knowledge is becoming dynamic and reusable within different business applications and processes. A systematic approach to creating, composing, delivering, maintaining and using documents such as technical manuals is an issue of vital importance in industry. The challenge lies in keeping track of documents, getting right resources and sub-resources for users, as well as in ensuring that information is presented in right contexts. Providing the right pieces of information are critical, for instance in trouble shooting situations or in service intervals of a technical object obeying certain maintenance schedules. Technical manuals are an important part of the technical product itself, whether it is a paper machine, a base station for mobile phones or a rock grinder.

Technical manuals and related value-added services are more and more delivered, used and managed through Internet. Technical manuals model the structure and functions of a technical object in different contexts by means of (a) content elements such as texts, technical illustrations, engineering drawings, tables, part lists, safety issues, animations, sounds and video, (b) structures between content elements such as user interfaces, linking and navigational features, and (c) delivery platforms such as Web based delivery or mobile devices.

Technical manuals are used in different contexts. Examples of contexts are user groups, tasks, situations, environmental aspects, delivery platforms and time, as well as any combination of these. Time is an essential context for instance in carrying out maintenance operations of a technical object according to given service schedules. The instructions related to these situations depend on time. Time contexts are defined by service intervals in the maintenance schedules and/or by user's requests. The context dependent parts of documents in a technical manual will be delivered to the user. For example in Figure 1 the user gives his/her context C = {U(technical staff), Ta(maintenance), E(on the ground), Ti(time-sensitive materials), S(maintenance schedules), D(Web)} as a request to the system which delivers maintenance schedules and links to related instructions to the user's device.

1.2 What is the problem?

Context dependent access will be the most relevant means of retrieving information from both the traditional and mobile Web. In service situations the maintenance crew onsite needs step-by-step instructions depending on the maintenance schedules for their devices. The maintenance procedure descriptions depend on the service intervals. For example, if the service interval is 50-hour $\leq t \leq$ 200-hour then the related document group $D_{g1} = (d_1, d_2, d_3)$ will be delivered to the user. If the service interval is 200-hour $\leq t \leq$ 400-hour then the deliverable document assembly will be $D_{g2} = (d_1, d_2, d_3, d_4)$. The problem is how to automatically manage the right document groups including text, technical illustrations and safety issues with associated time contexts. In our case study the contextual focus is on technical staff maintaining a rock grinder according to given maintenance schedules.

1.3 The theoretical and practical framework of our research

Relationship Navigation Analysis (RNA) [40], Allen's relations between time intervals [1, 2, 3] and extended links defined by the XML Linking Language (XLink) [37] provide a theoretical framework for our research. At the moment, XLink is the only formal linking language that is based on the international standard [23] and defines a multiple linking and linkbase mechanism on the Web. However, the concept of extended links can be adopted by any other XLink-like linking language on the future Web. Our implementation and demonstration environment is Mozilla based editors and browsers providing the practical framework for our research [14, 27].

1.4 Basic concepts

The basic concepts in our research are:

- Technical objects: Examples of technical objects are paper machines, base stations for mobile phones, rock grinders.
- Technical manuals: Technical manuals include structural, operational, maintenance and safety issues of a technical object.
- Contexts: User groups, tasks, situations, environmental aspects, delivery platforms and time define the circumstances of using technical manuals.
- Maintenance schedules: Service intervals of a technical object such as 8-hour maintenance, 400-hour maintenance and 2400-hour maintenance.

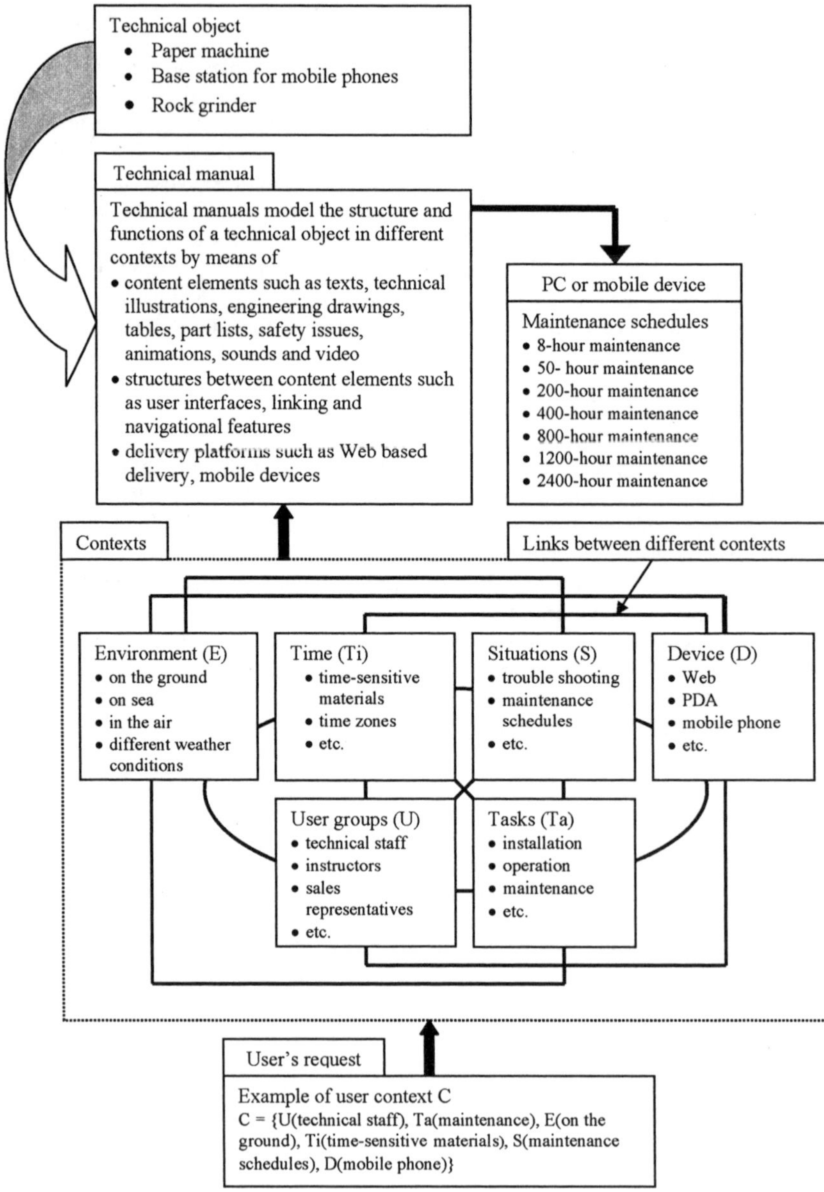

Figure 1. Technical manuals model the structure and functions of a technical object in different contexts.

- Relationship Navigation Analysis (RNA): RNA provides a systematic way of identifying useful relationships in application domains. RNA is based on generic relationship taxonomy. Application developers can implement each relationship as a link.
- XML Linking Language (XLink): XLink as defined by W3C allows link elements to be embedded into XML documents or inserted in separate XML-files in order to create and describe links between resources.
- Allen's relations: Allen has defined 13 relations between two time intervals. Time intervals are a set of constraints between two time points, a start and an end time.
- Time-sensitive relationship management: Time-sensitive relationships are functions of time, and they are activated when a context dependent temporal rule is valid. Time-sensitive relations can be implemented by means of temporal link traversals.

1.5 Our contribution

We introduce a method for time-sensitive relationship analysis and management. The method can be applied to time contextual composing, delivering and using document assemblies through traditional and mobile Internet. The method can be implemented by means of the XML Linking Language (XLink) specification. We have extended RNA's temporal relationship category with Allen's interval relations in order to formalize and describe time contexts and we have extended XLink's multiple linking mechanism with temporal linking facilities.

1.6 Overview of the paper

The remainder of the paper is organised as follows. Related work is reviewed in Section 2. Basic characteristics of technical manuals are described in Section 3. Time-sensitive relationship analysis and multiple link networks are introduced in Section 4. Implementation issues describing three different approaches, a rock grinder instruction manual as our test material and our demonstration are presented in Section 5. Section 6 is reserved for our conclusions and topics for further research.

2. Related work

Hypertext is a discipline that considers an application domain in terms of the relationships among its elements and processes. In hypertext applications relationships are implemented as external and/or internal links between contents and fragments of contents. According to hypertext research the solution to several link management problems is to separate the links from the content. Temporal linking has been widely studied in modelling multimedia synchronization, where time dependent media such as animation, sound and video is integrated with time independent media such as text, graphic and images [12, 18, 32, 34]. Nürnberg et al. [30] have described the development of conceptual architectures of hypermedia systems including various stages such as monolithic systems, open link service systems, open hyperbases, open hypermedia systems and component-based open hypermedia systems. The idea of a link service has existed since the days of Intermedia [17, 39]. From a user's point of view, the need for distributed link services has grown with the development of the World Wide Web. Chimera [4], Hyper-G [15] and Microcosm [10] are examples of the classic open hypermedia systems, where support for non-embedded links to

WWW pages has been developed. In Microcosm, SGML markup is used in the link databases and now some parts of the system have been modified to support some of the XLink-based linking facilities, such as controlling link behaviour and its presentation [9, 25].

Open hypermedia is an area that has been studied by the hypermedia community for several years and for which a number of systems have been implemented [7, 25, 29, 33]. In open hypermedia systems (OHS), links are stored and managed in linkbases. The idea of abstracting links from documents allows for a great deal of flexibility in link maintenance and re-use. Common usage of the Web involves embedding links within documents in the (X)HTML format. In this sense, the Web can be considered as a closed hypermedia system. However, the Web infrastructure allows hypertext links to be abstracted away from documents and managed separately, by using mechanisms defined in the XML Linking Language (XLink) specification and structure servers [26, 37]. XLink has a recommendation status in the World Wide Web Consortium (W3C). HLink aims to adopt some of XLink's features such as extended links to XHTML [35]. XLink- and HLink-like languages are developing the Web towards a world wide open hypermedia system and also towards a semantic web [5, 6, 35, 36].

There are already some extensions of XLink-language. XConnector is a project where XLink language has been extended to provide multimedia synchronization facilities for Web resources using links [28]. XLinkProxy is an implementation of an external link database, which is defined using XLink [11]. Christenssen and Hansen [13] have developed a direct mapping between the open hypermedia interchange format [16] and an XLink linkbase. Goate is an HTTP proxy that allows emulating XLink-like high-level linking with ordinary HTML browsers [24]. XML and XLink have also been investigated for modelling, browsing and re-using technical documents [8, 19, 31, 38, 41]. However, neither of these XLink-implementations solves the problem of modelling content adaptation according to time contexts.

The study presented in this paper is part of the research related to structured documents, link management and mobile work currently in progress in the Advanced Mutimedia Center at Tampere University of Technology, Pori [20, 21, 22].

3. Technical manuals

Technical manuals are created, managed and used during the life cycle of an artefact. Technical manuals describe the structure and function of a technical object such as a paper machine, a base station for mobile phones or a rock grinder. Manuals are used in various contexts. Environments, situations, tasks, time, user groups, users' devices and their combinations are examples of different contexts. There are many different user groups including designers, installation, operation, commissioning, maintenance and training personnel. All user groups have their own particular needs when searching for information. Moreover, the level of experience concerning a technical object will vary widely among the people in any user group. Different user groups, the level of expertise, as well as task and situation specific matters will affect the way people use technical manuals and how the fragments of information in technical manuals should be organised.

Technical manuals contain text, tables, and graphics, such as technical illustrations and engineering drawings, possibly also animation, sound and video sequences. Technical illustrations should not be confused with engineering drawings which can be difficult for non-technical people to interpret. Whereas a drawing will show every detail, technical illustrations depict exactly what is necessary and may even be slightly distorted to show various views. They can be used to show sectioned, cut-away and transparent views of equipment. They can also depict a process graphically. Generally, technical manuals are organized in chapters and subchapters. The different pieces of information can have many internal and external relationships. For example a specific subchapter or a drawing in the Installation chapter may also have an important role in the Maintenance chapter. Documents can include references to external information resources, such as standards and legal documents. Documents will need to be updated either often or now and then depending on the technical object it is related to. The basic components of technical manuals are presented in Table 1 [19].

Table 1. The basic components of technical manuals.

Component	**Examples of contents**
User guide	Guidelines for using a technical manual
Object independent general information	Company information, for instance links to home pages of the suppliers and vendors
Object dependent general information	Model or version number
Safety: notes, cautions, warnings	Remarks related to security from personal, environmental or equipment point of view
Structure of a technical object	A drawing
Functions of a technical object	Descriptions of the main functions of the technical object, may include links for example to procedural instructions related to trouble-shooting situations
Task and/or user group specific components • Installation • Operation • Commissioning • Maintenance • Training	Step by step procedures and related linkbases for installation, operation, commissioning and maintenance, user group specific training materials
Processes	Descriptions of the main processes, concerns especially more complex systems
Maintenance schedules	List of service interval and related maintenance instructions
Trouble-shooting situations	List of identified trouble-shooting situations and links to step by step instructions on how to solve them
Part lists	Parts lists including item number, description, quantity and part number linked to a numbered illustration. Even if the part numbers change the illustration remains valid.
Transportation	Regulations concerning transportation of a technical object or parts of it
Storage	Regulations concerning storage of a technical object or parts of it
De-installation and disposal	Regulations concerning de-installation and disposal of a technical object or parts of it
Environmental aspects	Regulations concerning the environmental aspects of a technical object or parts of it
Indexes and lists • Technical terms • Abbreviations • Tables • Technical illustrations • Engineering drawings • Animation, sound and video • List of links • Standards, other manuals	Linkbases related to technical terms, abbreviations, tables, technical illustrations, engineering drawings, animation, sound and video sequences, standards and other manuals related to the technical object
Users' notes	The user can add his/her comments on a special situation. This is an opportunity to share tacit information.

4. Time-sensitive relationship management

4.1 Time-sensitive relationship analysis

Relationship navigation analysis (RNA) provides a systematic way of identifying useful relationships in application domains [40]. RNA enhances system developers' understanding of application domains by broadening and deepening their conceptual model of the domain. RNA is based on generic relationship taxonomy and it can be applied to describe an application domain in terms of its relationships. It provides a systematic technique for determining the relationship structure of an application helping the designers to discover all potentially useful relationships in an application. According to Yoo and Bieber [40] RNA's generic relationship taxonomy contains 16 relationships. Temporal relationship is one of these, and it is defined as connecting an item of interest to temporally related items.

In our research we have extended RNA's temporal relationship analysis by means of Allen's relations between two time intervals. Time intervals are a set of constraints between two points, a start and an end time. Relationships between two time intervals (t_1, t_2) are defined by Allen's relations: before (t_1, t_2), equal (t_1, t_2), meets (t_1, t_2), overlaps (t_1, t_2), during (t_1, t_2), starts (t_1, t_2), finishes (t_1, t_2), and their inverse [1, 2, 3]. The time-sensitive relationship analysis point of view provides a unique starting point for identifying time dependent parts of an application domain, and for designing appropriate time contexts for different user groups and situations. We have applied time-sensitive relationship analysis to technical manuals. Examples of time-sensitive parts in technical manuals are instruction procedures in trouble-shooting situations and maintenance schedules obeying certain service intervals. In the first iteration, the designers can do the analysis. In the second iteration designers together with representatives from different user groups shall do the time-sensitive relationship analysis in order to understand the relationship structure and the users' needs more precisely. Time-sensitive relationship analysis comprises six main steps (Table 2). The analysis also helps to determine appropriate navigational structures on top of implemented links.

Table 2. Main steps of time-sensitive relationship analysis of a technical manual.

Step	Task
Analysis of users and/or user groups	Identifying individuals and/or groups using time-sensitive aspects of a technical manual.
Element analysis	Identifying different elements or element clusters of interest in a technical manual for each user and/or user group.
Time-sensitive relationship analysis	Identifying time-sensitive relationships for each element and/or element cluster of interest.
Navigation analysis	Identifying possible navigational structures for time-sensitive relationships.
Implementation analysis	Identifying implementation issues and evaluating existing technological methods and tools.
Cost/benefit analysis	Designers and users do an informal cost/benefit analysis based on iterative testing to decide which relationships would be useful and feasible enough to include in the final version of the application.

Time-sensitive relationships can be grouped into five major categories according to whether they present temporal orders, durations, clusters, iterations or conditionals. Orders are relationships that specify the order of occurrence of document fragments in a technical manual. Duration specifies relationships between the duration of intervals. Cluster relationships allow designers to group related documents and/or fragments of documents which can be used as a new entity in another temporal relationship. Iteration relationships specify that for instance a certain maintenance task should be carried out a specific number of times or until a certain condition is attained. Conditionals relationships are dependent on the state of the technical object. Conditional relationships could include a counting mechanism such as an hour meter of the technical object that allows the users to monitor system states and for-coming operations. Examples of temporal relationship categories are given in Table 3.

Table 3. Examples of temporal relationships in a technical manual.

Temporal relationship category	**Example: Maintenance**
Orders	The order of tasks in maintenance instructions.
Durations	Service intervals of a technical object.
Clusters	Textual maintenance instructions could be complemented with animation, sound and video sequences to obtain multimedia maintenance instruction procedures.
Iterations	A maintenance task should be repeated a specific number of times or until a certain condition is attained.
Conditionals	Maintenance schedules obey certain service intervals such as t=8 hours, t=50 hours and t=200 hours.

4.2 Multiple link networks

Time-sensitive relationships can be implemented as simple or extended links as defined by the XML Linking Language (XLink). Relationships with multiple endpoints connect not only two but a set of related nodes. In the XLink specification these are called extended links. An extended link is a link that associates an arbitrary number of resources and defines linking traversal rules between the resources. Traversal rules determine which participants are starting resources and which are ending resources in an extended link.

The traversal attributes define the desired traversal between resources participating in the multi-ended link. The resources are identified by their *label* attribute values. The *from* attribute defines the starting resources from which the traversal initiates. The *to* attribute defines resources that are traversed to, i.e. ending resources. The behaviour attributes specify how the ending resources will be shown to the user. For example *replace* means replacing windows on a screen. The semantic attributes describe the meaning of the link arc's ending resource relative to its starting resource. This contextual role can differ from the meaning of an ending resource when taken outside the context of this particular link arc. For example, a resource might generically represent a "technical object", but in the context of a particular link arc it might have the role of a "paper machine" and in the other context the role of a "rock grinder".

Time-sensitive links are functions of time, and they are activated when a context dependent temporal rule (TR) is valid. Allen's interval relations give us a base for describing temporal rules and creating time-sensitive linking and navigation support for time dependent Web applications. We have extended XLink's multi-ended link syntax with a *timerule* (TR) attribute and applied Allen's interval relations to formalize temporal rules related to the application and to describe them in a natural language. The *timerule* attributes can be embedded in the *xlink* element or they can be stored in separate *linkbase.xml* -files. We call this linking mechanism XLinkTime (Figure 2).

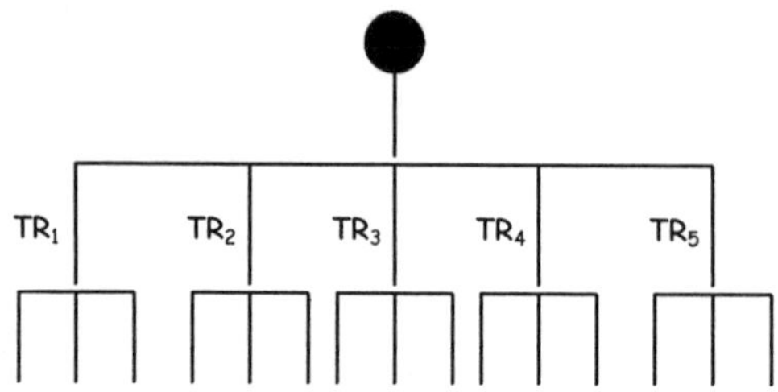

Figure 2. XLinkTime is a linking mechanism for time-sensitive relationship management. Examples of time rules TR_i, (i = 1, 2, 3, 4, 5) are $TR_1 = t_1$, $TR_2 = t_2$, $TR_3 = t_3$, $TR_4 < t_1$, $TR_5 > t_2$.

The implementation depends on the life-cycle of the application, on maintenance requirements, user needs and delivery platforms. The explicit syntax of the XLinkTime is always application dependent and the implementation requires a time analysis of the application concerned.

XLink's extended links can contain extended links as well, which makes multiple link networks possible. It is possible to create and maintain several context dependent link networks, which can be applied to a set of XML documents, in our case the maintenance instructions of a technical object. Maintenance procedure descriptions can be shown as a list of time dependent links. Each time dependent link points to a link network in a XML document base. According to the user input, which depends on the situation and on contextual time parameters, a certain link network with related step-by-step instructions will be delivered to the user's PC or mobile device (Figure 3).

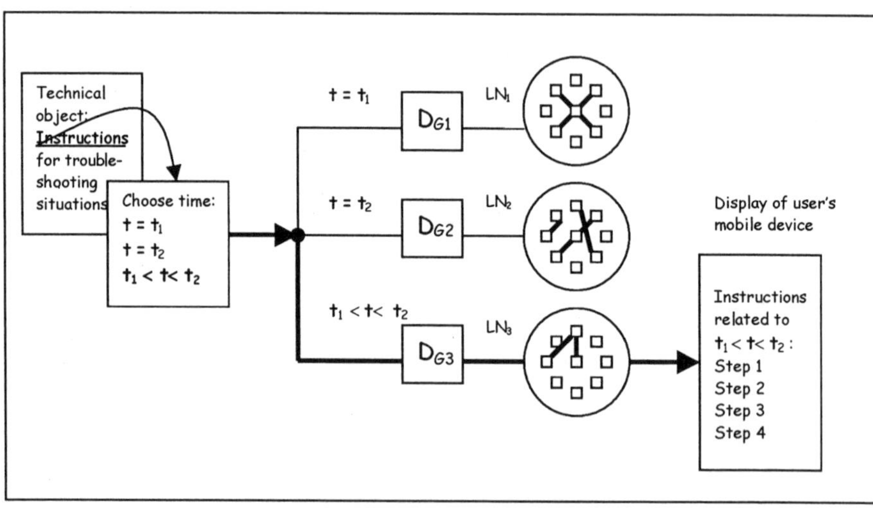

Figure 3. Time-sensitive linking mechanisms with multiple link networks (LN_1, LN_2, LN_3) and related document groups (D_{G1}, D_{G2}, D_{G3}). The user has given the time context $t_1 < t < t_2$ as an input to the system (bold line). Link network 3 and related document group D_{G3} are chosen and delivered to the user's device.

5. Implementation issues

5.1 Three approaches

When implementing time-sensitive linking mechanisms, there are several issues that come up. Is a) the time "now", b) the time a user decision, or is c) the time related to some other entity? It can be any of these and it should be stated somehow to the user. From the interval based point of view if the time is "now", an essential question is what is its interval? In this case, the interval is zero, meaning that the start time and the end time are equal and they have the same precision. Is the time interval absolute or relative? An example of an absolute time interval is a service interval of a technical object. Time zones are examples of relative time intervals. When defining a time context as "today", the time interval can be 0.00-23.59 or it can mean a date with a zero interval. If the time is a user decision and the application is used world wide, for example in internationally distributed enterprises, the application should include an automatic time transformation process for local and for central time. In our demonstration we use absolute time which is defined by the range of the use hours of a rock grinder.

There are basically three approaches for implementing time-sensitive links with current technology (Figure 4). The solutions usually differ only in the deployment of components. The solution to be chosen is a matter of application domain requirement specifications.

- Approach 1: The server transforms the data to HTML and there are no special needs for client applications. This approach is best suited for public Internet applications and can be used with legacy HTML-browsers. Time rules can be kept hidden in the service, or they can be brought to the client, and they are accessible by JavaScript. The service has an interface that uses Extensible Stylesheet Language Transformations (XSLT) or similar technology for rendering the view.

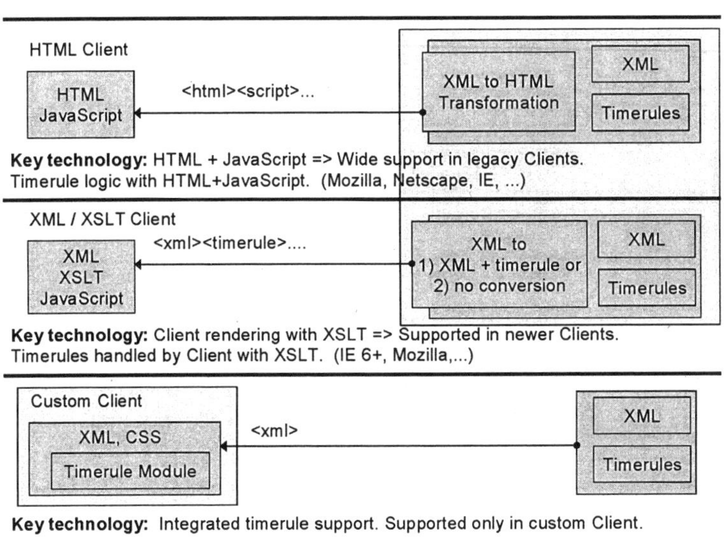

Figure 4. Three approaches for implementing time-sensitive linking. The requirements for the browser increase from top to bottom and the requirements for the server increase from bottom to top.

- Approach 2: The client transforms the data. This reduces the load on the server side and the clients are responsible for generating the views. Contemporary browsers support the XSLT transformations. This approach is best suited for public Internet applications.
- Approach 3: The server is a central repository for time rules. Client browsers are more sophisticated and have integrated support for time-sensitive linking. The approach is best suited for extranet and intranet applications.

When moving from the first approach to the third one several improvements are achieved. The server will be used only for data delivery, not for rendering. The time-sensitivity changes from emulation into integration. When time-sensitivity is integrated into a client application, the application can provide more link interface options to the user, such as menus and pop-up windows. We have used Approaches 2 and 3 in our demonstration.

Table 4. The structure of the rock grinder manual.

Level 1	Level 2	Level 3
Introduction	General	
General safety regulations	General safety regulations General remarks Start-up and operation General warranty conditions	The purpose of the warning signs A new shift
Process description	Work technique and material	General
Function	Function of the grinder Working methods Technical data Transmission Pressure settings Load holding valve	General General Electrical operation, Power transmission, Hydraulical system, Electrical system, Lubrication system, Framework, Hopper General , Reversible gear, Starting, Drive, Auto On Throttle non-return valves, Brake system, Pusher, Adjustment of operating pressure Counter knife bracket and spring tension
Operational instruction	Start and operation Starting a grinder Stopping a grinder Emergency stop	New shift, After work is over or during a temporary stop Before starting, Starting, Automatic operation Manual operation, Stopping When work is over, Emergency stop of grinder
Maintenance instructions	Safety measures **Maintenance** Replacing/grinding of grinder shaft teeth New counter knives Hard-facing Brake linings Central lubrication diagram Seal replacement, hydraulic cylinders	General, Welding operations, Working with oils and fuels **How to use the maintenance schedules**, Central lubrication, Oil grades recommended, Lubrication grease, Special notes on hydraulic oil, **8-h maintenance, 50-h maintenance, 200-h maintenance, 400-h maintenance, 800-h maintenance, 1200-h maintenance, 2400-h maintenance** General Alternation of counter knife shaft position, Hard-facing of grinder and counter knife teeth Welding material, Replacing brake lining Positions Replacement of seals
Trouble-shooting	Trouble-shooting	The engine stops, The conveyor stops, The brakes do not work, The pusher stops, The reversing valve goes backwards and forwards, Reverses too long, Does not reverse, The grinder shaft does not rotate

5.2 A rock grinder instruction manual as a test material

The instruction manual provides the user with fundamental data of the design and operation of the rock grinder that is essential to its correct handling and operation. In addition to three textual hierarchy levels the rock grinder instruction manual contains graphical elements such as technical illustrations and engineering drawings as well as safety regulation elements like Warning!, Note! and N.B.!. Warning! is used when there is a risk of injuries to personnel, Note! is used when there is a risk of break-down in the equipment and N.B.! is used to make information clearer. The structure of the manual is presented in Table 4.

5.3 Demonstration: Maintenance schedules

Our demonstration deals with the maintenance schedules on the following levels of the rock grinder manual: Maintenance instruction → Maintenance → How to use the maintenance schedules → TR-hour maintenance, where TR = 8, 50, 200, 400, 800, 1200 or 2400 (time rules TR are written in bold in Table 4). The grinder has seven service intervals: 8-hour maintenance, 50-hour maintenance, 200-hour maintenance, 400-hour maintenance, 800-hour maintenance, 1200-hour maintenance and 2400-hour maintenance. Each interval deals solely with those items that are relevant to it. A complete service comprises the previous service items. For example a 400-hour maintenance includes 8-hour + 50-hour + 200-hour + 400-hour services. The service interval instructions consist of between one and seven items to be taken into account. Each item includes between one and three tasks. The instructions are short texts like check lists containing technical illustrations and some Note! remarks. The maintenance schedules with related documents and fragments of documents can be presented by means of a multi-ended link with time-sensitive multiple link networks (Figure 5).

There are several time-sensitive contexts related to the service intervals of a rock grinder where the maintenance crew needs maintenance instructions. Examples of time-sensitive contexts (in hours) from the user's point of view are:

a) $t = t_i$, $t_i = (8, 50, 200, 400, 800, 1200, 2400)$ represent the service intervals when certain maintenance operations must be done. The maintenance instructions related to each t_i can be delivered automatically to the user's device when a certain t_i is achieved or according to the user's request.
b) t = 200, the user wants to check which operations are included in the 400-hour maintenance.
c) t = 300, the user wants to check which maintenance operations have been performed during the last service intervals.
d) t = 600, a trouble-shooting situation may occur and the user wants to check which maintenance operations should have been carried out during the earlier service intervals in order to exclude possible neglects.
e) t = 600, a trouble-shooting situation may occur and the user wants to check which operations should have been carried out during the previous service interval in order to exclude possible neglects and which operations are to be carried out during the following service interval.
f) t = 790, the 800-hour maintenance check point is approaching and the user wants to check which operations have to be performed during the next working day.
g) t = 450, the user wants to check which operations should have been performed at the 400-hour maintenance.

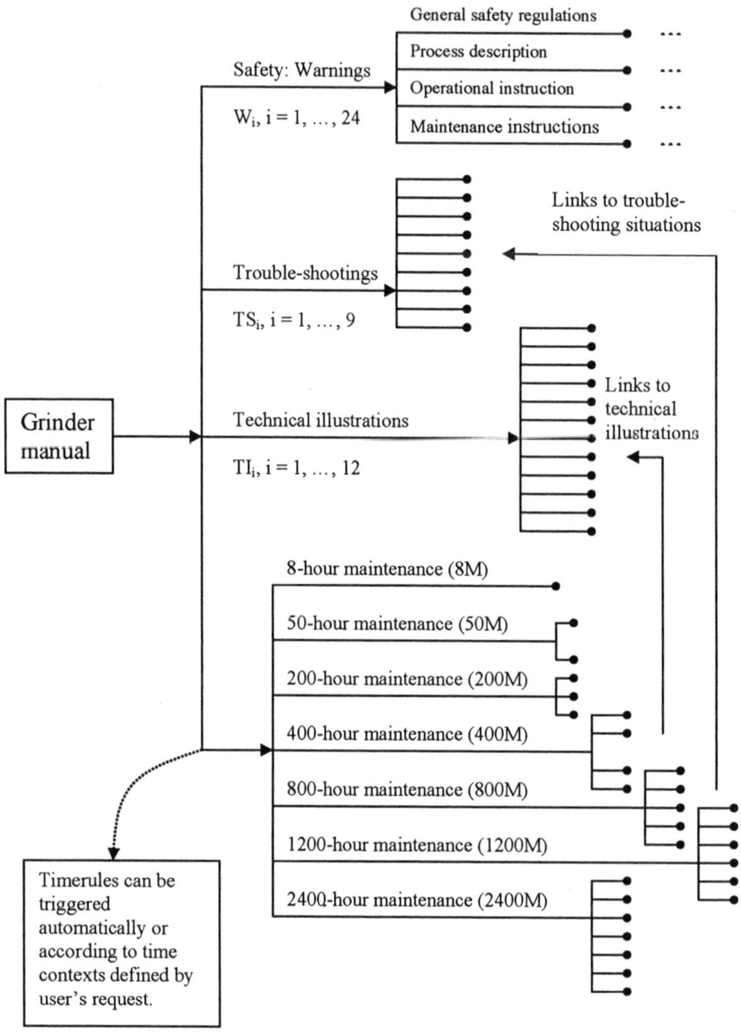

Figure 5. A multi-ended link with time-sensitive multiple link networks describing maintenance schedules of a grinder manual together with related documents.

h) $t = 900$, the user wants to check which operations are to be done.
i) $400 \leq t \leq 800$, there are two rock grinders in use, one that has been bought as new and has now been used 1200 hours, and another one that has been bought as second-hand (400 hours). The user wants to compare their operation during the time interval $400 \leq t \leq 800$.
j) $1200 \leq t \leq 2400$, the user wants to compare the operations of the two rock grinders in example i) before their annual 2400-maintenance.

Correspondences between time contexts related to service intervals listed above and Allen's interval relations are given in Table 5.

Table 5. Examples of time contexts from the user's point of view corresponding to Allen's interval relations.

Allen's interval relations	**Example of time-sensitive contexts**
Right-before	Just before a certain time instant; example f.
Right-after	Just after a certain time instant; example g.
Before	Before a certain time instant; example c.
After	After a certain time instant; example h.
Meets	Consecutive service intervals; example b.
During	Comparing the operation of two rock grinders; example i.
Equal	8-, 50-, 200-, 400-, 800-, 1200-, 2400-hour maintenances; example a.
Overlaps	Two intervals $0 \leq t \leq 600$ and $400 \leq t \leq 1200$ overlap; example e.
Start	Intervals $0 \leq t \leq 400$ and $0 \leq t \leq 600$ start at the same time; example d.
Finish	Comparing the operation of two rock grinders before their annual maintenance; example j.

We have constructed a demonstration in Mozilla-environment (Figure 6). The demonstration consists of four main files. Baseloader.xul loads an external xml-file that contain xlink-links. Situations.xul produces a tree-structure on the left frame. The contents are read from the linkbase document that has been loaded by baseloader.xul. Xlink.js is a Java script which builds up a datamodel from xlinks. Simgrinder.xml is a linkbase for maintenance schedules. A user selects a linkbase from a linkbase frame. Linkbase documents are loaded and a tree of documents is generated and updated. The tree view consists of nodes that group maintenance functions according to the service intervals. The user can choose between different service intervals and related tasks according to the time conditions which are at hand.

6. Conclusions and topics for further research

We have discussed time-sensitive Web-based application domains from the relationship management point of view. The focus of the paper has been on analysis and implementation issues of temporal relationships. Relationships have been implemented as time-sensitive links which are activated when a certain time condition is valid. Our general application domain has been technical manuals of technical objects. This paper has dealt with the instruction manual of a rock grinder. Time conditions are defined by the maintenance schedule of the rock grinder, which includes seven service intervals. Each service interval consists of certain maintenance procedures and tasks. Related time contexts are defined by the users, i.e. the maintenance crew and/or by the technical object itself.

We have extended a temporal relationship category of generic relationship taxonomy by means of Allen's relations between two time intervals. The implementation of time-sensitive relationships is based on the XML Linking Language (XLink) specification. We have introduced a time-sensitive linking mechanism, XLinkTime, by extending XLink's

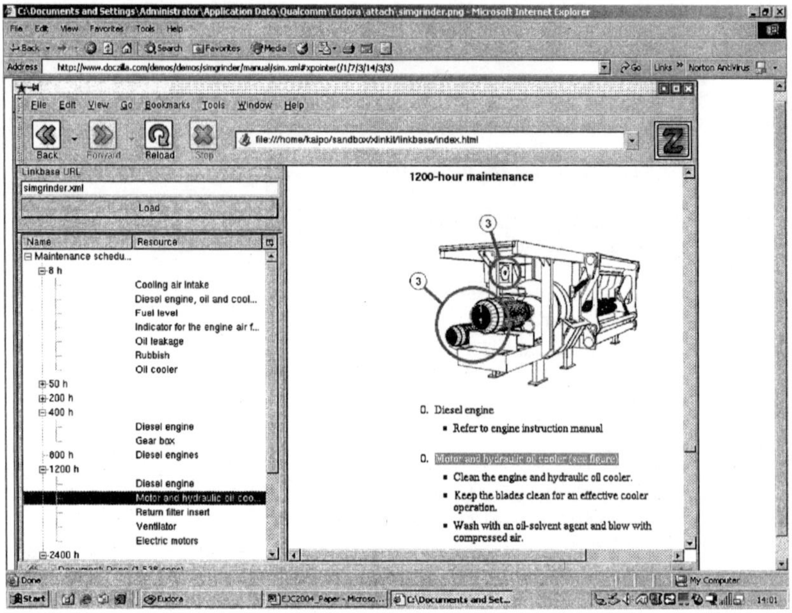


```
<?xml version="1.0" ?>
<loadbase>
<startrsrc xlink:type="arc"xlink:arcrole="http://www.w3.org/1999/xlink/properties/linkbase"
xlink:label="ms" xlink:href="sim.xml/>
<linkbase xlink:label="linkbase" xlink:href="simgrinder.xml" />
<load xlink:from="ms" xlink:to="linkbase" actuate="onLoad" />
</loadbase>
...
<links xmlns:timerule=http://www.tut.fi/amc/2003/05/timerule xmlns:xlink="http://www.w3.org/1999/xlink"
xlink:type="extended">
<local xlink:type="resource" xlink:title="Maintenance schedules" xlink:label="ms" />
<!-- Maintenance Schedules -->
<!-- 1200 h -->
<remote xlink:type="locator" timerule:start="0 h" timerule:end="1200 h" timerule:title="1200 h"
xlink:label="ms" xlink:title="Diesel engine"
xlink:href=http://www.doczilla.com/demos/demos/simgrinder/manual/sim.xml#xpointer(/1/7/3/14/3/1)
<remote xlink:type="locator" timerule:start="0 h" timerule:end="1200 h" timerule:title="1200 xlink:label="ms"
xlink:title="Motor and hydraulic oil cooler"
xlink:href="http://www.doczilla.com/demos/demos/simgrinder/manual/sim.xml#xpointer(/1/7/3/14/3/3)" />
<remote xlink:type="locator" timerule:start="0 h" timerule:end="1200 h" timerule:title="1200 h" xlink:label="ms"
xlink:title="Return filter insert"
xlink:href="http://www.doczilla.com/demos/demos/simgrinder/manual/sim.xml#xpointer(/1/7/3/14/3/5)" />
<remote xlink:type="locator" timerule:start="0 h" timerule:end="1200 h" timerule:title="1200 h" xlink:label="ms"
xlink:title="Ventilator"
xlink:href="http://www.doczilla.com/demos/demos/simgrinder/manual/sim.xml#xpointer(/1/7/3/14/3/7)" />
<remote xlink:type="locator" timerule:start="0 h" timerule:end="1200 h" timerule:title="1200 h" xlink:label="ms"
xlink:title="Electric motors"
xlink:href="http://www.doczilla.com/demos/demos/simgrinder/manual/sim.xml#xpointer(/1/7/3/14/3/9)" />
<arc xlink:type="arc" xlink:from="ms" xlink:to="ms" />
</links>
```

Figure 6. Time-sensitive links applied to maintenance schedules of the rock grinder.

multi-ended link and multiple link network structures with a *timerule* attribute. XLinkTime provides temporal ruling facilities for choosing, assembling and delivering documents or fragments of documents from the time-based point of view. We have demonstrated XLinkTime by embedding the time rules in the XML document's *xlink* element and by separating them from the XML documents and storing them into a separate *linkbase.xml* file.

In the next phase, we will construct a mobile demonstration with a 3D time-sensitive link user interface based on Scalable Vector Graphics (SVG). We are also interested in applying XLinkTime to other possible application domains such as distributed project management. In a theoretical sense, granularity of temporal linking rules and RDF-based (Resource Description Framework) management of time-sensitive linking structures are challenging issues for further research.

References

[1] Allen, J. F. 1983. Maintaining knowledge about temporal intervals. Communications of the ACM, Vol. 26, No. 11, pp. 832 – 843.

[2] Allen, J. F. 1984. Towards a general theory of action and time. Artificial Intelligence, Vol. 23, No. 2, pp. 123 – 154.

[3] Allen, J. F. 1991. Time and time again: The many ways to represent time. International Journal of Intelligent Systems, Vol. 6, No. 4, pp. 341 –355.

[4] Anderson, K. M., Taylor, R. N. and Whitehead, E. J. 1994. Chimera: Hypertext for heterogeneous software environments. In Proceedings of the ACM European Conference on Hypermedia Technology (ECHT '94), September, 18 – 23, Edinburgh, Scotland, UK. New York: ACM Press. Pp. 94 – 197.

[5] Berners-Lee, T. 1999. Weaving the Web. New York: Harper SanFrancisco. 226 p.

[6] Berners-Lee, T., Hendler, J. and Lassila, O. 2001. The Semantic Web. Scientific American, vol. 284, No. 5, pp. 28 – 37.

[7] Bieber, M., Vitali, F., Ashman, H., Balasubramanian, V. and Oinas-Kukkonen, H. 1997. Fourth generation hypermedia: some missing links for the World Wide Web. International Journal of Human-Computer Studies, Vol. 47, No. 1, pp. 31 - 65.

[8] Bry, F. and Kraus, M. 2002. Advanced modelling and browsing of technical documents. In Proceedings of the 2002 ACM Symposium on Applied Computing. New York: ACM Press. Pp. 520 - 524.

[9] Carr, L., Bechhofer, S. Goble, C. and Hall, W. 2001. Conceptual linking: Ontology-based open hypermedia. In: Proceedings of the 10th World Wide Web Conference (referred 31.5.2001) <URL: http//www10.org>.

[10] Carr, L. A., Hall, W. and Hitchcock, S. 1998. Link services or link agents? In: Grønbæk, K., Mylonas, E. and Shipman, F. M. (eds.) Proceedings of the Ninth ACM Conference on Hypertext and Hypermedia, Pittsburgh, PA 20 – 24 June 1998, USA. The Association for Computing Machinery. Pp.113 – 122.

[11] Ciancarini, P., Folli, F., Rossi, D. and Vitali, F. 2002. XLinkProxy: External linkbases with XLink. The New review of Hypermedia and Multimedia, Vol. 8, 2002, pp. 27 – 37.

[12] Courtiat, J.–P. and De Oliveira, R. C. 1996. Proving temporal consistency in a new multimedia synchronization model. In Proceedings: ACM Multimedia 96, Boston, USA. USA. The Association for Computing Machinery. Pp.141 - 152.

[13] Christensen, B. G. and Hansen, F. A. 2002. XLink – Linking the Web and open hypermedia. In Millard, D., Haake, J. M. and Reich, S. (eds.). Proceedings of the International Workshop on Open Hypermedia Systems, Core Concepts & Research Directions. Pre-Conference Workshop at the ACM 13th International Conference on Hypertext and Hypermedia, June 12th, 2002, Maryland, USA. Hagen: FernUniversität. Pp. 9 – 18.

[14] DocZilla 2003. DocZilla Homepage (referred 20.11.2003) <URL: http://www.doczilla.com/>.

[15] Grønbæk, K. and Trigg, R. H. 1999. From Web to workplace: Designing open hypermedia systems. The MIT Press Books, Boston, MA. 424 pp.

[16] Grønbæk, K., Sloth, L. and Bouvin, N. O. 2000. Open hypermedia as user controlled meta data for the web. In Proceedings of the 9th International World Wide Web Conference, May 2000, Amsterdam, Holland. Pp. 553 – 566.

[17] Halasz, F. and Schwartz, M. 1994. The Dexter hypertext reference model. Communications of the ACM, Vol. 37, No. 2, pp. 30 – 39.

[18] Hardman, L., Schmitz, P., van Ossenbruggen, J., ten Kate, W. and Rutledge, L. 2000. The link vs. the event: activating and deactivating elements in time-based hypermedia. The New Review of Hypermedia and Multimedia, Vol. 6, pp. 89 - 109.

[19] Heimbürger, A. 1999. Link management in technical customer documentation. In: Käkölä, T. (ed). The Proceedings of the 22nd Information Systems Research Seminar in Scandinavia (IRIS 22), Enterprise architecture for virtual organisations, August 7 - 10, 1999, Keuruu, Finland. University of Jyväskylä, Department of Computer Science and Information Systems, Technical Reports TR-21, Vol. 2. Pp. 13 - 26.

[20] Heimbürger, A. 2003. Modelling time-sensitive linking mechanisms. In: Jaakkola, H., Kangassalo, H, Kawaguchi, E. and Thalheim, B. (eds.). Frontiers in Artificial Intelligence and Applications, Vol. 94, Information Modelling and Knowledge Bases XIV. Amsterdam: IOS Press. Pp. 26 - 42.

[21] Heimbürger, A. 2003. Contextual multi-ended linking: Developing a method for content adaptation. In Proceedings of the International Workshop on Interactive Computer Aided Learning (ICL 2003), Learning Object and Reusability of Content, September 24 – 26, 2003, Villach, Austria. Pp. 1 – 25.

[22] Heimbürger, A. 2003. Developing time-sensitive hypertext linking and navigation support. In: Proceedings of the International Federation for Information Processing (IFIP) Open Working Conference, E-Training Practices for Professional Organisations (eTrain 2003), July 7 – 11, 2003, Pori, Finland (to appear).

[23] ISO 8879. 1986. Information processing – Text and office systems – Standard generalized markup language (SGML). Geneva: International Organization for Standardization. 155 p.

[24] Martin, D. and Ashman, H. 2002. Goate: An infrastructure for new Web linking. In Millard, D., Haake, J. M. and Reich, S. (eds.). Proceedings of the International Workshop on Open Hypermedia Systems, Core Concepts & Research Directions. Pre-Conference Workshop at the ACM 13th International Conference on Hypertext and Hypermedia, June 12th, 2002, Maryland, USA. Hagen: FernUniversität. Pp. 19 - 25.

[25] Miles-Board, T., Kampa, S., Carr, L. and Hall, W. 2001. Hypertext in the Semantic Web. In: Davis, H., Douglas, Y. and Durand, D. G. (eds.) 2001. ACM Hypertext 2001, Proceeding of the 12th ACM Conference on Hypertext and Hypermedia. Aarhus, Denmark, 14 – 18 August, 2001. New York: The Association for Computing Machinery. Pp. 237 - 238. ACM ISBN 1-59113-420-7.

[26] Millard, D. et al. AHA! Meet Auld Linky: Integrating designed and free-formed hypertext systems. In Proceedings of the 14th ACM Conference on Hypertext and Hypermedia, August 26 – 30, 2003, Nottingham, UK. New York: ACM Press. Pp. 161 – 169.

[27] Mozilla 2003. Mozilla Homepage (referred 20.11.2003) <URL: http://www.mozilla.org/>.

[28] Muchaluat-Saade, D. C. and Soares, L. F. G. 2002. XConnector and XTemplate: improving the expressiveness and reuse in web authoring languages. The New review of Hypermedia and Multimedia, Vol. 8, 2002, pp. 139 -169.

[29] Neto, R. B. et al. 2002. An open linking service supporting the authoring of Web documents. In Proceedings of the 2002 ACM Symposium on Document Engineering (DocEng '02), November, 8 – 9, 2002, McLean, Virginia, USA. New York: ACM Press. Pp. 66 – 73.

[30] Nürnberg, P. J., Leggett, J. J. and Wiil, U. K. 1998. An agenda for open hypermedia research. In: Grønbæk, K., Mylonas, E. and Shipman, F. M. (eds.) Proceedings of the Ninth ACM Conference on Hypertext and Hypermedia, Pittsburgh, PA 20 – 24 June 1998, USA. The Association for Computing Machinery. Pp.198 – 206.

[31] Paganelli, C. and Mounier, E. 2003. Information retrieval in technical documents: from user's query to the information-unit tagging. In Proceedings of the 2003 ACM 21st Annual Conference on Documentation, October, 12 - 15, 2003, San Francisco, CA, USA. New York: ACM Press. Pp. 133 - 139.

[32] Pérez-Luque, M. J. and Little, T. D. C. 1996. A temporal reference framework for multimedia syncronization. IEEE Journal on Selected Areas in Communications, Vol. 14, no. 1, pp. 36 – 51.

[33] Reich , S. and Anderson, K. M. (eds.) 2000. Open hypermedia systems and structural computing. In Proceedings of 6th International Workshop, Ohs-6, 2nd International Workshop, Sc-2, San Antonio, Texas, USA, May/June, 2000. Berlin: Springer. 183 p. (Lecture Notes in Computer Science 1903)

[34] Vazirgiannis, M. 1999. Interactive multimedia documents. Modelling, authoring, and implementation experiences. Lecture Notes in Computer Science 1564. Berlin: Springer-Verlag. 161 p.

[35] W3C 2003a. The World Wide Web Consortium: HLink: Link recognition for the XHTML (referred 20.11.2003) <URL: http://www.w3.org/TR/2002/WD-hlink-20020913/>.

[36] W3C. 2003b. The World Wide Web Consortium: Semantic Web Activity, (referred 20.11.2003) <URL: http://www.w3.org/2001/sw/>.

[37] W3C. 2003c. The World Wide Web Consortium: XML Linking Language (XLink) Version 1.0 W3C Recommendation 27 June 2001, (referred 20.11.2003) <URL: http://www.w3.org/TR/xlink/>.

[38] Wang, W. and Rada, R. 1998. Structured hypertext with domain semantics. ACM Transactions on Information Systems, Vol. 16, No. 4, pp. 372 – 412.

[39] Yankelovich, N., Haan, B. J., Meyrowitz, N. K. and Drucker, S. M. 1988. Intermedia: the concept and construction of a seamless information environment. IEEE Computer, Vol. 21, No. 1, pp. 81 – 96.
[40] Yoo, J. and Bieber, M. 2000 A systematic relationship analysis for modelling information domains. In: Rossi, M. and Siau, K. (eds.) Information modelling in the new millennium. Hershey, USA: Idea Group Publishing. Pp. 485 – 501.
[41] Zhuge, H. 2003. Active e-document framework ADF: model and tool. Information Management, Vol. 41, No. 1, pp. 87 – 97.

Information Modelling and Knowledge Bases XVI
Y. Kiyoki et al. (Eds.)
IOS Press, 2005

A *Kansei* Metadata Generation Method for Music Data Dealing with Dramatic Interpretation

Asako IJICHI†and Yasushi KIYOKI††
†*Graduate School of Media and Governance, Keio University*
††Faculty of Environmental Information, Keio University
Endoh 5322, Fujisawa, Kanagawa, 252-8520, Japan
{aco,kiyoki}@sfc.keio.ac.jp

Abstract. In this paper, we propose a "Kansei" metadata generation method for music data dealing with dramatic interpretation. The concept of "Kansei" includes several meanings on *sensitive recognition*, such as human senses, feelings, sensitivity and psychological reactions. The aim of our research is to generate metadata for transitional media data. Our metadata generation method computes correlations between music data and impression words by reflecting the degree of a gap of two adjacent impressions in temporal transition of music. This method consists of the following processes: 1) setting a tree structured music granularity 2) metadata generation dealing with dramatic interpretation 3) extraction of primitive impression from music data. This mechanism makes it possible to create metadata which reflect natural human senses. This paper shows several experimental results of metadata generation to clarify the feasibility and effectiveness of our method.

1 Introduction

In this paper, we propose a metadata generation method for music data dealing with dramatic interpretation. In the design of the multimedia database systems, there are two important issues to generate metadata from media data. The first issue is how to deal with "Kansei" of human beings. The second issue is how to represent a metadata dealing with "temporal transition and its effects" of music data.

As discussed in [2], the field of "Kansei" was originally introduced as the word *aesthetics* by Baumgrarten in 1750. The aesthetic of Baumgrarten had been established and succeeded by Kant with his ideological aesthetics. The concept of "Kansei" includes several meanings on *sensitive recognition*, such as human senses, feelings, sensitivity and psychological reactions. The "Kansei" information plays an essential role in human's understanding and memory of media data. By introducing kansei information into the media retrieval technology, we can realize the natual and meaningful music retrieval.

One of the most significant features of music data is *temporal transition*. We focus an attention to transitional "Kansei" information in music data. Since music data give impression with temporal response, metadata of the music should express a concept of temporal response. An important feature of transitional media data is that the temporal transition affects interpretation of the "Kansei" information in media data. When users listen to music, the current part of "Kansei" impression felt by users is affected by the previous part of "Kansei" impression. To generate appropriate metadata for transitional media data, it is important to deal with transition of "Kansei" information. We introduce a method for calculating "effects of transitional media data (dramatic interpretation)".

Transitional media data consists of several parts, and each part has its unique role, such as a theme part and an insertion part. Dramatic interpretation leads to a theme impressively by using the characteristic of phenomenon of temporal transition.

As an instance in movies, a scene expresses happiness. Then subsequent scene expresses tragic episode. This is the example of dramatic interpretation to emphasis the tragic episode by the gap between "happiness" and "tragic" episode.

Dramatic interpretation is commonly recognized. Effects of dramatic interpretation can be shared as common perception. Human beings are affected by dramatic interpretation from temporal transition of "Kansei" information. To realize media data retrieval by "Kansei", it is indispensable to deal with temporal transition and its effects of "Kansei" information. That is, it is possible to generate "Kansei" metadata with calculating dramatic interpretation and reflecting it to each metadata.

In this paper, we propose a metadata generation method for music data dealing with dramatic interpretation. The main objective of this method is to realize a music retrieval environment based on "Kansei". Our metadata generation method computes correlations between music data and impression words by reflecting the degree of a gap between two adjacent parts of impression in temporal transition of music. This method makes it possible to generate metadata with the most impressive part emphasized.

There are several challenges and approaches to retrieve music data[1][9]. Related research can be summarized as follows: 1)music retrieval by attribute information, 2)music retrieval by kansei information and 3)music retrieval by transitional raw music data. ID3-Tag information of MP3 provides attribute information that is used to retrieve music by incidental information of music, such as composers and players. A framework for media data retrieval by kansei was introduced in [6]. In the framework, "Kansei" metadata are created without temporal transitions. This framework is suitable for calculating correlations between adjectives and music. A query-by-humming system for a music database was introduce in [10]. The target data of the system are transitional raw music. A direct retrieval for transitional music data is realizes by calculating correlations between a query and raw music data. The semantic associative search system[7][8] is used to media data retrieval by receiving keywords representing the user's impression and the media data. In [5], we have proposed our framework for music retrieval- by-impression.

Our method makes it possible to provide natural and meaningful media retrieval dealing with temporal transition.

2 An impression-metadata generation method for music data retrieval-by-impression dealing with impression transition

We propose a metadata generation method for music data dealing with dramatic interpretation. Our method computes correlations between music data and the impression words by reflecting the degree of a gap between two adjacent impressions in temporal transition of music. The method generates metadata with the most impressive part emphasized. Moreover, According to music structures, the generation method is applied recursively to match the granularity(size of a part) of music corresponds to the user's query.

The method realizes to generate metadata dealing with dramatic interpretation. The method is applied to any level granularity.

The method consists of following processes:

1. setting a tree structured music granularity(applied recursively)
2. metadata generation dealing with dramatic interpretation
3. extraction of primitive impression from music data

The outline of our method is shown in Figure1.

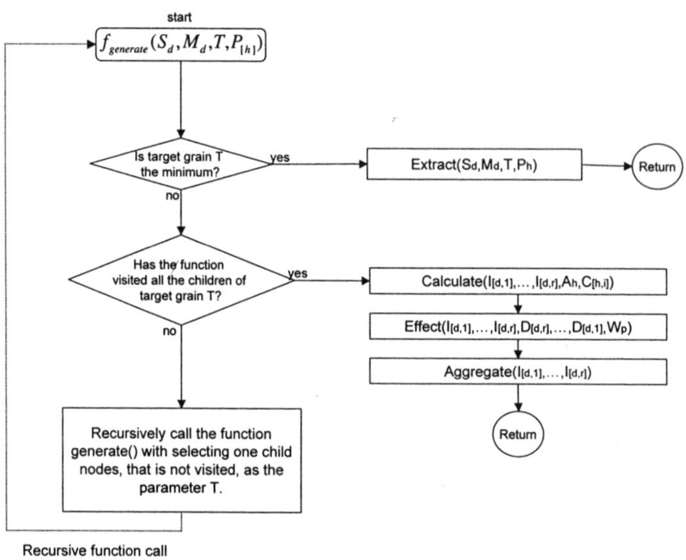

Figure 1: The outline of our metadata generation method

2.1 *Setting the tree structured music granularity*

This process sets the tree structured music granularity to generate metadata. This process transacts recursively until the result corresponds with the target granularity.

When we listen to a piece of music, we do not understand the whole of music at once. We divide the music into certain parts, understand the meaning of each part and memorize them. We call the part of music *"music grain (music section)"*. Each music grain gives its own impression.

There are several kinds of partitions to divide the music into grains. We call the criteria of partition of music *"music granularity"*. Music granularity is represented in the tree structure as shown in Figure2.

A composer writes a piece of music in a certain musical from. Each music form has its own tree structured music granularity. Therefore, each music has its own tree structure, length of a grain and an inclusive relation between two adjacent granularities. For example, a section consists of several sentences, a piece of music consists of several sections, and a CD album consists of several pieces of music. To create metadata for a grain in the granularity setting, it is necessary to deal with the metadata of the child nodes grains. For example, to create metadata of a piece of music, it is necessary to deal with sections, sentences, phrases and motif in music. As another example, to create metadata of a CD album, it is necessary to deal with music, sections, sentences, phrases and motif. We understand a piece of music, pay attention to a certain music granularity, and recognize the impression of each music grain. Accordingly, the granularity setting depends on the attention of users. Grains in different granularity setting have different meanings. A metadata generation method to retrieve music data should deal with music granularity in users' query. The function for setting granularity of a tree structured music is defined as:

Generating metadata

$$f_{generate}(S_d, M_d, T_s, P_{[h]}) \rightarrow \{I_{[d,1]}, ..., I_{[d,r]}\} \quad (1)$$

S_d=Granularity structure($d = 1, n$)(n=the maximum number of pieces of music)

M_d= Music data

T_s=Granularity required by users($s = 1, t$)(s=the number of granularities)

$P_{[h]}$=impression extraction method ($h = 1, e$)(e=the number of methods)

$I_{[d,j]}$=impression($j = 1, u$)(u=the number of grains)

2.2 *Metadata generation dealing with dramatic interpretation*

This metadata generation process calculates the degree of a gap between two adjacent impressions in temporal transition. As described in 2.1, we divide music into several grains, recognize the impression of each music grain and store them. We call the difference of impressions between two adjacent music grains *"transitional gap"*. This process calculates a transitional gap automatically and extracts metadata of transitional media data by reflecting transitional gap.

Music is one of the transitional media data. The impression of each music grain is individual in temporal transition and namely each impression is not mingled with other impressions. Each impression in temporal transition affects only the value of the subsequent grain's impressions. This method calculates a transitional gap and is used to emphasize impressions of the subsequent grain impressively. Functions of the metadata generation mechanisms are defined as:

Calculating the degree of a gap

$$f_{calculate}(I_{[d,1]}, ..., I_{[d,r]}, P_h, C_{[h,i]}) \rightarrow \{D_{[d,1]}, ..., D_{[d,i]}\} \quad (2)$$

$C_{[h,i]}$=Calculating method of the degree of two impressions($h = 1, g$)(g =the number of the methods)

$D_{[d,1]}$=Degree of a gap($d = 1, k$)(k=the number of gaps)

Weighting the gap

$$f_{effect}(I_{[d,1]}, ..., I_{[d,m]}, D_{[d,1]}, ..., D_{[d,i]}, W_p) \rightarrow \{I_{[d,1]}, ..., I_{[d,m]}\} \quad (3)$$

W_p=Weighting method($p = 1, q$)(q=the number of methods)

Aggregating metadata

$$f_{aggregate}(I_{[d,1]}, ..., I_{[d,m]}) \rightarrow \{I_{[d,1]}, ..., I_{[d,j]}\} \quad (4)$$

2.3 Extracting impression from music data

This function is applied to the minimum level granularity of music data, and represents the impression of the music grains by adjectives and its correlation value. The minimum level granularity is the lowest level of granularity structure of each music. The lowest-level of granularity corresponds to the minimum size of music grain, such as a motif, for express impression. The function for extracting impression from music data is defined as:

Extracting impression

$$f_{extract}(S_d, M_d, P_h) \rightarrow \{I_{[d,1]}, ..., I_{[d,m]}\} \quad (5)$$

The data structure to analyze impression of music data is shown in Figure2.

3 The Implementation Method

In this section, we show an implementation method of metadata generation. We have implemented an experimental system for classical music data.

3.1 Setting the tree structured music granularity

This process sets the music granularity for structuring music data. To generate metadata of music in the granularity setting, it is necessary to deal with details of temporal transition of music effects of dramatic interpretation.

3.2 Metadata generation dealing with dramatic interpretation

In this implementation, we apply a cyclic adjective list to calculate a transitional gap of two adjacent impressions. The cyclic adjective list is shown in Figure3. In the study of the paper[3], the gap of impression on the cycle is defined as follows: 1)the angle on the cycle

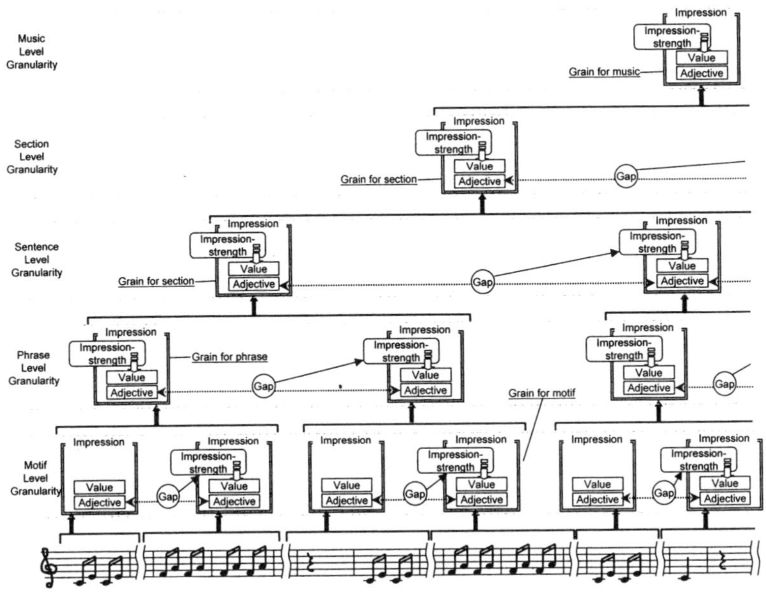

Figure 2: The data structure to analyze impression of music data

corresponds to the distance of impression, and 2)the maximum distance corresponds to the inverse impression. The parameters for transitional gaps are categorized as follows:

1)The maximum gap(parameter=2)
2)The minimum gap(parameter=0)
3)The other gap(parameter=1)

3.3 *Extraction of primitive impressions from music data*

This process extracts primitive impressions from music data. This process is based on the study of psychological experiments, where a lot of association tests for impression in music have been done on relation between classical music and adjectives[3][4]. The procedure of this process consists of the two steps:

1. Extract six important elements of music: rhythm, tempo, pitch, melody, harmony and tone.

2. Calculate correlation values of those six elements of music with the cyclic adjective list which is arranged in eight related groups.

The outline of the extraction method is shown in Figure4.

By applying this procedure to music data, our automatic metadata extraction method measures correlations between music data and adjective words, and obtains impression words corresponding to the music data.

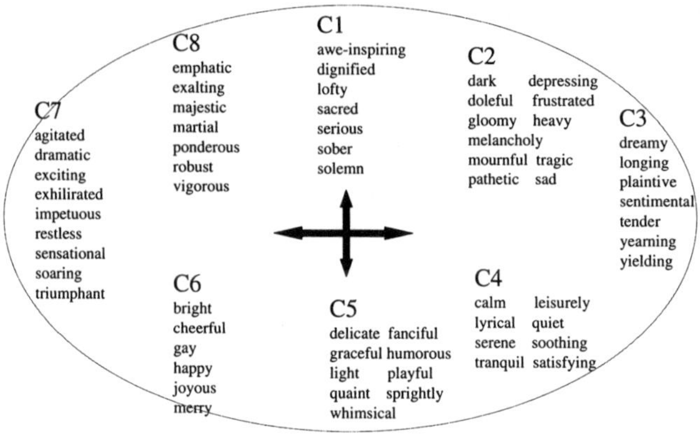

Figure 3: The cyclic adjective list

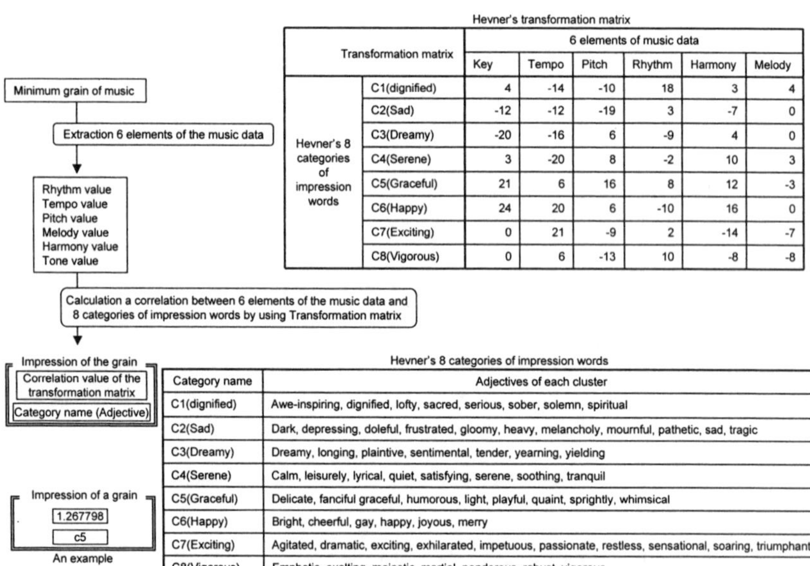

Hevner's transformation matrix

Transformation matrix		6 elements of music data					
		Key	Tempo	Pitch	Rhythm	Harmony	Melody
Hevner's 8 categories of impression words	C1(dignified)	4	-14	-10	18	3	4
	C2(Sad)	-12	-12	-19	3	-7	0
	C3(Dreamy)	-20	-16	6	-9	4	0
	C4(Serene)	3	-20	8	-2	10	3
	C5(Graceful)	21	6	16	8	12	-3
	C6(Happy)	24	20	6	-10	16	0
	C7(Exciting)	0	21	-9	2	-14	-7
	C8(Vigorous)	0	6	-13	10	-8	-8

Hevner's 8 categories of impression words

Category name	Adjectives of each cluster
C1(dignified)	Awe-inspiring, dignified, lofty, sacred, serious, sober, solemn, spiritual
C2(Sad)	Dark, depressing, doleful, frustrated, gloomy, heavy, melancholy, mournful, pathetic, sad, tragic
C3(Dreamy)	Dreamy, longing, plaintive, sentimental, tender, yearning, yielding
C4(Serene)	Calm, leisurely, lyrical, quiet, satisfying, serene, soothing, tranquil
C5(Graceful)	Delicate, fanciful graceful, humorous, light, playful, quaint, sprightly, whimsical
C6(Happy)	Bright, cheerful, gay, happy, joyous, merry
C7(Exciting)	Agitated, dramatic, exciting, exhilarated, impetuous, passionate, restless, sensational, soaring, triumphant
C8(Vigorous)	Emphatic, exalting, majestic, martial, ponderous, robust, vigorous

Figure 4: Extraction of primitive impression from music data

4 Experimental Study

We have implemented an experimental system realizing our proposed method. We have performed two experiments.

The first experiment is related to impression gaps occurring with temporal transition of impression of music. This experiment is for investigating how people feel the gap of two adjacent impressions to reflect the emphasis as the dramatic interpretation. The second experiment is an example of metadata generation for music data.

4.1 Experiment1

This experiment is for investigating how people feel the gap between two adjacent impressions between grains of music data and how the gap influences impression of the grain. In this experiment, we have used following objects.

10 test subjects to make the example of the humans' impression for grains.

5 pieces of music (each music has 3 grains)

Individual-impression (Humans' impression of each grain which is listened to separately.) Test subjects listen to one grain and then give articulation for the grain to refresh an impression of previous grain, and repeat the same action to every grain.

Sequential-impression (Humans' impression of a grain which is listened to sequentially.) The test subjects listen to the music from beginning to end, and then they give articulation for the grain of music to refresh impression of music, and repeat the same action to every music.

Test subjects give Individual-impression of the grain by choosing an impression word group from the cyclic adjective list (C1, C2, C3, C4, C5, C6, C7 and C8) shown in Figure[3]. They also evaluate the strength of its impression (impression-strength) in ten grades (0.1-1.0) for each grain. The strength of each grain is specified by a test subject. A part of results are shown in Table1.

Table 1: The answer of Individual-impression and its strength of each grain in Test Subject1

Grain ID	Individual impression	Impression strength
1	C5 (Graceful)	0.2
2	C1 (Dignified)	0.2
3	C6 (Happy)	0.3
...	...	...

Test subjects give Sequential-impression of the grain by choosing an impression word group from the cyclic adjective list. Then, they also evaluate the impression-strength in ten grades (0.1-1.0) for each grain. The strength of each grain is specified by the test subject. A part of results are shown in Table2.

We compare the results of Individual-impression (Table1) and Sequential-impression (Table2) to evaluate two viewpoints, *impression-viewpoint* and *value-viewpoint*. As *impression-viewpoint*,

Table 2: The answer of Sequential-impression and its strength of each grain in test subject-1

Grain ID	Sequential impression	Impression strength
1	C5 (Graceful)	0.4
2	C1 (Dignified)	0.5
3	C6 (Happy)	0.7
...	...	...

we evaluate whether the impression of each grain is equivalent between Individual-impression and Sequential-impression. As *value-viewpoint*, we evaluate how different the impression-strength in each same grain between Individual-impression and Sequential- impression.

As an example of *impression-viewpoint*, the results of the test subject-1 are shown in Table3. The test subject-1 has the same impression of a grain listened to separately and sequentially.

Table 3: Results of the test subject-1 (*impression-viewpoint*)

Music id	grain id	Individual impression	Sequential impression
1	11	C5	C5
	12	C1	C1
	13	C6	C6
2	21	C7	C7
	22	C3	C3
	23	C7	C7
3	31	C1	C1
	32	C3	C3
	33	C5	C5
5	51	C1	C1
	52	C5	C5
	53	C1	C1
6	61	C1	C1
	62	C3	C3
	63	C7	C7

As shown in Table 4, all test subjects (10 test subjects) evaluate *impression-viewpoints* for 5 music data, and 9 test subjects have the same impression between Individual and Sequential impressions, in each of 5 pieces of music.

We have selected those 9 test subjects for the following experiments, and we use their results in the *value-viewpoint*.

To compare the degrees of impression-strength between impressions of a grain in the cases for Individual and Sequential impression, we show the strength of impressions for each grain in Tables5 and 6, respectively. The strength is expressed as the average of impression-strength values among the 9 test subjects.

Experiment1 (Tables 1 and 2) has shown that the impression-strength to each grain is different between Individual impression and Sequential impression. That is, the order of grains influences the impression-strength for each grain. Moreover, as shown in Tables 5 and 6, the impression-strength values of the first grain are similar between Individual impression and Sequential impression. The differences in the subsequent grains are expressed as follows.

Table 4: Similarity(%)between Individual and Sequential impressions

Test subject ID	Similarity(%) between Individual and Sequential impressions
1	100%
2	100%
3	100%
4	100%
5	100%
6	100%
7	100%
8	100%
9	100%
10	40%

Table 5: Average of impression-strength (Individual impression) among 9 test subjects

grain ID	Average of impression-strength	Impression-metadata
11	0.4	c5
12	0.3	c1
13	0.3	c6
21	0.2	c7
22	0.2	c3
23	0.2	c7
31	0.3	c1
32	0.3	c3
33	0.4	c5
51	0.2	c1
52	0.2	c5
53	0.2	c1
61	0.4	c1
62	0.4	c3
63	0.3	c7

Table 6: Average of impression-strength (Sequential impression) among 9 test subjects

grain ID	Average of impression-strength	Impression-metadata
11	0.4	C5
12	0.5	C1
13	0.7	C6
21	0.2	C7
22	0.4	C3
23	0.8	C7
31	0.3	C1
32	0.4	C3
33	0.7	C5
51	0.2	C1
52	0.4	C5
53	0.8	C1
61	0.4	C1
62	0.5	C3
63	0.9	C7

The impression-strength values of the second grain are different. The difference on impression-strength is expressed as follows.
(Individual impression strength in the second grain) $\leq$ (Sequential impression strength in the second grain)

The strength values of the impression of the third grain are different. The difference in impression-strength is expressed as follows.
(Individual impression strength in the third grain) $\leq$ (Sequential impression strength in the third grain)

The strength values of the impression of the second and third grains in Sequential impression are different. The difference is expressed as follows:
(Sequential impression strength in the second grain) $\leq$ (Sequential impression strength in the third grain)

These results mean that the impression for each grain is influenced by the sequence of previous grains. Experiment1 (Tables 1 and 2) has shown that the test subjects feel the gap between previous and following grains, and the gap influences the impression strength of the current grain. These experimental results show that the direction of our approach for reflecting impression transition occurring as time passes in music is reasonable.

4.2 *Experiment2*

The second experiment is the example to generate metadata dealing with the dramatic interpretation in various music granularity shown in Figure 5.

This example shows that our metadata generation method is flexible and feasible to any level of music granularity.

5 Conclusion

In this paper, we have proposed the "Kansei" metadata generation method for music data dealing with dramatic interpretation. In the experimental study, we have shown two experimental results for the metadata generation method, and those experimental results have shown that our metadata generation method is feasible and effective for extracting impressions of music and for applying it to music data retrieval based on human's impression. As our future work, we will extend and apply our method to different types of media data dealing with dramatic interpretation. Furthermore, we will study methodology for integrating media data within the framework of "impression" in human beings.

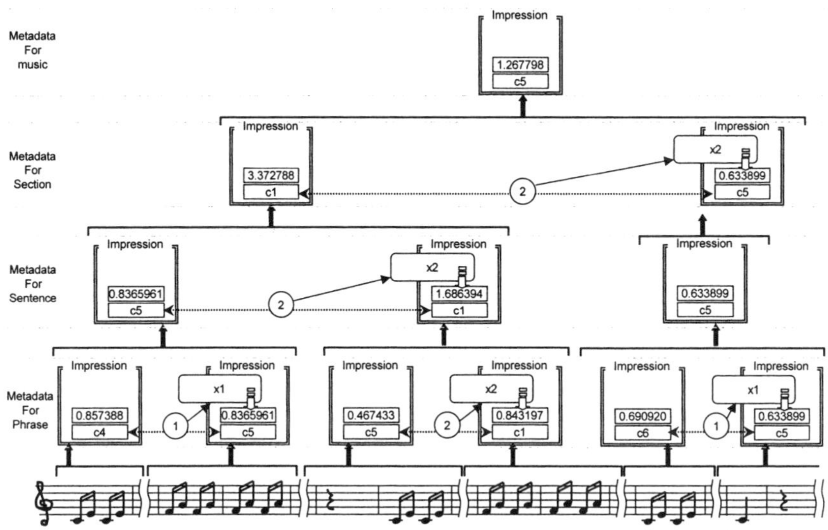

Metadata for music	C5 1.267798					
Metadata for section	C1 3.372788				C5 1.267798	
Metadata for sentence	C5 0.8365961		C1 3.372788		C5 0.633899	
Metadata for phrase	C4 0.857388	c5 0.8365961	c5 0.467433	c1 1.686394	c6 0.690920	c5 0.633899

Figure 5: An example of metadata generation

References

[1] A. Sheth, W. Klaus(eds.): "Multimedia Data Management - Using Metadata to Integrate and Apply Digital Media", McGraw-Hill, 1997.

[2] A. Harada(eds.): "The framework of Kansei engineering", Report of Modelling the Evaluation Structure of Kansei, pp49-55, 1997

[3] K. Hevner: "Experimental Studies of the Elements of Expression in Music", American Journal of Psychology, Vol.48, pp.246-268, 1936.

[4] K. Hevner: "Expression in Music: A Discussion of Experimental Studies and Theories", Psychological Review, Vol.42, pp.186-204, 1935.

[5] A. Ijichi, Y. Kiyoki: "An Automatic Metadata Generation Method for Music Retrieval-by-Impression Dealing with Impression-Transition", Proceedings of the 7th IASTED International Conference on Internet and Multimedia System and Applications, pp281-288, 2003.

[6] T. Kitagawa and Y. Kiyoki, "Fundamental framework for media data retrieval systems using media-lexico transformation operator", Information Modeling and Knowledge Bases, IOS Press, Vol.12, pp. 316-326, May 2001.

[7] Y. Kiyoki, T. Kitagawa and T. Hayama: "A Metadatabase System for Semantic Image Search by a Mathematical Model of Meaning", Multimedia Data Management Using Metadata to Integrate and Apply Digital Media, McGraw-Hill, Chapter 7, 1998.

[8] Y. Kiyoki, T. Kitagawa, and T. Hayama, "A Metadatabase System for Semantic Image Search by a Mathematical Model of Meaning", ACM SIGMOD Record (Special issue on metadata for digital media), W. Klaus, A. Sheth, eds., Vol.23, No. 4, pp.34-41, Dec. 1994.

[9] W. Klaus, A. Sheth(eds.): "Special issue on metadata for digital media", ACM SIGMOD Record, Vol.23, No. 4, Dec. 1994

[10] N. Kosugi et al: "A Practical Query-By-Humming System for a Large Music Database", ACM Multimedia 2000, pp.333-342, 2000.

Information Modelling and Knowledge Bases XVI
Y. Kiyoki et al. (Eds.)
IOS Press, 2005

Turn Taking in Database Semantics

Roland Hausser

Universität Erlangen-Nürnberg
Abteilung Computerlinguistik (CLUE)
rrh@linguistik.uni-erlangen.de

Abstract

One of the most basic structural phenomena of natural language communication is turn taking, i.e. the communicating agents taking turns in being speaker and hearer. Without turn taking, there is only one-sided monologue as the limiting case.

The linguistic framework called Database Semantics (AIJ'01) treats the agent's switching back and forth between the speaker and the hearer mode as a well-defined and well-motivated computational problem. Its solution is a precondition for modeling natural language communication by programming suitable machines (talking robots). All syntactic and semantic analysis of natural language must be integrated into the turn taking underlying communication.

Turn taking is modeled as the interaction of three kinds of LA-grammars, called LA-hear, LA-think, and LA-speak. LA-hear interprets natural language signs by reading their content into the hearer's database. LA-think provides the content to be verbalized by autonomously navigating through the speaker's database content (conceptualization). LA-speak maps the content traversed by LA-think into natural language surfaces.

After a general introduction to the time-linear algorithm of LA-grammar, the rule and derivation format of the three grammars modeling turn taking are explained. The examples are based on a Java implementation currently under development by Mr. Arkadius Kycia in cooperation with the author.[1]

1 Basic Algorithm of LA-grammar

Today, there is a choice between two full-blown grammar formalisms complete with algebraic definition, hierarchy of language classes, complexity classes, and standard methods of computational implementations in different programming languages, namely PS-grammar (PSG) and LA-grammar (LAG). The most basic difference between these two frameworks is that the derivation order of LA-grammar is time-linear while that of PS-grammar is not.

Time-linear means linear like time and in the direction of time. LA-grammar models time-linearity in that it takes a sequence, for example the words a b c d e ..., as input and always combines what has been analyzed so far, called sentence start (*ss*), with the current next word (*nw*) into a new sentence start (*ss'*), for example a and b into (a b), (a b) and c into ((a b) c), ((a b) c) and d into (((a b) c) d), etc. This kind of combination is called *left-associative*.

Database Semantics (dbs) chooses LA-grammar as its algorithm because the structure of natural language is inherently time-linear. This is reflected not only in the structural order of the natural language signs, but also in the procedures of language interpretation (hearer mode) and language production (speaker mode). In addition, the complexity properties of

[1]This paper benefited from comments by Prof. Meyer-Wegener, Department of Computer Science at the Universität Erlangen-Nürnberg. Naturally, he is not responsible for any weaknesses or mistakes.

LA-grammar are much more suitable for the structures found in natural language than those of PS-grammar. Also, LA-grammar is type transparent[2] while PS-grammar is not.

LA-grammar as exemplified by NEWCAT'86, CoL'89, and TCS'92 was developed prior to dbs (FoCL'99, AIJ'01) and is a conventional system in that it takes unanalyzed language expressions as input and renders grammatically analyzed language expressions as output. Such LA-grammars consist of (i) a lexicon, (ii) a set of start states, (iii) a set of rules, and (iv) a set of final states.

As an example, consider the following definition of an LA-grammar for the formal language $a^k b^k c^k$, consisting of expressions like abc, aabbcc, aaabbbccc, etc.

1.1 LA-GRAMMAR FOR $a^k b^k c^k$

LX $=_{def}$ {[a (A)], [b (B)], [c (C)]}
$ST_S =_{def}$ {[(A) $\{r_1, r_2\}$]}
r_1: (X) (A) $\Rightarrow$ (AX) $\{r_1, r_2\}$
r_2: (AX) (B) $\Rightarrow$ (XB) $\{r_2, r_3\}$
r_3: (BX) (C) $\Rightarrow$ (X) $\{r_3\}$
$ST_F =_{def}$ {[ε rp_3]}.

The lexicon LX is defined as a list of analyzed words. The analysis of a word consists of the surface, also called lemma, and some kind of lexical characterization (e.g., the syntactic category, the meaning, etc.). The lexicon defined in 1.1 consists of three analyzed words, namely [a (A)], [b (B)], and [c (C)]. The word [a (A)], for example, consists of the surface a and the category (A).[3]

The set of start states ST_S specifies which words (represented by their category) can start the parsing of a sentence, and which specific rules should be tried to combine the first word with the next. In our example, the set of start states has only one element, namely [(A) $\{r_1, r_2\}$]. Like any state in LA-grammar, the start state is defined as an ordered pair consisting of a category, here (A), and a rule package, here $\{r_1, r_2\}$.

An LA-grammar may have several start states and start computing on various kinds of input. In 1.1, however, the set named ST_S contains only one start state. This LA-grammar is like a little machine which will start to compute if and only if it gets a sign with the category (A) on its input tape as initial input.

The set of rules in 1.1 contains three elements, called r-1, r-2, and r-3. Their general structure may be characterized as the following schema:

1.2 SCHEMA OF LEFT-ASSOCIATIVE RULES IN LA-GRAMMAR

r_i: *ss nw* $\Rightarrow$ *ss'* rp_i

A LAG rule consists (i) of a rule name, e.g. r_i, (ii) a pattern for the *ss* (sentence start, first input), (iii) a pattern for the *nw* (next word, second input), (iv) a pattern for the output called *ss'* (resulting, or new, sentence start), and (v) a rule package, e.g. rp_i.

[2] According to Berwick & Weinberg 1984, a grammar is type transparent if the declarative linguistic analysis and the derivation procedure on the computer are merely different realizations of the same algorithm. For further explanation see FoCL'99, p. 170 f.

[3] In LA-grammars for natural languages, with their very large number of words and wordforms, the lexicon is part of a component for automatic word form analysis. For small lexica, like the one needed for $a^k b^k c^k$, it suffices to give the explicit list.

The combination of rule names and rule packages, e.g. r_i and rp_i, in LAG constitute a finite state transition network. The combination of an *ss*, *nw*, and *ss'* pattern in a LAG rule constitutes the rule's categorial operation.[4]

A left-associative combination step requires a current state $(ss\ rp_i)$ and a next word *nw*:

(i) $(ss\ rp_i) + nw$

The combination step is computed by applying the rules of the current state's rule package rp_i to the current sentence start and the current next word:

(ii) rp_i: *(ss nw)*

Applying a rule successfully to the input *(ss nw)* results in a new state consisting of a new sentence start *ss'* (represented formally by its category) and a new rule package rp_j. By adding a new next word to this new state provides the ingredients for the next left-associative combination step:

(i) $(ss'\ rp_j) + nw'$

Again, the system applies the rules of the rule package to the input pair:

(ii) rp_j: *(ss' nw')*

This process continues until all the next words in the input have been consumed. If more than one rule is successful in a combination step, the resulting branches are pursued in parallel (cf. 2.5). If none of the rules is successful in a combination step, the derivation fails.

The set of final states S_F specifies which expressions are *complete* final expressions, like aaabbbccc of $a^k b^k c^k$, in contradistinction to incomplete but nevertheless legal intermediate expressions like aaabbbcc. In our example, the set of final states contains only one element, namely $[\varepsilon\ rp_3]$. This means that a parsed expression is a legal complete expression of the formal language $a^k b^k c^k$ if its final category is ε,[5] and its final rule package has been activated by r-3.

Because of the absolute type transparency of LA-grammar, the declarative linguistic analysis and the derivation procedure on the computer are merely different realizations of the same left-associative algorithm. For example, the LAG parser NEWCAT implemented in Lisp takes an LA-grammar and a string as input and generates an analysis like the following. It is based on the LA-grammar defined in 1.1 and the input string aaabbbccc:

1.3 NEWCAT PARSING aaabbbccc USING LIST-BASED NOTATION

```
*START-0
1
    (A) a
    (A) a
*RULE-1
2
    (A A) a a
    (A) a
*RULE-1
3
    (A A A) a a a
    (B) b
*RULE-2
4
    (A A B) a a a b
    (B) b
*RULE-2
5
```

[4]The LA-grammars considered here belong to the subclass of C-LAGs. In C-LAGs the pattern of the sentence start is restricted to patterns like (aX), (Xa), or (aXb), while patterns like (XaY) are excluded – whereby a, b are constants and X, Y are variables. Because of this restriction, a rule application may be taken as the primitive operation of C-LAGs for complexity analysis. For further discussion see FoCL'99, p. 207 f.

[5]Also called 'empty category', corresponding to 'empty stack' in automata theory.

```
    (A B B) a a a b b
    (B) b
*RULE-2
6
    (B B B) a a a b b b
    (C) c
*RULE-3
7
    (B B) a a a b b b c
    (C) c
*RULE-3
8
    (B) a a a b b b c c
    (C) c
*RULE-3
9
    (NIL) a a a b b b c c c
```

This LA-grammar analysis is generated directly as the formatted protocol (trace) of the LA-parser NEWCAT .[6] Note that LA-grammar parses context-sensitive $a^k b^k c^k$ in linear time.[7]

NEWCAT'89 presents a Lisp implementation complete with source code for LA-parsing 221 sentence structures of German and 114 sentence structures of English. The syntactic analysis provided by the LA-parser is designed to handle agreement, word order, and valency, and is derived in a strictly time-linear fashion.

2 Feature Structure Notation of JLAG Implementation

In the long run, the list-based format of early LA-grammar as described in the previous Section proved to be too cumbersome for natural language analysis. This motivated a transition from the original list-based notation to an equivalent feature structure notation. The following example illustrates the two notations, using rule r-1 of the LA-grammar defined in 1.1.

2.1 Comparing List Notation with Feature Structure Notation

original list-based notation:

r_1: (X) (A) $\Rightarrow$ (AX) $\{r_1, r_2\}$

current feature structure notation:

r_1: [cat: X] [cat: A] leftcopy A ss-cat copy_ss $\{r_1, r_2\}$

While the list-based notation represents the ss-patterns as the list (X), the feature structure notation represents it as the attribute/value-pair [cat: X], and similarly for the nw-pattern.[8] Furthermore, the categorial operation coded in the list-based notation by means of the result pattern (aX) is coded in the feature structure notation by specifying the operations explicitly. For example, the operation leftcopy A ss-cat in the 2.1 means that the constant A is to be added to the left of the current value of the cat-attribute in the ss.[9]

[6] To obtain a better comparison of the input categories, surface and category are printed in inverse order, e.g., (A) a instead of a (A).

[7] For a complexity-theoretical analysis of LA-grammar see TCS'92 and FoCL'99, Chapters 11 and 12.

[8] The attributes of a feature structure are also called features. We use the term 'attribute' in order to avoid confusion with the term 'feature' as used in psychology, for example in the comparison of features and templates.

[9] The operations copy_ss or copy_nw derive from the fact that the feature structure version of LA-grammars for natural language treat the sentence start as a set of proplets and allow to specify whether or not the next word is to be added to this set.

Feature structures consisting of attritutes with constants or variables as values are descriptively more powerful than lists or list patterns. It is therefore easy to translate from the old notation into the new, as illustrated by the following LA-grammar definition of $a^k b^k c^k$:

2.2 LA-GRAMMAR FOR $a^k b^k c^k$ IN FEATURE STRUCTURE NOTATION

$$LX =_{def} \{ \begin{bmatrix} \text{sur: a} \\ \text{cat: A} \end{bmatrix}, \begin{bmatrix} \text{sur: b} \\ \text{cat: B} \end{bmatrix}, \begin{bmatrix} \text{sur: c} \\ \text{cat: C} \end{bmatrix} \}$$

$$ST_S =_{def} \{[\ [\text{cat: A}]\ \{r_1, r_2\}]\}$$

r_1:	[cat: X]	[cat: A]	leftcopy A ss-cat copy_ss	$\{r_1, r_2\}$
r_2:	[cat: AX]	[cat: B]	leftremove A ss-cat rightcopy B ss-cat copy_ss	$\{r_2, r_3\}$
r_3:	[cat: BX]	[cat: B]	leftremove B ss-cat copy_ss	$\{r_3\}$

$$ST_F =_{def} \{[\ [\text{cat: } \varepsilon]\ rp_3]\}.$$

This LA-grammar is equivalent – in the sense of a notational variant – to that defined in 1.1.

The new notation of the data and the LA-grammar rules was developed as part of an implementation of Database Semantics in Java, currently under development. There also resulted a new display format for the syntactic derivation:

2.3 COMPARING LIST-BASED WITH FEATURE STRUCTURE DISPLAY

list-based NEWCAT display (see also 1.3)

```
1
    (A) a
    (A) a
*RULE-1
2
    (A A) a a
```

feature structure JLAG display (see also 2.4)

```
2.1
r-1: [cat: X]  [cat: A]  leftcopy      rp:{r-1, r-2}    #rule level
                         copy_ss

     [sur: a]  [sur: a]                                 #input level
     [cat: A]  [cat: A]
```

While the NEWCAT format displays only the input, the rule name, and the output, the JLAG format shows the complete rule in conjunction with the input, whereby the patterns of the rule level are aligned to be directly above the input. The output (not shown in 2.3) is displayed as the *ss* of the next combination step. Consider the following derivation:

2.4 JLAG PARSING aabbcc USING FEATURE STRUCTURE NOTATION

```
rrh@venus:~/java/JLAG/bin> java anbncn.ANBNCN

#  Enter a sentence or '-help'.

anbncn.ANBNCN-HEAR> a a b b c c
# Hearer mode #

1.
Start state, NW:         [cat: A]               rp:{r-1, r-2}

                         [sur: a]
                         [cat: A]

2.1
r-1: [cat: X]   [cat: A]   leftcopy        rp:{r-1, r-2}
                           copy_ss

     [sur: a]   [sur: a]
     [cat: A]   [cat: A]

3.12
r-2: [cat: AX]   [cat: B]   leftremove       rp:{r-2, r-3}
                            rightcopy
                            copy_ss

     [sur: aa]   [sur: b]
     [cat: AA]   [cat: B]

4.121
r-2: [cat: AX ]   [cat: B]   leftremove       rp:{r-2, r-3}
                             rightcopy
                             copy_ss

     [sur: aab]   [sur: b]
     [cat: AB ]   [cat: B]

5.1212
r-3: [cat: X    ]   [cat: C]   leftremove       rp:{r-3}
                               copy_ss

     [sur: aabb]   [sur: c]
     [cat: BB  ]   [cat: C]

6.12121
r-3: [cat: X     ]   [cat: C]   leftremove       rp:{r-3}
                                copy_ss

     [sur: aabbc]   [sur: c]
     [cat: B    ]   [cat: C]

########################  Final State: [Branch: 12121]

  [sur: aabbcc]
  [cat:       ]
----------------------------------------
### Parsing finished successfully.
```

Each combination step has a number, for example 5.1212. The digit preceding the full stop refers to the word number in the input, here the fifth word (for which reason the start state has the number 1 and the final state has no number). The digits after the full stop specify which rule in the rule package of a combination was successful (whereby a 1 represents the first rule name in the package, a 2 the second rule, etc.). For example, the combination number 3.12 indicates the addition of the third word in the input, based on a successful application of the first rule of the rule package in the first combination and the second rule of the rule package of the second combination.

If more than one rule in a rule package is successful, the derivation is ambiguous. Ambiguities are handled by LA-parsers as parallel paths. Consider for example the following derivation of the context-sensitive language WW:[10]

2.5 JLAG FOR WW PARSING abab

```
ww.WW-HEAR> a b a b
# Hearer mode #

1.
Start state, NW:         [cat: A]              rp:{r-2, r-1}

                         [sur: a]
                         [cat: A]

2.2
r-1: [cat: X]   [cat: seg]   rightcopy       rp:{r-2, r-1}
                             copy_ss

     [sur: a]   [sur: b  ]
     [cat: A]   [cat: B  ]

3.21
r-2: [cat: seg X]   [cat: seg]   leftremove       rp:{r-2}
                                 copy_ss

     [sur: ab    ]  [sur: a  ]
     [cat: AB    ]  [cat: A  ]

3.22
r-1: [cat: X ]   [cat: seg]   rightcopy       rp:{r-2, r-1}
                              copy_ss

     [sur: ab]   [sur: a  ]
     [cat: AB]   [cat: A  ]

4.211
r-2: [cat: seg X]   [cat: seg]   leftremove       rp:{r-2}
                                 copy_ss

     [sur: aba   ]  [sur: b  ]
     [cat: B     ]  [cat: B  ]

4.222
r-1: [cat: X  ]   [cat: seg]   rightcopy       rp:{r-2, r-1}
                               copy_ss

     [sur: aba]   [sur: b  ]
     [cat: ABA]   [cat: B  ]

########################  Final State: [Branch: 211]

  [sur: abab]
  [cat:     ]
----------------------------------------
### Parsing finished successfully.
```

The time-linear ambiguity arises with the addition of word 3: in 3.21, rule r-2 treats word 3 as the first word of the second W, while in 3.22, rule r-1 treats word 3 as an addition to the first W. The ambiguity is continued in the addition of word 4. However, as shown by the branch number 211 of the Final State, only the hypothesis represented by the first of the two parallel path results in a correct complete expression of WW.

The new notation based on feature structures and specified operations does not change the mathematical properties of LA-grammar as compared to the old one. This is because (i) the rules are restricted to left-associative compositions, and the set of operations of each rule is (ii) finite in number and (iii) non-recursive. LA-grammar parses context-sensitive WW in low polynomial time, namely n^2.

[10]See FoCL, example 11.5.6 on p. 218, for the declarative specification of WW in the list-based format.

3 *Turn Taking* as a Software Engineering Task

The feature structures, illustrated above with formal languages, are called *proplets* in applications of LA-grammar to natural language. Proplets represent the functor-argument structure of a proposition (intrapropositional relations) and the connections between propositions (extrapropositional relations) in terms of attribute values. Proplets are well-suited for the storage and retrieval required for modeling natural language communication.

Based on proplets, the time-linear algorithm of LA-grammar is used to realize the following procedures: (i) natural language interpretation during input, handled by **LA-hear**, (ii) inferencing and the sorting of content during thought, handled by **LA-think**, and (iii) language production during output, handled by **LA-speak.**[11]

In communication, these three LA-grammars cooperate as follows:

3.1 THE CYCLE OF NATURAL LANGUAGE COMMUNICATION

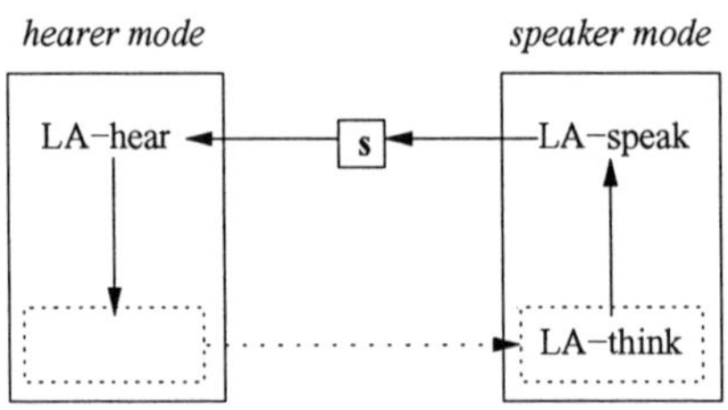

This schema[12] represents the process of *turn taking*, i.e., the communicating agents taking turns in being speaker and hearer. The small box containing **s** represents the sign being transmitted. Turn taking has the following aspects:[13]

3.2 TWO ASPECTS OF TURN TAKING

1. *From the outside:*

   Two communicating agents are observed as they are taking turns. This is represented by 3.1 when the two boxes are taken to be two different agents, one in the hearer and the other in the speaker mode.

2. *From the inside:*

   One communicating agent is observed as it switches between being the speaker and the hearer. This is represented by 3.1 when the two boxes are taken to be the hearer and the speaker mode of the same agent (dotted right hand arrow).

Before taking a closer look at the rules and derivations of LA-hear, LA-think, and LA-speak in Sections 4, 5, and 6, respectively, let us illustrate our conceptual solution to turn taking with a simple example. It shows the processing of the sentence the girl sang in the hearer and the speaker mode.

[11]The names of these LA-grammars are motivated by the most basic variants of the hearer and the speaker mode. They are not intended to exclude, for example, written signs as input to LA-hear and as output of LA-speak. Instead, the system of Database Semantics presented here is based on modality-independent sign representations. The part of peripheral cognition is omitted for simplicity.

[12]In schema 3.1 contextual recognition and action are omitted for simplicity.

[13]For a study of turn taking with a much wider scope than the present paper see Thórisson 2002.

3.3 A FORMAL ILLUSTRATION OF TURN TAKING

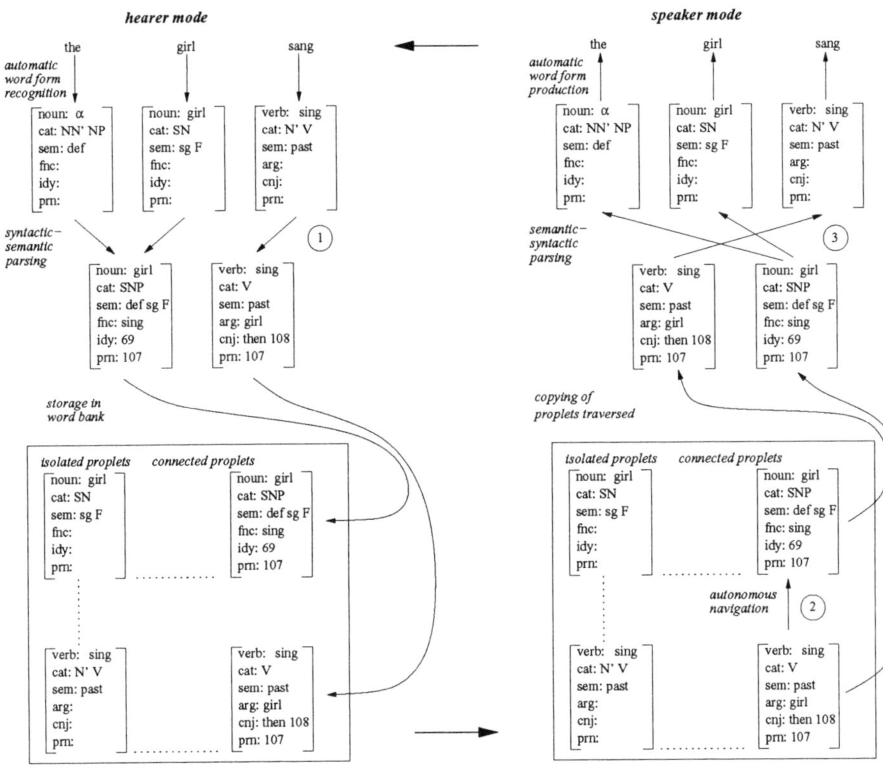

Starting with the hearer mode on the left hand side, we see the input sequence in the top line: the girl sang. Next below is automatic word form recognition. It provides each of the three input surfaces with a corresponding isolated proplet retrieved from the lexicon.

These lexically analyzed word forms are the input to LA-hear, represented as ①. This interpretation parser produces two connected proplets as output, whereby the noun girl has been absorbed into the determiner the. The connected proplets *girl* and *sing* are stored at the end of their respective token lines in the database,

Then the agent switches into the speaker mode, as represented by the bottom right arrow. Inside the database on the right, language production starts with an autonomous navigation serving as conceptualization. This navigation is powered by LA-think, represented as ②.

The proplets traversed are copied and used as input to LA-speak, represented as ③. This production parser is for handling (i) the language-specific word order and (ii) function word precipitation (which is the counterpart to absorption during interpretation). In other words, the language-independent proplet sequence *sing girl* is mapped by LA-speak into the language-dependent proplet sequence *DET girl sing*.

Using the values of certain attributes, for example, sing and past, these proplets are mapped by automatic word form production into appropriate language-dependent surfaces, here sang. When these surfaces have been produced as an external sign, as indicated by the top left arrow, the system switches again into the hearer mode.[14] In this way, the same content may be

[14]We may assume that the agent utters the sentence *The girl sang.* (speaker mode), thereby hearing this utterance (hearer mode). After interpretation and storage of the content, the hearer turns into a speaker, remembers

sent back and forth ad infinitum.

For a model of natural language communication, this artificial, absolutely minimal example of turn taking must be extended systematically to more realistic exchanges. First, instead of one agent blindly processing the same sentence by switching between the hearer and the speaker mode, there should be at least two agents exchanging different sentences. Second, the expressions used in the dialog should include all syntactic moods, i.e., not only declarative, but also interrogative and imperative. Third, the interpretation of a sentence in the hearer mode should initiate the production of a *meaningful* response in the speaker mode.

To get an idea of what is required, consider the following examples:

3.4 Examples of Turn Taking Using Different Syntactic Moods

1. Two declarative sentences:

   A1S: The girl sang. :A2H
   A2S: That is good. :A1H

2. Yes/no-interrogative and yes/no-answer:

   A1S: Did the girl sing? :A2H
   A2S: Yes. :A1H

3. Wh-interrogative and wh-answer:

   A1S: Who sang? :A2H
   A2S: The girl. :A1H

4. Imperative and responding action:

   A1S: Sing! :A2H
   A2S: performs action requested.

A1S stands for agent 1 in the speaker mode, A1H for agent 1 in the hearer mode, and correspondingly for A2S and A2H. The sign passed from the speaker, e.g. A2S, to the hearer, e.g. A1H, is placed between the two agents involved, for example A2S: That is good. :A1H. In the case of the imperative, the most appropriate response by the hearer, e.g. robot, is not the production of another sentence, but rather the performance of the action requested.

In order to extend the minimal system illustrated in 3.3 to the examples of 3.4, the syntactic-semantic parsing of LA-hear and the semantic-syntactic parsing of LA-speak must be upscaled to handle the constructions in question. Equally importantly, however, LA-think must be extended to control the navigation underlying language production in a meaningful way.

These tasks are of an empirical nature. In other words, once the functioning of natural language communication has been solved in principle, the addition of new words to the lexicon, of new constructions to LA-hear and LA-speak, and of new inferences and behavior patterns to LA-think[15] is just a matter of doing the descriptive work, both at the level of the declarative specification and the software implementation.

the content, and produces the utterance again, etc.

[15]In order to ensure long term upscalability of the system, the contextual database housing LA-think should be complemented with contextual recognition and action as soon as technically possible.

4 LA-Hear

We begin the process of incremental upscaling by describing the rule and derivation format of LA-think. Consider the rule DET+NN (determiner plus noun), defined as part of LA-hear.2 in Hausser 2002.

4.1 LA-HEAR RULE **DET+NN**

$$\textbf{DET+NN}\ \begin{bmatrix}\text{noun: } ① \\ \text{cat: } n'\ x \\ \text{sem: } y\end{bmatrix} \begin{bmatrix}\text{noun: } \alpha \\ \text{cat: } n \\ \text{sem: } z\end{bmatrix} \quad \begin{array}{l}\text{delete } n' \text{ ss-cat} \\ \text{nw-sem} \xrightarrow{\boxed{a}} \text{ss-sem} \\ \text{nw-noun} \xrightarrow{\boxed{r}} ① \\ \text{copy}_{ss}\end{array} \quad \{\text{NOM+FV, FV+NP, S+IP}\}$$

Like the rules for the formal language $a^kb^kc^k$ defined in 2.2, this LA-hear rule consists of (i) the rule name DET+NN, (ii) patterns for the sentence start and the next word, defined as feature structures with variables and constants as values, (iii) a set of operations, and (iv) a rule package.

Conceptually, the difference between the rules for formal languages and for natural languages is most visible in their respective operations. In formal languages, the operations are mostly for adding, e.g. `leftcopy`, or subtracting e.g. `leftremove`, category segments in the cat-attribute of the sentence start. In natural language, in contrast, the operations are for canceling valency positions, e.g. `delete n' ss-cat`, and for copying values between proplets, e.g. `nw-sem` $\xrightarrow{\boxed{a}}$ `ss-sem`.

As an example of a complete LA-hear derivation consider the syntactic-semantic parsing of the girl sang.

4.2 LA-HEAR DERIVATION OF The girl sang.

```
la1.LA1-HEAR> the girl sang.
# Hearer mode #
1.
Start state, NW:        [cat: NN' NP]        rp:{DET+NN, DET+ADN, NOM+FV}

                        [sur: the    ]
                        [noun: _n1   ]
                        [cat: NN' NP]
                        [sem: def    ]
                        [mdr:        ]
                        [fnc:        ]
                        [idy: +1     ]
                        [prn:        ]
2.1
DET+NN: [cat: NN' NP]  [cat: NN     ]  delete_n'_ss-cat    rp:{NOM+FV,
        [noun: 1    ]  [noun: alpha]  nw_noun-|r|->_n1       FV+NP, S+IP}
        [sem: y     ]  [sem: z      ]  nw_sem-|a|->ss_sem
                                       copy_ss

        [sur: the   ]  [sur: girl  ]
        [noun: _n1  ]  [noun: girl ]
        [cat: NN' NP]  [cat: SN    ]
        [sem: def   ]  [sem: sg F  ]
        [mdr:       ]  [mdr:       ]
        [fnc:       ]  [fnc:       ]
        [idy: +1    ]  [idy: +1    ]
        [prn:       ]  [prn:       ]
3.11
NOM+FV: [noun: alpha  ]  [verb: beta ]  nw_verb-|e|->ss_fnc    rp:{AUX+NFV,
        [cat: NP      ]  [cat: NS3' V]  delete_np'_nw-cat      S+IP, FV+NP}
        [fnc:         ]  [arg:       ]  ss_noun-|a|->nw_arg
                         [ctn:       ]  ps_verb-|a|->nw_ctp
                                        ps_prn-|a|->nw_ctp
                                        nw_verb-|a|->ps_ctn
```

```
                                            copy_nw

         [sur:            ]  [sur: sang  ]
         [noun: girl      ]  [verb: sing ]
         [cat: SNP        ]  [cat: N' V  ]
         [sem: def sg F]     [sem: past  ]
         [mdr:            ]  [mdr:       ]
         [fnc:            ]  [arg:       ]
         [idy: 3          ]  [ctn:       ]
         [prn: 4          ]  [ctp:       ]
         [wrdn: 1         ]  [prn:       ]
4.112
S+IP: [cat: V      ]   [cat: V' DECL]   delete_vt_ss-cat        rp:{IP+START}
      [verb: alpha]                     delete_vt'_nw-cat
                                        nw_cat-|e|->ss_cat
                                        copy_ss

      [sur:        ]   [sur: .      ]
      [verb: sing ]    [cat: V' DECL]
      [cat: V      ]
      [sem: past  ]
      [mdr:        ]
      [arg: girl  ]
      [ctn:        ]
      [ctp:        ]
      [prn: 4      ]
      [wrdn: 3     ]

########################  Final State: [Branch: 112]
[sur:       ]
[verb: sing]
[cat: DECL ]
[sem: past ]
[mdr:       ]
[arg: girl ]
[ctn:       ]
[ctp:       ]
[prn: 4     ]
[wrdn: 3    ]

--------------------------------------
### Parsing finished successfully.
Proplets Added to Wordbank.

[sur:            ] [sur:        ]
[noun: girl      ] [verb: sing]
[cat: SNP        ] [cat: DECL ]
[sem: def sg F]    [sem: past ]
[mdr:            ] [mdr:       ]
[fnc: sing       ] [arg: girl ]
[idy: 3          ] [ctn:       ]
[prn: 4          ] [ctp:       ]
[wrdn: 1         ] [prn: 4     ]
                   [wrdn: 3    ]
```

The derivation begins with the start state pattern `[cat: NN' NP]` matching the lexical lookup of the first word, the. In composition 2.1, the first rule in the rule package of the start state, namely the familiar DET+NN, is applied to the lexical proplets *the* and *girl*, resulting in a more complete *girl*-proplet.

This proplet is shown as the sentence start of composition 3.11, in which the rule NOM+FV adds the third word of the sentence, sang as the *nw*, copying the value `girl` into the arg-attribute of *sing* and the value `sing` into the fnc-attribute of *girl*. Finally, in composition 4.112 the full stop is added by the rule S+IP (sentence plus interpunctuation), changing the value of the cat-attribute of *sing* to DECL (declarative).

The natural language derivation 4.2 resembles those for the formal languages $a^kb^kc^k$ in 2.4 and WW in 2.5 in that the explicit rule patterns are aligned above the input proplets. The result of the derivation is a set of two proplets, *girl* and *sing*, which are ready to be sorted into the word bank.

This LA-hear derivation may seem like a conventional linguistic analysis (and associated parser) insofar as both take language expressions as input and render some analysis as output. They differ, however, in that the object of a conventional linguistic analysis is the language sign, independent of the process of natural language communication, while the object of LA-hear is a computational model of the hearer mode.

From this important difference in their goals there follow several technical differences between the two approaches:

4.3 COMPARING LA-HEAR WITH A CONVENTIONAL ANALYSIS

1. *Lexical semantics of content words*

   Conventional linguistics offers various different methods to characterize word meaning, ranging from semantic primitives, oppositions, metalanguage definitions, truth functions, etc., none of which are integrated into a computationally viable theory of cognition.

   Database Semantics defines lexical semantics in terms of (i) isolated proplets (feature structures) and (ii) the concepts serving as the values of their part-of-speech attributes. The concepts are procedurally defined and originate in the recognition and action of peripheral cognition.

2. *Compositional semantics*

   Conventional linguistics characterizes 'what belongs together semantically' in terms of constituent structures represented as trees and motivated by movement and substitution tests. In categorial grammar, constituent structures double as functor-argument structures. The semantics analysis does not go beyond the limits of the sentence.

   Database Semantics characterizes the functor-argument structure in terms of attribute values rather than trees. Compositional semantics comprises all the values provided in addition to lexical semantics – resulting from the copying operations of the LA-hear rules. The semantic analysis includes the concatenation of sentences in terms of conjunction and identity.

3. *Relation between syntax and semantics*

   Conventional linguistics treats syntax and semantics as two separate levels, whereby the semantics is usually defined for a preexisting autonomous syntax.

   Database Semantics treats syntactic and semantic parsing simultaneously in terms of connecting proplets via value copying, without a separate or autonomous level of syntax. Semantically uninterpreted languages simply have fewer and different operations than interpreted ones (compare 2.5 and 4.2).

4. *Derivation order and derivation results*

   Conventional linguistics uses a derivation order based on the principle of possible substitutions, resulting in hierarchies represented as trees.

   Database Semantics uses a time-linear derivation order based on the principle of possible continuations, resulting in unordered sets of proplets.

Other differences are surface compositionality, type transparency, and the functionality of reference based on internal matching. All these properties of Database Semantics in general and LA-hear in particular are necessary for achieving a computationally viable, functional model of the hearer mode.

5 LA-Think

The second kind of LA-grammar used in Database Semantics is LA-think. While the task of LA-hear is filling the database with content by interpreting natural language, the task of LA-think is navigating autonomously through the content of the database. This navigation models the process of thought and is used for (i) inferencing and (ii) the speaker's conceptualization.

For simplicity, let us consider a word bank into which LA-hear has read the content of the sentence The girl sang., as described in the previous section:

5.1 PROPLETS REPRESENTING The girl sang. IN THE WORD BANK

```
Owner Proplets  |-------- token line ---------->
[noun: girl] [sur:          ]
             [noun: girl    ]
             [cat: SNP      ]
             [sem: def sg F]
             [mdr:          ]
             [fnc: sing     ]
             [idy: 3        ]
             [prn: 4        ]
             [wrdn: 1       ]
             [trc: 2 6      ]

Owner Proplets  |-------- token line ---------->
[verb: sing] [sur:          ]
             [verb: sing    ]
             [cat: DECL     ]
             [sem: past     ]
             [mdr:          ]
             [arg: girl     ]
             [ctn:          ]
             [ctp:          ]
             [prn: 4        ]
             [wrdn: 3       ]
             [trc:          ]
```

This display of the database content shows two isolated proplets, [noun: girl] and [verb: sing], and two corresponding connected proplets (underneath the terms 'token line'). The isolated (lexical) proplets are simplified to show only the part of speech attributes and their values (concepts). The word bank 5.1 differs from the result of derivation 4.2 in that the proplets in 5.1 are ordered, using the concepts as the key.

Next consider the definition of an LA-think rule for navigating through this word bank, namely **V_N_V** of LA-think.2 defined in Hausser 2002.

5.2 LA-THINK RULE **V_N_V**

$$\textbf{V\_N\_V}\ \begin{bmatrix}\text{verb: }\beta\\ \text{arg: \#x }\alpha\text{ y}\\ \text{prn: m}\end{bmatrix}\begin{bmatrix}\text{noun: }\alpha\\ \text{func: }\beta\\ \text{mdr: NIL}\\ \text{prn: m}\end{bmatrix}\quad \begin{matrix}\text{mark }\alpha\text{ in ss-arg}\\ \text{output position ss}\\ \text{activate LA-speak.2}\end{matrix}\quad \{\text{3 V\_N\_V, 4 V\_N\_N, 5 V\_V\_V}\}$$

Like the LA-hear rule 4.1, this LA-think rule consists of (i) the rule name V_N_V (for verb-noun-verb), (ii) two patterns, one for the current proplet (serving as the 'sentence start') and one for the next proplet (serving as the 'next word'), (iii) a set of operations, and (iv) a rule package.

As an example of an LA-think navigation consider traversal of the proposition the girl sang, represented by the proplets *sing girl.*

5.3 LA-THINK NAVIGATION TRAVERSING *sing girl*

```
la1.LA1-THINK> sing
# Think mode #

1.
Start state, NW:        [verb: sing]              rp:{V_N_V, V_N_N}

                        [sur:       ]
                        [verb: sing]
                        [cat: DECL ]
                        [sem: past ]
                        [mdr:       ]
                        [arg: girl ]
                        [ctn:       ]
                        [ctp:       ]
                        [prn: 5     ]
                        [wrdn: 3    ]
2.
V_N_V: [verb: alpha]  [noun: beta   ]  mark alpha in ss-arg   rp:{V_N_V,
       [arg: beta  ]  [fnc: alpha   ]  output pos ss          V_N_N, V_V_V}
       [prn: m     ]  [prn: m       ]  activate LA-sp
                      [mdr:         ]

       [sur:       ]  [sur:         ]
       [verb: sing ]  [noun: girl   ]
       [cat: DECL  ]  [cat: SNP     ]
       [sem: past  ]  [sem: def sg F]
       [mdr:       ]  [mdr:         ]
       [arg: girl  ]  [fnc: sing    ]
       [ctn:       ]  [idy: 4       ]
       [ctp:       ]  [prn: 5       ]
       [prn: 5     ]  [wrdn: 1      ]
       [wrdn: 3    ]
------------------------------------
#####  End of navigation

Activated sequence of proplets:

[sur:       ] [sur:         ]
[verb: sing] [noun: girl    ]
[cat: DECL ] [cat: SNP      ]
[sem: past ] [sem: def sg F]
[mdr:       ] [mdr:         ]
[arg: #girl] [fnc: sing     ]
[ctn:       ] [idy: 4       ]
[ctp:       ] [prn: 5       ]
[prn: 5     ] [wrdn: 1      ]
[wrdn: 3    ] [trc: 10      ]
[trc: 9     ]
```

LA-think is activated by the input of a first concept which starts the navigation, here *sing*. This concept is usually provided automatically by autonomous recognition. For example, recognizing a feeling of hunger may start a navigation through recent hunger satisfaction routines, recognizing the Empire States Building may start the agent to reminisce about a recent trip to New York, recognizing dirty dishes may start dish washing activity, etc., depending on the agent's current set of priorities. In our very simple example, the concept starting the LA-think navigation is provided by the user typing `sing` after the top line prompt `la1.LA1-THINK>`.

If the word bank contained several connected *sing* proplets, they would be presented to the user with a request to choose one. In our example, however, this step is omitted because the word bank contains only one such proplet. In derivation step 1, this proplet is matched with the start state of LA-think.

Next applies the rule V_N_V contained in the rule package of the start state. By matching the first pattern with the proplet *sing*, the variables `alpha`, `beta`, and `m` are bound to the values `sing`, `girl`, and `5`, respectively. These bound variables reoccur in the second pattern, thus providing all the information necessary to retrieve a legitimate successor proplet, *girl*. The navigation ends after the application of V_N_V because the word bank in our example contains only two proplets.

To prevent redundant traversal of the same verb noun sequence, the argument currently traversed is marked with # (operation `mark alpha in ss-arg`). This is necessary because in propositions with more than one argument, e.g. *give Mary man book*, the rule V_N_V calls itself in its rule package, activating the following proplet sequence in the word bank:

V_N_V operations	activated sequence
give_Mary_give	*give Mary*
give_man_give	*give Mary man*
give_book_give	*give Mary man book*[16]

In such a derivation, the `arg` attribute of the V proplet will start out with the values `Mary man book`. After traversing the first N proplet, the mark operation of V_N_V changes these values to `#Mary man book`, after the second N proplet to `#Mary #man book`, and after the third to `#Mary #man #book`.

In order to allow repeated but non-redundant application of the rule V_N_V, the variable `alpha` in the *ss* pattern is bound to `Mary` in the value sequence `Mary man book`, to `man` in the value sequence `#Mary man book`, and to `book` in the `#Mary #man book`. Relative to the value sequence `#Mary #man #book`, the value assignment to the variable `alpha` fails, aborting this latest attempt at applying the rule.

The # marking of values by LA-think rules (only in LA-think.2 and higher) is of a temporary nature and does not change the content of the database. After completion of an intrapropositional navigation, the markers are automatically removed.

The second rule operation in 5.2, i.e., `output position ss`, specifies that the navigation returns to the *ss* proplet. This ensures that the current *ss* proplet, i.e. the verb, will be the new *ss* in the next rule application.

The third rule operation in 5.2, i.e., `activate LA-speak.2`, switches the system from LA-think.2 to LA-speak.2 for the realization of the surfaces derived from the proplet traversed. The rules of LA-speak.2 have complementary operations which switch the system back to LA-think.2 for continuation of the navigation.

6 LA-Speak

The third type of LA-grammar used in Database Semantics is called LA-speak. LA-speak is driven by an LA-think navigation.[17] As an example of an LA-speak rule consider **-DET**.

[16]Next, the navigation may continue extrapropositionally from *give* (see left column) to the verb of another proposition, provided the current verb's ctp (connective to previous) or ctn (connective to next) attribute has a value.

[17]For simplicity this process has been described in 3.3 as copying the proplets traversed by the LA-think navigation and using the copies as input to LA-speak. In a more parsimonious solution, LA-speak merely matches the proplets activated by the LA-think navigation and derives the desired output without any copying of database proplets.

This rule is for precipitating the determiner of a noun phrase, and is defined in Hausser 2002 as part of LA-speak.2.

6.1 LA-speak rule -DET

$$\textbf{-DET} \begin{bmatrix} \text{verb: } \beta \\ \text{arg: \#x \#}\alpha\text{ y} \\ \text{prn: m} \end{bmatrix} \begin{bmatrix} \text{noun: } \alpha \\ \text{mdr: z} \\ \text{func: } \beta \\ \text{prn: m} \end{bmatrix} \quad \begin{matrix} \text{lex-d } [\text{noun: } \alpha] \\ \text{if z} \neq \text{NIL activate LA-think.2} \end{matrix} \quad \{\text{-ADN, -NOUN}\}$$

Like the LA-hear rule 4.1 and the LA-think rule 5.2, this LA-speak rule consists of (i) the rule name -DET, (ii) two patterns, (iii) a set of operations, and (iv) a rule package.

The precipitation of the determiner from the noun proplet is handled by the -DET operation lex-d, which is defined as follows:

6.2 The function lex-d for realizing determiners

If one of the following patterns matches an activated proplet N, then lex-d produces the associated surface:

pattern	surface	pattern	surface
[noun: α, cat: SNP, sem: indef sg]	a(n)	[noun: α, cat: SNP, sem: pl exh]	every
[noun: α, cat: NP, sem: sel x]	some	[noun: α, cat: PNP, sem: pl exh]	all
[noun: α, cat: NP, sem: def x]	the		

Other lex-operations of LA-speak are lex-n for realizing names or pronouns, lex-nn for nouns, lex-fv for finite verb forms, lex-ax for auxiliaries, lex-nv for non-finite verb forms, lex-adn for adnominal forms, and lex-p for interpunctuation signs. The method illustrated in 6.2 is the production counterpart to automatic word form recognition.

As an example of an LA-speak derivation consider production of the surface the girl sang. from the activated database proplets *sing girl*.

6.3 LA-speak producing *The girl sang.*

```
##### LA-SPEAK
Start state, NW:          [verb: alpha]            rp:{-DET, -NoP}

                          [sur:        ]
                          [verb: sing ]
                          [cat: DECL  ]
                          [sem: past  ]
                          [mdr:       ]
                          [arg: #girl ]
                          [ctn:       ]
                          [ctp:       ]
                          [prn: 4     ]
                          [wrdn: 3    ]
                          [trc: 9     ]
```

```
1.
-DET: [verb: beta]  [noun: alpha  ]   lex-d        rp:{-ADN, -NOUN}
      [arg: alpha]  [mdr: z       ]
      [prn: m    ]  [fnc: beta    ]
                    [prn: m       ]

      [sur:      ]  [sur:         ]
      [verb: sing]  [noun: girl   ]
      [cat: DECL ]  [cat: SNP     ]
      [sem: past ]  [sem: def sg F]
      [mdr:      ]  [mdr:         ]
      [arg: #girl]  [fnc: sing    ]
      [ctn:      ]  [idy: 3       ]
      [ctp:      ]  [prn: 4       ]
      [prn: 4    ]  [wrdn: 1      ]
      [wrdn: 3   ]  [trc: 10      ]
      [trc: 9    ]
-DET produced: the
2.
-NOUN: [verb: beta]  [noun: alpha  ]   lex-n        rp:{-FVERB, -DET, STOP}
       [arg: alpha]  [mdr: z       ]
       [prn: m    ]  [idy: n       ]
                     [prn: m       ]

       [sur:      ]  [sur:         ]
       [verb: sing]  [noun: girl   ]
       [cat: DECL ]  [cat: SNP     ]
       [sem: past ]  [sem: def sg F]
       [mdr:      ]  [mdr:         ]
       [arg: #girl]  [fnc: sing    ]
       [ctn:      ]  [idy: 3       ]
       [ctp:      ]  [prn: 4       ]
       [prn: 4    ]  [wrdn: 1      ]
       [wrdn: 3   ]  [trc: 10      ]
       [trc: 9    ]
-NOUN produced: girl
3.
-FVERB: [verb: beta]  [cat: X       ]   lex-fv       rp:{-DET, -STOP}
        [cat: DECL ]
        [arg: alpha]
        [prn: m    ]

        [sur:      ]  [sur:         ]
        [verb: sing]  [noun: girl   ]
        [cat: DECL ]  [cat: SNP     ]
        [sem: past ]  [sem: def sg F]
        [mdr:      ]  [mdr:         ]
        [arg: #girl]  [fnc: sing    ]
        [ctn:      ]  [idy: 3       ]
        [ctp:      ]  [prn: 4       ]
        [prn: 4    ]  [wrdn: 1      ]
        [wrdn: 3   ]  [trc: 10      ]
        [trc: 9    ]
-FVERB produced: sang
4.
-STOP: [verb: beta]  [cat: X       ]   lex-p        rp:{-DET, -NOUN}
        [cat: DECL ]
        [arg: alpha]
        [prn: m    ]

        [sur:      ]  [sur:         ]
        [verb: sing]  [noun: girl   ]
        [cat: DECL ]  [cat: SNP     ]
        [sem: past ]  [sem: def sg F]
        [mdr:      ]  [mdr:         ]
        [arg: #girl]  [fnc: sing    ]
        [ctn:      ]  [idy: 3       ]
        [ctp:      ]  [prn: 4       ]
        [prn: 4    ]  [wrdn: 1      ]
        [wrdn: 3   ]  [trc: 10      ]
        [trc: 9    ]
-STOP produced: .

Generated words: the girl sang .
```

LA-THINK

The LA-speak rules apply in sequence, using their lex-functions to extract the surfaces in question from the two proplets activated by the LA-think navigation.

In propositions consisting of two proplets, like the above example, the LA-think navigation traverses the proposition in question completely, then switches to LA-speak, produces the surface completely, and finally switches back to LA-think (for example, to initiate traversal of the next proposition). In propositions consisting of more than two proplets, however, the actions of LA-think and LA-speak interact incrementally.

For example, the sentence The heavy old car hit the beautiful tree. is based on the proplet sequence *hit car heavy old tree beautiful*, represented abstractly as VNAANA. After LA-think has traversed the first two proplets *hit car*, the system switches to LA-speak to produce the surface the. Then it switches to LA-think to traverse *heavy*, switches to LA-speak to produce heavy, switches to LA-think to traverse *old*, switches to LA-speak to produce old, car, and hit, switches to LA-think to traverse *tree*, switches to LA-speak to produce the, switches to LA-think to traverse *beautiful*, and finally switches to LA-speak to produce beautiful, tree and the full-stop.

This incremental interaction between LA-think and LA-speak is motivated not only by psychological considerations, but also by computational efficiency. The reason is that realizing the surface of proplets as soon as possible (before the navigation has reached the end of the proposition) results in a more restricted set of candidates for matching by the LA-speak rules than the activated sequence for a complete proposition.[18]

7 Conclusion

Our computational model of turn taking is based on two basic innovations: (i) the algorithm of LA-grammar (TCS'92) and (ii) the data structure of a word bank (AIJ'01). The algorithm of LA-grammar is based on a strictly time-linear derivation order. The data structure of a word bank represents content in the form of concatenated propositions, defined as (unordered) sets of feature structures called proplets.

Proplets code the intrapropositional functor-argument structure and the extrapropositional relations of identity and conjunction as attribute values, independent of any graphical restrictions. For this reason, proplets are optimal for the time-linear storage and retrieval as required for the theoretical reconstruction of natural language communication, including turn taking.

Communication is modeled by three LA-grammars applying in sequence. LA-hear interprets natural language as a set of proplets which are sorted into the word bank. LA-think activates proplets in the word bank by navigating from one proplet to the next, serving as conceptualization. LA-speak maps the proplets traversed by the autonomous navigation into corresponding natural language surfaces.

These LA-grammars are explicitly defined for a substantial fragment of English and serve as the declarative specification of an initial version of the overall system. They are supported by an implementation in Java, which will be demoed as part of the talk.

---

[18]This holds especially at the beginning of a derivation. Later, as more proplets are added to the activated sequence, the earlier proplets have been systematically marked in past LA-think rule applications – which reduces the set of matching candidates from the other end. As a consequence, a maximally incremental system can guide the intended matching with much simpler and more efficient pattern definitions than one which activates future proplets earlier than necessary.

Bibliography

Berwick, R.C. & A.S. Weinberg (1984) *The Grammatical Basis of Linguistic Performance: Language Use and Acquisition*, Cambridge, Mass.: MIT Press.

Hausser, R. (2002) *Human-Computer Communication in Natural Language*, unpublished manuscript, p. 140. Available at: http://hakata.mt.cs.cmu.edu/11-792/hausser3.pdf.

Kycia, A. (2004) *Implementierung der Datenbanksemantik in Java*, MA-thesis in progress, Universität Erlangen-Nürnberg.

Thórisson, K. (2002) "Natural Turn-Taking needs no Manual: Computational Theory and Model, from Perception to Action," in Granström, B., D. House, & Karlsson (eds.) *Multimodality in Language and Speech System*, p. 173–207, Dordrecht: Kluwer.

Abbreviations:

NEWCAT'86 = Hausser, R. (1986) *NEWCAT: Natural Language Parsing Using Left-associative Grammar*. (Lecture Notes in Computer Science 231), pp. 540, Berlin, New York: Springer-Verlag.

CoL'89 = Hausser, R. (1989) *Computation of Language, An Essay on Syntax, Semantics and Pragmatics in Natural Man-Machine Communication*, Symbolic Computation: Artificial Intelligence, pp. 425, Berlin, New York: Springer-Verlag.

TCS'92 = Hausser, R. (1992) "Complexity in Left-Associative Grammar," *Theoretical Computer Science*, Vol. 106.2:283-308, Dordrecht: Elsevier.

FoCL'99 = Hausser, R. (1999/2001) *Foundations of Computational Linguistics, Human-Computer Communication in Natural Language*. 2nd Edition 2001, pp. 578, Berlin, New York: Springer-Verlag.

AIJ'01 = Hausser, R. (2001) "Database Semantics for Natural Language." *Artificial Intelligence*, Vol. 130.1:27–74, Dordrecht: Elsevier.

Information Modelling and Knowledge Bases XVI
Y. Kiyoki et al. (Eds.)
IOS Press, 2005

A Model of Unforgeable Digital Certificate Document System

Koichi NOZAKI*, Hideki NODA**, Eiji KAWAGUCHI** and Richard EASON***
**Nagasaki University,* ***Kyushu Institute of Technology,* ****University of Maine*
(nozaki@net.nagasaki-u.ac.jp,* ***{noda, kawaguch}@know.comp.kyutech.ac.jp)*

Abstract. Certificate documents today are either written or printed matter on a paper surface or plastic card. No digital data are accepted as a certificate document. People always asked to send an original certificate documents by traditional post mail even in this developing information age. The present paper proposes an unforgeable certificate document system wherein each document has a forgery-protective capability by itself. It does not need any communication with an authentication bureau. The system has several variations according to the key usage and additionally incorporated data. If this system is implemented, all certificate documents may be digitalized.

1. Introduction

Almost all documents nowadays are handled in a digital manner except for one very important case. It is the certificate document. Today, an authentic certificate document, that is original, cannot take a digital form. Certificate documents must be either written or printed on a paper surface or on a plastic card. For example, a driving permit or a passport is always a printed matter. No digital data work as a driver's license or a passport. All current certificates consist of information data and a base substance on which the information data is located. The base substance itself does not have any meaning. It only carries the information.

The reason why a digital document is not accepted as a certificate is very obvious. All digital documents currently used are easily forged without leaving any clues for forgery detection. No one has any means to tell whether a given digital document is authentic or not.

Printed matter is believed to be much more secure or more difficult to forge than digital data because the printed information is generally difficult to erase and to change, and the state of the art printing techniques to defy attempts to forge documents have been developed every day. Handwritten matter, e.g., a signature, is also regarded as a difficult-to-forge document because mimicking someone's hand writing skill and habits is not an easy task for an ordinary person. So, a certificate document often carries a signature made by an authorized person in such a way that it is difficult to separate it from the document information.

In the meantime, with the advent of the information age there is a surge of demand among people in the world that the certificate documents can be also handled in a digital manner. People hope to send and receive certificates through the Internet. The reality, however, is that people still send and receive paper certificates by way of traditional post mail with high cost.

In this situation, an invention of an unforgeable digital document system is long awaited. If invented, digital data could work as a certificate document by itself, and transmitted to any place in the world through the Internet. Currently, there are "signature certification" systems that use a public-and-private key system [1]. However, these systems need time and cost on the user's side. This makes them unsatisfactory for everyday use.

2. Document forgery

Document forgery is classified into two following categories.

(1) Type-1 forgery: Altering a part of an authentic certificate that is original

(2) Type-2 forgery: Producing a new fake certificate with false information

Type-1 forgery occurs when some part of the original document is altered to benefit someone who is not benefited by the original document. In this case the base substance, such as a paper sheet or plastic card, is legal and valid, but the information contained is forged. For example, a stolen photo credit card may be altered by replacing the photo picture on the card, the expiration date on a drivers license could be altered to make it still valid after the original license expires, the grading scores on a student report card may be changed from C's and D's to all A's, etc.

Type-2 forgery is the case when both the base substance and the information data are false, but it is hard to tell if it is genuine or fake because the base substance and the style of the document look very authentic. Most passport forgeries belong to this category. Type-2 forgeries include attempts to create authentic looking, but false, information data.

The most important aspect of the Type-2 forgery is as follows. Forging is always targeted at a certificate document that is issued by a high authority, such as the government of a country, a public organization, world known big business, or a very famous person in the world. While, a certificate issued by an average citizen, or un-trusted company, are seldom forged. This is because people regard a high authority as simply trustworthy without any authentication. Once a high authority have announced or published something, people trust it as it is. The high authority itself automatically serves as an authentication bureau.

An "unforgeable document" should satisfy the following conditions.

- C-1 A forged document has some difference from an authentic original document in some way
- C-2 The forgery detection can be done without referring to the authentic original document
- C-3 There is a concrete verification method that detects a forged document without communicating with an authentication bureau.

If a system can produce such a digital document that satisfies all these conditions, the system is titled as an unforgeable digital document system. An "unforgeable digital certificate document system" is an unforgeable digital document system that is used for a certification system.

3. Information structure in the current digital document

Documents consist of characters, digits, tables, illustrations, photo pictures, etc. Its digital version, i.e., digital document, is a computer file. Digital documents are designed to produces visual information on the computer monitor. Therefore, every digital document can take an image form rather than a combination of the different file types (e.g., one "BMP" file instead of the combination of "WORD", "JPEG" and "EXCEL" files) if the overall visual information is the same.

A current digital document has a single-layered information structure in the sense that all information is just as it looks. No hidden information is included in the document. So, it is quite easy to alter the original document using a text editor or an image editor. It is also easy to produce an entirely fake document that looks very authentic by using computers. Nobody can tell whether a given digital document is authentic or not.

Therefore, an unforgeable digital-document, if ever exists, cannot take a single-layered information structure. Instead, it should have "a multi-layered" information structure having the following properties.

(A) The information data on both the external and internal layer are tightly linked.
(B) The internal information is not extracted without a special means to extract.
(C) The internal information is not changeable as attempted.

Fig. 1 illustrates a concept of the single-layered document and the corresponding multi-layered document.

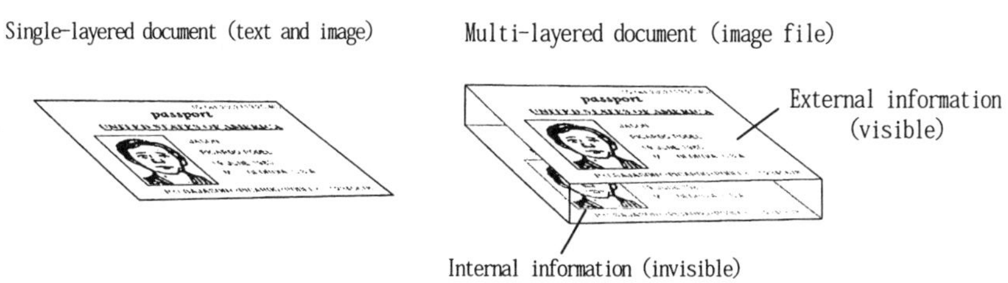

Fig. 1 A single-layered and a multi-layered document

In this example the multi-layered document is an image data, and the external and the internal information is the same as the original single-layered document information. Once someone has altered the external layer by using an image editor, there occurs some discrepancy between the two layers. Or, the internal information is significantly disordered.

4. How to implement a multi-layered information structure

We know that a steganographic technique can implement a multi-layered information structure in an image data by using an embedding program. The external layer in that case is the visual information of the original document, which is the cover image in steganography, and the internal layer is the embedded information. If we embed the original document in the internal layer, we will get a multi-layered information structure as is shown in Fig. 1.

A BPCS-Steganography [2] [3] is a good method for this purpose because the embedding capacity is very large and the embedded data is quite fragile. Fig. 2 shows a data embedding procedure by steganography in general. An embedding program is used for embedding, an extracting program is used for extracting the embedded information.

Fig. 3 illustrates a scheme to create a multi-layered digital document from an original document. In this procedure the Document Image data becomes larger than the original document in size.

As the internal information is not changeable as attempted, any alterations on the external layer can be detected by comparing with the internal layer. If both meet, the document is authentic otherwise the document is forged.

Fig. 2 Data embedding by steganography

Secret data is embedded in the cover image to get the stego image. There is no visible degradation in the stego image. Extracting the embedded data is executed by an extracting program with/without a key.

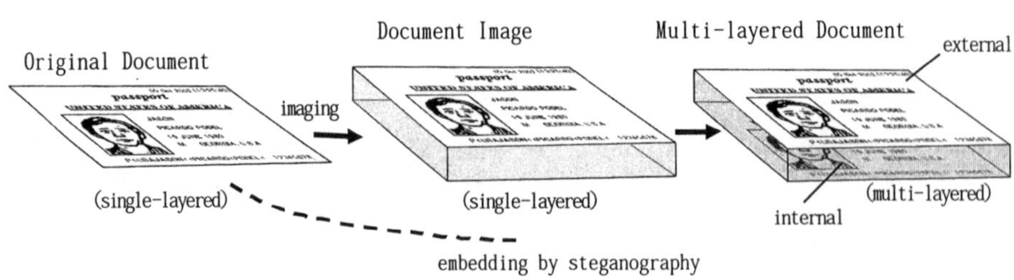

Fig. 3 Multi-layered document production procedure

5. Unforgeable digital certificate systems

As we have shown in Fig. 3, the multi-layered digital image is the basic structure of the unforgeable digital document. It has some variations according to the functions in the document producer and the key in the system as well as other incorporated data.

5.1 Type-1 system: A Type-1 forgery protective system

Type-1 system consists of the Document Producer and the Document Verifier. The system can only detect whether the document was altered or not after it was created. It does not matter if the document is a certificate or a general document.

Fig. 4 (a) illustrates the flow diagram of the Document Producer of Type-1. The input to the system is an Original Digital Document (Doc_{org}). The output is the Multi-Layered Digital Document consisting of the external and the internal information, i.e., $Doc=(Doc_{ext}, Doc_{int})$. We call a Type-1 multi-layered document a Type-1 document.

While, Fig. 4 (b) shows the flow of the Type-1 Document Verifier. A Type-1 forgery on a Type-1 multi-layered document is easily detected by this Document Verifier.

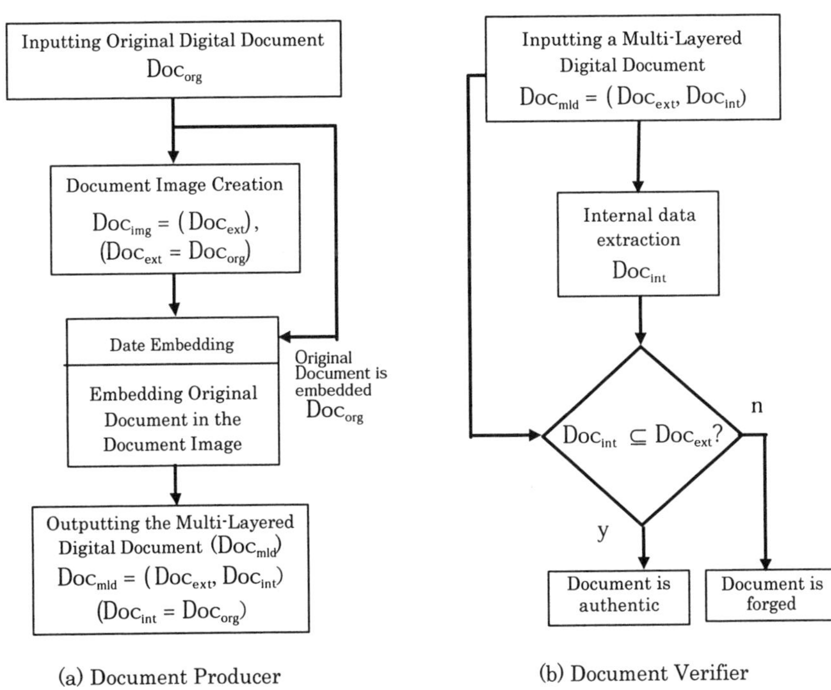

Fig. 4 Type-1 digital unforgeable document system

5.2 Type-2 system: A Type-2 forgery protective system

A type-2 forgery is attempted on a certificate document issued by a high authority, which needs no authentication by a third party. Passport forgery is a typical example of this forgery type.

A Type-2 system consists of a Type-2 Document Producer and an associated Document Verifier. A Type-1 system is extended to a Type-2 system by incorporating a pair of keys, namely, a Document Production Key (Kp) and a Document Verification Key (Kv). Kv is produced from Kp according to a one-way scrambling function in an unseen manner within the system.

A Type-2 Document Producer works as follows.

First, a document production key (Kp) is arbitrarily selected. It is inputted to the system and scrambled one-way to create a document verification key (Kv). This Kv is outputted from the system, and opened public. It is used when the document is verified in the Document Verifier. The document production key (Kp) must be kept secret. It is a type of a private key.

Actually, Kv is confidentially used for embedding the Original Document (Doc_{org}) in the Document Image (Doc_{img}) to produce the multi-layered document (Doc_{mlt}).

However, the key scrambling and the document embedding operations in this Document Producer are unseen from the outside. The key-scrambling and the document embedding are made inseparable in the program. The system is made in a reverse-engineering-protective manner.

Therefore, Kp works as if it is the key to embed the document, even if Kv is actually used. We call a Type-2 multi-layered document a Type-2 document.

A Type-2 Document Verifier works as follows.

A Type-2 multi-layered document data is first inputted to the system. The document verification key (Kv) is also provided to the system. Then the internal information is extracted in the system and compared with the external information. If both meet, the document is authentic, otherwise is forged.

The reason why a Type-2 certificate document is unforgeable is obvious. Someone who tries to forge a Type-2 document must create a document that can be verified by the Kv that has been opened public by the certificate issuing authority. The only way the forger can succeed is that he must steal the Kp. However, it is securely kept secret by the authority. Therefore, as long as Kp is at the certificate issuer's hand, the certificate can never be forged.

The important point in the Type-2 system is that the system uses a pair of keys (Kp and Kv). The basic idea is the same as the traditional "public-key and private-key" system. However, The system here does not need any authentication bureau. The difficulty of detecting Kp from Kv is based on both the reverse-engineering protectiveness of the system and the one-way scrambling function. Therefore, the technique to make the system difficult to reverse-engineer (e.g., obfuscation of the program source code) is quite important. The theoretical evaluation of the system security is difficult, but it is practically very secure, or we can make it very safe by many programming techniques.

5.3 Type-3 system: An extension of the Type-2 system

A Type-2 system is upgraded by incorporating other information data. Some document in some case needs to be verified the time of document production. In some other case it might be better or necessary to carry information about the geographical location of the document production. Such additional information must not be changed after the document was created.

The time data is acquired by a Network Time Protocol Server (NTP Server) through the Internet. It is quite accurate timing system. We describe this data by TS. In some situation the time data is also obtained from the Radio Standard Time Broadcasting system that is commonly available in many countries today.

As for the geographical location data, the Global Positioning System (GPS) gives us very detailed Latitude-and-Longitude data all over the world. We designate such positioning data by GP.

The document production time and place data (TS and GP) is the data that are unique to each document. No two documents can have the same data if they are included in the document. We call a Type-2 system with TS and/or GP data a Type-3 system. TS and GP are added to the original document. They are placed both external and internal layers of the document image. Fig. 5 (a) illustrates the flow of a Type-3 Document Producer. While, Fig. 5 (b) is the corresponding Document Verifier.

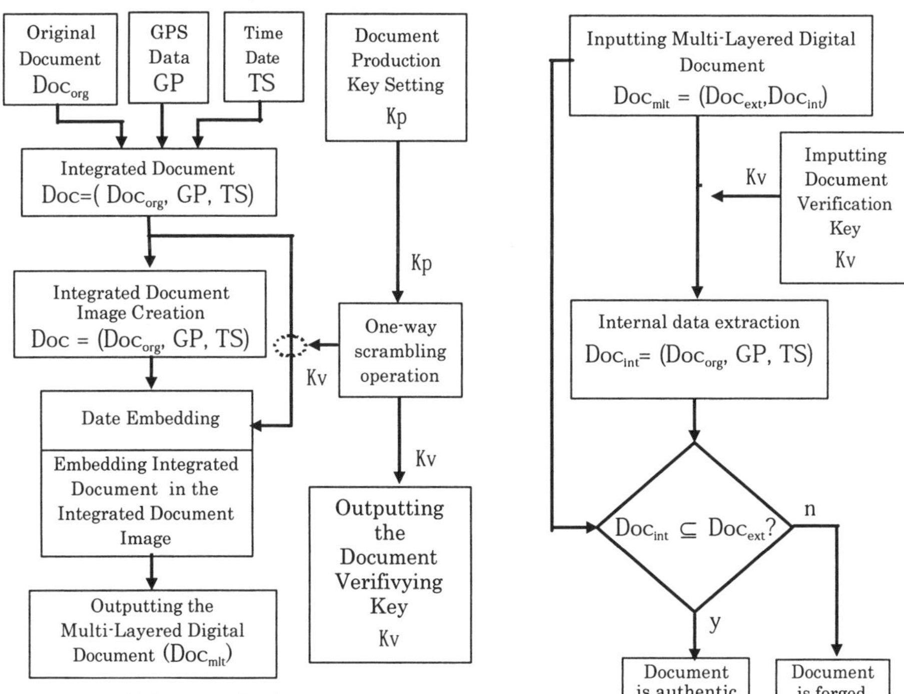

(a) Document Producer

(b) Document Verifier

Fig. 5 Type-3 digital unforgeable document system

6. Practical applications of the unforgeable certificate system

We will show two example applications here.

6.1 Digital passport system

Fig. 6 shows a model of a digital passport. The unforgeable passport data is stored in the IC memory that is mounted on the passport card. The printed information on the card shows the passport information in the IC, but is not the primary data. Rather, it is secondary data just for human eyes. Forgery detection is executed locally at the airport immigration inspection booth in an off line manner. The data in the IC memory can be personally used for sending authorized passport information to remote place through the Internet.

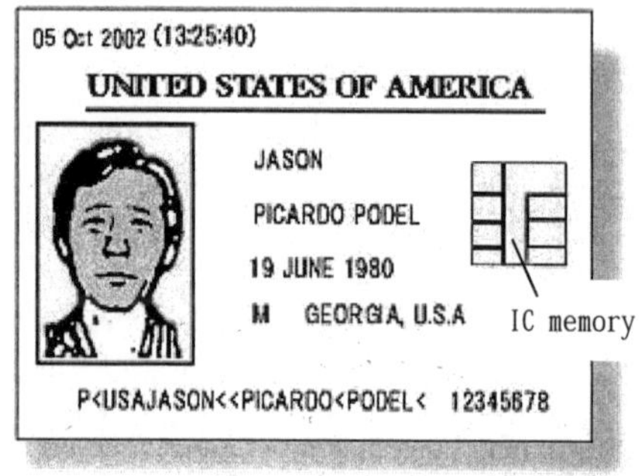

Fig. 6 Digital passport

6.2 Digital camera system

Fig. 7 is illustrates a digital photo picture with the TS and GP data at the top and the camera model date at the bottom.

The photo is verified by the key that is provided by the camera maker (or located at the "Danon's home page"). If the internal picture is the same as the external, it is an original picture, otherwise a tampered picture. No one can modify this picture without causing damage on the internal picture. This may be very good for many forensic purposes.

Fig.7 Unforgeable digital photo

7. Conclusions

We showed an unforgeable digital certificate document system based on the steganographic technique. If this system is realized, digital sending and receiving of certificate document become very easy and secure. The authors have started working on implementing a prototype system.

References

[1] http://www.murdoch.edu.au/elaw/issues/v9n3/lim93_text.html

[2] Koichi Nozaki, Michiharu Niimi, et al "A large capacity steganography using color BMP images", Proceedings of ACCV'98-Third Asian Conference on Computer Vision, pp.16-23, 1998.

[3] Eiji Kawaguchi and Richard Eason, "Principle and application of BPCS-Steganography", Proceedings of SPIE Multimedia Ststems Applications, Vol.3528, pp.464-473, 1998.

Information Modelling and Knowledge Bases XVI
Y. Kiyoki et al. (Eds.)
IOS Press, 2005

An Empirical Study of Context Knowledge in Engineering Design Practice

Muhammad Ikhwan JAMBAK and Joris. S. M. VERGEEST
Department of Design Engineering
Faculty of Industrial Design Engineering
Delft University of Technology
Landbergstraat 15, 2628 CE Delft, The Netherlands

Abstract. Knowledge gathering and information retrieval is known as a challenging field of research, especially in product design. This research is aimed to develop a method and tool to support designers to gather and retrieve information during conceptual product design. As part of the research, the existence and the role of context in engineering design practices were studied. The design process and the context knowledge appeared to be representable in a formal framework.

Introduction

Large amounts of knowledge and information are involved throughout the product design process [1]. Designers need knowledge in order to accomplish the design task in a sophisticated manner. Poor or inadequate knowledge and information during the design process could result in unsuccessful product designs. Achieving a method and tool to support designers to gather knowledge and to retrieve information during product design has become an important issue nowadays. Although knowledge and information are abundantly available and surrounding the designer, to get access to the most relevant knowledge or information for a particular design process is not necessarily an easy task. The challenge of this issue not only came from the difficulties to utilize the knowledge and information coming from various locations, domains and disciplines. It is also a challenge because in new product design, the design concept is often vague and incomplete. In addition the amount of information or knowledge to be handled is too large, too broad and often not manageable in a well-structured system.

One of the potential solutions to the above problems is offered by the "context" concept. This concept is being explored in the field of artificial intelligent [2][3] and knowledge management [4], and natural language processing [5]. A comprehensive survey of context has been done by [5]. Although the term context is far from a consensus it has to do with the relationship among knowledge or information. Indeed, [4] said that knowledge does never stand for its own; it is always used in a context. For this research, a "context knowledge" terminology was introduced referring to a subset of all knowledge that identifies the context of specific knowledge. Our hypothesis towards a method and tool of knowledge gathering and information retrieval during product design is that if the designer can communicate context knowledge, only then the relevant knowledge or information can be gathered from knowledge or information provider.

With the purpose of supporting designers, the upcoming system should meet the designer's expectation and should be really helpful during the product design. Therefore, we propose a framework facilitating us to describe more detail about the aim and the scope of the project and give us a general ideal of steps towards the solution. Another reason is that it acts as a model of the design process that explicitly includes knowledge and information handling. More than that, this framework will play an important role in a step-by-step implementation planning and as a basic architecture of the system. For more details about the framework, we refer to [6].

In order to confirm that the upcoming system can be helpful and useful for the designer, we did some study on the data from an experiment, reflecting the practice of product engineering design. If we understand context from practice, then there is more chance to have a helpful and useful support system. For this purpose, the data from the so-called D_1 experiment of the Desys project [7] were used in this study. This paper reports the result of the study. In the first section, the Desys project, D_1 experiment, and its data will be described. In the second section, some study indicating the context and the way in which it being treated was studied. The third section reports about the role of context in information transaction. Having learned these phenomena, we formalize the design process and context knowledge in section four. The conclusion was drawn in section five.

1. The Desys Project

Desys was a research project conducted at the Faculty of Industrial Design Engineering of the Delft University of Technology. It investigated the role of information in the early phase of product development process [7]. The Desys project aimed to develop tools to support conceptual design activities. In order to have a computerized conceptual supporting tool, the requirements of such computer system were gathered through empirical observation of the design process.

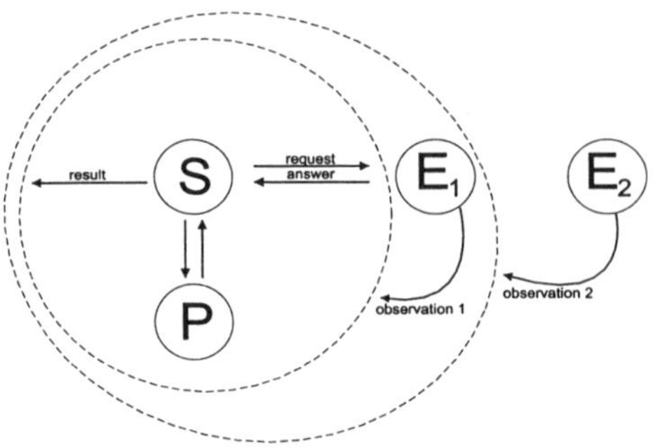

Figure 1: Structure of D_1 Experiment

As part of the Desys project, D_1 experiment was aimed to study information handling provided by means of design support and identifying any opportunities to do so. This experiment as depicted in Figure 1 [8], was based on a conceptual design task given to a designer. The designer *S* performing the design process *P*, was located in a observation room, where he was assisted by an experimenter E_1. E_1 was playing role as an expert information provider and meanwhile E_1 also observing *S*. *S* and E_1 were sitting at the same table in a room with audio and video equipment. In the same time E_2 was located in a separate room controlling the equipment and observing the *S*. Both experimenters E_1 and E_2 observed the requests, answers and results, and also noted the activities of *S*. However, since E_1 had to answer questions from *S* therefore his/her observation was considered as incomplete.

The task was chosen as one that reflected an industrial problem, addressing a mixture of generating a concept together with its technical specifications in both the mechanical and electrical domain. *S* had to design a bicycle lighting system that still powers the bulbs when the bicycle speed is low or the bicycle is standing still for a short time. *S* had to generate a brief description of the concept to show that it fulfills the requirements.

E_1 and E_2 had to make real time notes into an observation list as shown in figure 2. This list contains the start and end of the task, observed even (activities, request, answer, and result) that represented by its time, description and indication. At the end of each session E_1 and E_2 made a first version of transcript by combining their notes and observing through the video and audio recording if necessary. This transcript file called *f-file* (figure 2) was structured as follows:

- File identification
- Header
- Start time of the task
- Transcription of the events
- End time of the task

```
# file dls001.f4
#
# this file contains the data list f of d1.20.06.94.1
# ...
# ...
# 13:08:25 start_of_assignment
13:09:40 activity (marking the assignment on page 1)
13:12:35 activity (marking the assignment on page 2)
13:13:00 activity (marking notes about the requirements; 42)
# sample 1
13:13:30 request  (What are the different types of rear-wheels on which
                   the generator has to fit, how are they integrated in
                   their environment, how does the internal of the hub
                   look like such as chain gear-case, overcoat guard,
                   pedal break, caliper break and drum brake)
13:14:30 activity (collecting information on the technical environment)
# end sample 1
13:15:00 activity (sketching; 42 centre)
13:15:30 answer   (information sheet 'hub';1)
13:15:45 request  (doest the bicycle has a drum brake)
13:15:45 answer   (a drum brake and a five speed gear hub)
13:15:45 activity (writing down these data; 42)
13:16:00 activity (reading sheet 'hub';1)
13:16:25 activity (reading the assignment)
13:16:35 activity (marking the assignment)
13:16:40 answer   (the rear wheel has a diameter of 26 or 28 inch;42)
13:17:00 request  (is it allowed to change the cup around the hub)
13:17:00 answer   (no it is not allowed to change the cup)
13:18:00 activity (looking for a turning and non turning part near the
                   hub; 42)
# sample 2
13:18:30 activity (considering an alternative outside axle)
13:18:45 ......

.
# end sample 27

.
# 15:41:45 end_of_assignment
```

Figure 2: An example of typical f-file

2. Indication of the Existence of Context in a Design Process

2.1. Indicating the Existence of the Context by the Emergence of New Issues

Only if there is enough initial knowledge, one can manipulate data [9]. We argue that this condition also applies to the situation where one has to gather additional knowledge or to retrieve information. The context should be sufficient to get the relevant knowledge or information from the knowledge/information provider. Along with the iteration process of searching-receiving knowledge/information, the context becomes more and more clear, as illustrated in figure 3.

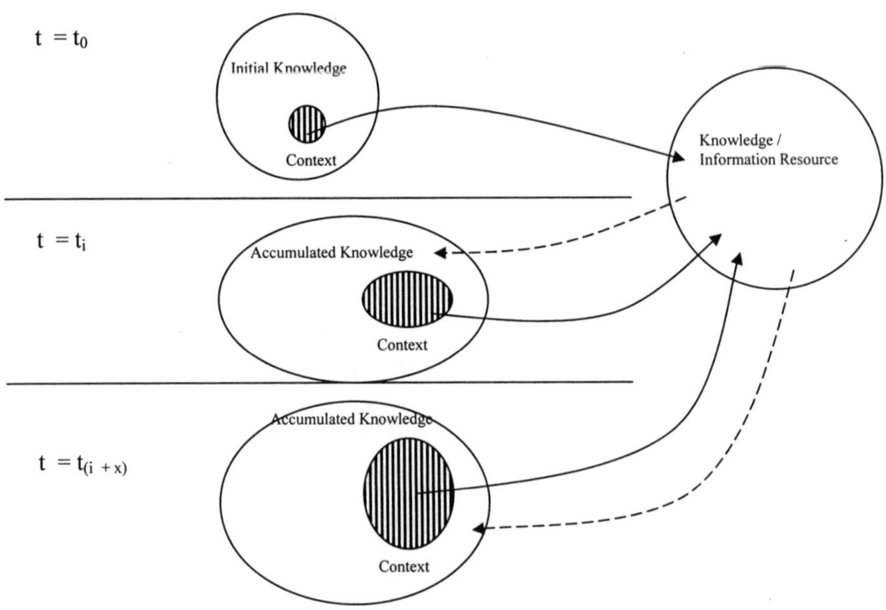

Figure 3: Knowledge growth and information exchange

2.1.1. Hypothesis and Expectation

The emergence of new issue can indicate the growth of context. With the assumption that the designer's knowledge is growing when he/she receives the incoming knowledge or information from the information provider. As shown in figure 3, if the context is subset of the accumulated knowledge, the capability to identify the necessary knowledge is also increased. From this study, we expect to find new design issues during the design process.

2.1.2. Method

The data from the transcript file (*f-file*) was read sequentially, and divided into blocks. For each block (lines within a sample plus lines before it started) in the *f-file*, every issue found (keywords or terms of information requested by designer, dialogues between designer and expert, and also documents studied by the designer) was extracted. At the end of each block, the most dominating issue tackled by the designer is considered as a designer's "Focus" in that particular stage. However, when a designer tackled all the issues equally, then the focus will not be seen. When a new issue appeared it was tagged as "New". In addition, when new issues were inherited from the previous connected terminology, a tag "Sub" was included for its next appearance. Furthermore, when two issues can be seen to have a connection to each other, a "Connect" tag was given.

1st block

Issues:

a. Rear-wheels
b. Generator
c. Environment
d. Hub
e. Chain, new: gear-case
f. Overcoat guard
g. Pedal brake
h. Caliper break
i. Drum brake

Focus: Environment

2nd Block

Issues:

a. Hub, New: Five speed gear, cup, turning and non-turning part, construct outside, driving cup
b. Drum brake
c. Rear-wheels, New: diameter
d. New: Axle
e. New: Inertia

Focus: Hub

Figure 4: Observed new design issues

2.1.3. Results

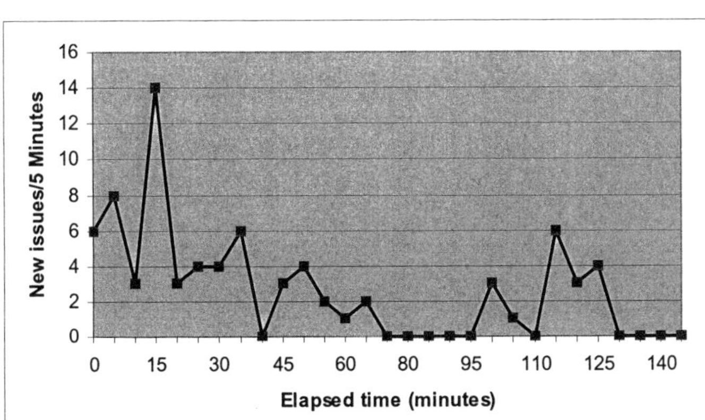

Figure 5: The emergence new design issues

After studying the design requirements, the designer requests some information (tagged as 6 initial issues at time 13:30:30) as shown in figure 2. This can be regarded as a starting point of the context-knowledge growth. Every new issue arisen from the starting point until the end of the design assignment was noticed within 5 minutes time intervals. New issues have been frequently noticed at the first period of 15 minutes where the highest number (14) occurred in the period between minute 10 and 20. There were no new issues in the period between minute 35 and 45, also in between minute 65 and 100, as well as after minute 125 until the end of the design assignment. Figure 6 shows the accumulation of new issues from the start until the end of the design process.

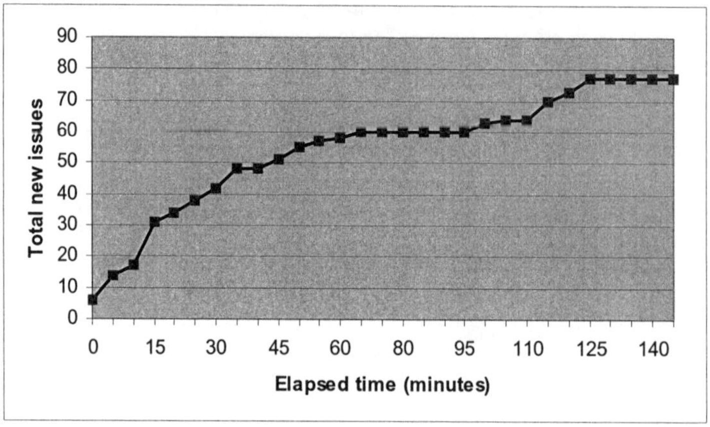

Figure 6: Total new issues

2.1.4. Concluding remark

This study confirms that the designer has the initial context knowledge as he/she can make some information request. This analysis also shows that an issue might influence a new knowledge/information discovery. In the periods where there were no new issues, it can be observed that designer was busy with the elaboration or dealing with the previous issues. Indeed, it can be concluded that at the end period of the design assignment, where the idea/concept is practically almost concrete, no new issues arose anymore.

2.2. Indicating The Existence of The Context By Answering Time

The setting of D_1 experiment as described in figure 1, give the opportunities to the expert (E_1) to develop his/her context on the design process, since the expert is within the scope of the process. Our assumption, since the expert was the only knowledge/information provider that could develop the context and the others (book, physical model, picture, and sheet) cannot, the experts must be the quicker knowledge/information provider to the designer.

2.2.1. Method

Using the difference between the time the subject got an answer and the time when he or she asked (or looked for) information, for each type of answering medium the delay time of answer delivery can be found.

2.2.2. Results

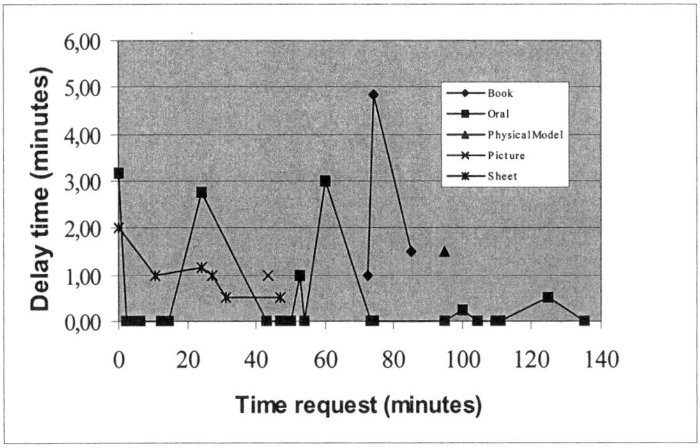

Figure 7: Answering time per channel of communication

Figure 7 shows the results of the study. Books imply the longest delay time and oral replies are given in the shortest time. To make the result more clearly to be understood, we make the frequency distribution of the above result as shown in figure 8. Most of the oral channel communication gives the answer in less than one minute, while nineteen of them were directly replied by the knowledge/information provider.

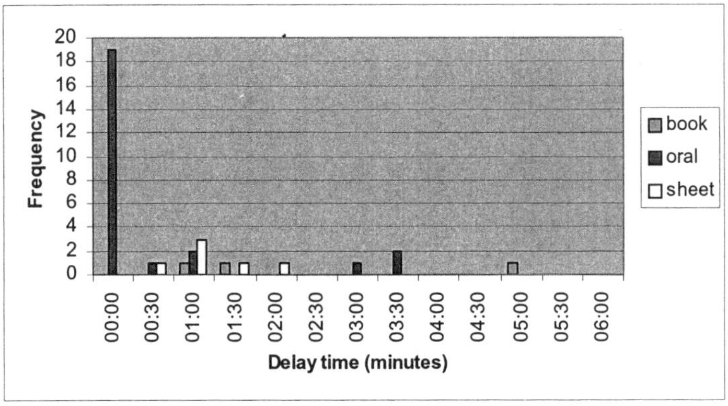

Figure 8: Frequency distribution of answering times

2.2.3. Concluding remark

This study confirms the important role of the context within a communication of knowledge gathering or information retrieving between the designer and the information provider.

3. Context role in communication

In addition to the result of the study as described in subsection 2.2., a study of the role of the context in a communication between designer and knowledge/information provider, was conducted.

3.1. Hypothesis and expectation

Our hypothesis is that every question to an information provider can and needs to be accompanied by context(s). A richer context will give a better chance to the information provider to provide a right and relevant answer at the right time.

We analyzed the questions made by the designer (subject) in session 1 and 2 of the D1 experiment. This analysis aimed to figure out the relationship between the questions made by the subject and the context surrounding the question-answer process. Therefore, this analysis focused on the following points:

1. Concreteness of the question
2. Relevance of the question

Our expectations at the beginning of this analysis are:

1. We will find some cases that the questions are not concrete or considered as an implicit question but the information provider can manage to understand the questions.
2. We will find some cases where the question is related to the previous questions.
3. We will see how contexts transfer/dialogue occur within the question-answer process

3.2 Analysis

Some terminologies are introduced for the analysis; they are as described as follows:

1. Subject is a designer who asking questions to expert (as information provider)
2. Observer is the one who is observes the transcript file (*f-file*)
3. Time is the time stamp in the protocol when the question occurred
4. Concreteness of question, the question is explicit if the observer can understand it without checking the previous question-answer process. "Y" means the question is explicit and "N" means the question is implicit.
5. Channel of communication, is how the question or context transacted between the subject and the information provider. "S" means the question and or context is transacted literally. "T" means the question or context is transacted via a written

text. It can happen that the question is transacted literally but the context is transacted via a text. In the D1 protocol, session 2, at time 14:29:00, for example, the subject asking a question "Does this formula give the effective voltage?" In this case, the subject was assumed showing a written formula to the information provider. "P" means the question or context is transacted via a picture/drawing sheet. "O" means the question or context is transacted via a physical object.

6. Question related to previous question, "Y" if the observer assumed that the question has a relationship with the question-answer process or request activities.
7. Question related to very earlier question, "Y" if the question is related to an earlier question or request activity.
8. Confirmation need, "Y" if the information provider asking the subject to confirm his/her understanding of the question or to redirect his context.
9. Context word(s) need, "Y" if the question contains one or more words that can point out to the previous question-answer process or request activities.

The Procedure of the analysis of this analysis is as follows:

1. The observer should read the *f-file* sequentially
2. The observer have to find the question in the activity "request" of *f-file*
3. The observer have to notice the time stamp when the request happen
4. The request that is not in a question form should be excluded from the table but should be notify (because it involved in context build)
5. The observer should check the question whether he/she can understand the question at first.
6. The observer should assume how the question is transacted, either literally, or need an explanation or additional information using texts, pictures or physical objects.
7. The observer should check whether the question has relationship with the previous requests and answers activities of the *f-file*
8. The observer should check whether the question has relationship with the earlier question (not the whole requests and answers activities)
9. The observer should check the answer activity after related to the request activity, whether it has a question form or not
10. The observer should check whether the question contains one or more words that can relate to other requests or answers activities of the *f-file*

3.3. Results

Table 1: context role in knowledge and information transaction

	Time Elapsed (t)	Concreteness of question	Channel of communication	Question related to previous questions/activities	Question related to very earlier question/activities	Confirmation need	Context word(s) need
Session 1							
	13:15:45	Y	S				
	13:19:30	Y	S	Y	Y		Y

13:26:00	Y	O				
13:56:30	Y	S				Y
14:00:30	Y	S	Y	Y		Y
14:02:00	Y	S				Y
14:06:00	Y	S	Y	Y		Y
14:07:30	Y		Y	Y		Y
14:48:30	Y	S	Y	Y		Y
14:58:30	Y	S	Y	Y		Y
15:40:00	Y	S	Y	Y		Y
Session 2						
13:35:00	Y	S				
13:36:00	Y	O	Y	Y		Y
13:42:20	Y	S	Y	Y		Y
13:47:40	Y	S	Y	Y	Y	Y
13:51:00	Y	S			Y	
13:51:00	Y	S	Y	Y		Y
13:55:00	Y	S	Y	Y		Y
13:58:50	Y	S	Y	Y		
14:04:05	Y	S	Y	Y		
14:07:50	Y	S	Y			
14:12:00	Y	S				
14:23:40	Y	S	Y	Y		
14:26:30	Y	P	Y	Y		
14:29:00	Y	T	Y	Y		Y
14:35:05		S	Y	Y		
14:39:40	Y	S	Y			Y
14:43:50	Y	S	Y	Y		Y
14:51:30	Y	S	Y			Y
14:51:40	Y	S	Y	Y		
15:00:00	Y	S	Y			Y
15:00:00		S	Y	Y		

3.3.4. Discussion

3.3.4.1. Concreteness of question

From the result above, most of the questions can be considered as explicit. In session one, none of eleven questions is implicit. And only two out of twenty-two are implicit in session two. All of these implicit questions are related to the previous questions. In the contrary, not all the explicit questions have relationship with the previous questions.

This result shows us that the implicit question happens if the question has a strong relationship with the previous questions. For instance, in session two at 14:35:05 the question "are they equally expensive?" are not explicit to the observer. However, after the observer sees the previous request at 14:33:40 about the current of existing batteries, it is clear to the observer that questions should be as follow: "are these existing batteries equally expensive?" Here, we can see that the information provider already has a context, and therefore it is not a problem to understand the implicit questions.

3.3.4.2. *Relevance of question*

Both of sessions one and two, show that most of the question has relationship with the previous question or request activities. Indeed, most of the questions have relationship with the earlier question or request activity. This result shows that the context has been develop during the question-answer process and the even during design process as whole. Few of the questions have not relationship both to the whole request-answer process nor the earlier question or request. This result gives the indication that a new context has been built up. On the other hand, the questions at 14:51:30, 14:51:30 and 15:00:00 of session two have relationship with the previous request-answer process but has nothing to do with the earlier question. This could be the indication of the subject shift back to the specific focus of design problem that he or she left it before.

The result also shows that most of questions contain word(s) that can point out to the previous request-answer process or request activities. All the questions that contain context word(s) also have relationship with the previous request-answer process or request activities.

3.3.4.3. *Context transfer / dialogue*

Most of the cases the context transfer are using speech as channel of communication. Or the questions are explained literally during the dialogue between the subject and the information provider. However, it happens twice that the contexts are transferred via (or refer to) a physical object and once in using a text.

Another evidence found in this result was that most of the cases, the initiative of context building are come from the subject who has the question, but also the context could be requested by the information provider in order to deliver an answer. In this case, the information provider will reply a question from subject with another question. Or in other words, a confirmation needed by the information provider in order to be in the right context.

3.3.5. *Concluding remark*

The results of this analysis are consistent with some of our expectations. We found some cases where the information provider manages to understand the implicit question. We also found that most of the questions have relationship with the previous question-answer process or request activities that give the indication of a specific/local context has been developed. We also found some possible channel of communication to transfer the context. However, this analysis is not possible to measure quality of the answer because the absence of the feedback from the subject regarding to the incoming data/information from the information provider. The other reason is that these question-answer processes were occurred as a part of the design process where the result of the design process itself is also in progress.

4. The proposed formalism

Guided by the results of this study, we proposed a formalism of the design process and the context-knowledge. This formalism is being used as the foundation of the development of a knowledge gathering and information retrieving support tool during the product design.

4.1. Design process formalism

Let a design process be denoted as D. This process is started at t_{start} and ended at t_{end}. In this process, designer(s) denoted as S is creating models $M_{(t)}$ that can be changed from time to time until the end of design process at t_{end}. The knowledge provider(s) R such as databases, books, experts, etc., provide a set of items of information $J = \{d_1^j, d_2^j, d_2^j ... d_m^j\}$ at $T^J = \{t_1^j, t_2^j, t_3^j ... t_m^j\}$, where d_i is an information item provided by R at time t_i^j. If all the received information items $d_1^j, d_2^j, d_2^j ... d_m^j \in J$ and I is the set of all information, then: $J \subset I$. In order to receive the necessary information, S produces and passes $Q = \{q_1, q_2, q_3 ... q_m\}$ at $T^Q = \{t_1^q, t_2^q, t_3^q ... t_m^q\}$ to R, where q_i is a query item passed by S to R at time t_i^q. Therefore, the design process D can be described as follows:

$$D = (S, R, M_{(t)}, t_{start}, t_{end}, J, Q, T^J, T^Q)$$

We assume that before $t = t_{start}$ the designer S has accumulated pieces of knowledge $k_1^i, k_2^i, k_3^i ... k_m^i$ in a set of initial knowledge K_i. Tacit knowledge such designer experience and or any knowledge he or she knows about before $t = t_{start}$ are included in this initial knowledge. These all pieces of knowledge denoted as $k_1^i, k_2^i, k_3^i ... k_m^i$ are members of K_i.

We assume that all of the received information items of J can be interpreted or transformed into knowledge by the designer in a process P, then, the process P, can be seen as a mapping from I to K, thus: $P : I \rightarrow K$, i.e., If $P(d_m^j) = k_m$.

If all of incoming information in J can be transformed into knowledge and assumed nothing lost during the process and can be stored, then this knowledge are accumulated in designer's set of accumulated knowledge $K_a(t)$ as pieces of knowledge that denoted as $k_1^a, k_2^a, k_3^a ... k_m^a$. The accumulated knowledge (K_a) at time t, then can be expressed as follow:

$$K_a(t) = K_i \bigcup_{m=1,n} P(d_m^j), t_m^j \leq t$$

The accumulation knowledge $K_a(t)$ is a union of the initial knowledge and the knowledge from the interpretation or transformation process of all received information items. At the beginning of design process $K_a(t_{start}) = K_i$ because there is no interpretation or transformation from information to knowledge yet. At the end where $t = t_{end}$, the knowledge accumulated at the completion of D can be seen as:

$$P(J) = K_a(t_{end})$$

As a note, if K is the set of all knowledge, then $K_i \subset K$ as well as $K_a(t) \subset K$ also.

4.2. Context knowledge formalism

As mentioned in previous section, context knowledge is a subset of all knowledge accumulated by the user and has been or will be transferred or shared to the knowledge provider. Therefore, at time t, the context knowledge $K_c(t)$ can be expressed as $K_c(t) \subset K_a(t)$.

However, in order to share the context knowledge, the user S must express it as understandable information to the knowledge provider R. Let E as a process to express the knowledge into a set of context information C. if $d_1^c, d_2^c, d_3^c ... d_m^c$ are pieces of information deducted form K_c in a process of $E(K_c)$, then $d_1^c, d_2^c, d_3^c ... d_m^c \in C$ and therefore also true that $d_1^c, d_2^c, d_3^c ... d_m^c \in I$.

If it is also true, that the designer can also express his/her initial knowledge K_i into a set of information V that containing information pieces $d_1^v, d_2^v, d_3^v ... d_m^v$ in a $E(K_i)$ process. Therefore, for a query q_m that is passed by the designer S at time $t_m^j < t_m^q \leq t$, a set of additional context information C_m can be delivered along this query, where $C_m \subset \{V \cup J\}$ and $d_i^c \in C_m : d_i^c$ related to q_m.

5. Conclusion

The results of this study have provided indications and evidences of the existence of the context during the design process. The evidences showed the important role of two types of contexts. The first type of the context, which is increasing along with the growth of the accumulated knowledge, is very important in knowledge and information transaction (subsection 2.1). The other type is very helpful for the communication between the designer and the knowledge/information provider (subsections 2.2 and section 3). Although both of them are important this research will be focus on the first type the context. For this reason, a formalism of design process and context knowledge has been provided, and will be the base for a context management system to support product design.

6. References

[1] Rodgers, P.A. et al: *The management concept of design knowledge in modern product development organizations.* J. Computer Integrated Manufacturing, **14** (1), 2001, p. 108-115.

[2] Akman, V. and Surav, M.: *Steps Toward Formalizing Context,* AI Magazine **17**(3), 1996, p. 55-72

[3] Buvač, S., Buvač, V, and Mason, I.A., *Metamathematics of contexts,* Fundamenta Informaticae, **23** (3), 1995, p. 263-301

[4] Schwotzer, T.: *Context driven spontaneous knowledge exchange.* Proceeding of 1st German Workshop on Experience Management: Sharing Experience about Sharing of Experience, Minor, M. and Staab, S. (Eds), Berlin, March 7-8, 2002, p. 131-138

[5] Brézillon, P.: *Context in Problem solving: A Survey.* The Knowledge Engineering Review, **14** (1), 1999, p. 1-34.

[6] Jambak, M.I. and Vergeest, J.S. M.: *Framework towards contextual information handling during product conceptualization,* Digital Proceeding of Advance Engineering Design, Prague, June 1-4, 2003,

[7] van Breement, E.J.J.: *Empirical results of observing product development processes.* K-memo 329 - The Desys D1 experiment. Delft: Faculty of Industrial Design Engineering, Section Engineering Design, Delft University of Technology, 1996.

[8] Knoop, W.G.: *The empirical research discussion platform.* Report K366. Delft: Faculty of Industrial Design Engineering, Delft University of Technology, 1997

[9] Ullman, D.G.: *Robust Decision-making for engineering design.* J. Eng. Design, 2001, **12** (1) p. 3-13

Information Modelling and Knowledge Bases XVI
Y. Kiyoki et al. (Eds.)
IOS Press, 2005

A context-dependent geocoding method for document databases

Yoshihide HOSOKAWA † and Naohisa TAKAHASHI †
Department of Electrical & Computer Engineering, Nagoya Institute of Technology

Abstract. Location-based services for car or human navigation are seen as one of the main applications of spatial computation technology. Such location-sensing technology as GPS has increased the popularity of those services. However, the information provided by them has been limited because no mechanism has been established for the automatic extraction of geocodes from information resources. A 'geocode' shows the latitude and longitude of a landmark on Earth. 'Geocoding' indicates a process for extracting landmarks (and their geocodes) related to given text information.

In this paper, we present a context-dependent geocoding method for document databases. The main feature of our method is that it translates location names included in given document data into corresponding geocodes by recognizing the context of the document data. We define 'context' as a set of terms in the document data that captures the meaning (the single landmark) of a location name in the document data. Consequently, it is possible to apply geocoding to document data that do not include such detailed geographical information as address, zip code, or phone number.

We clarify the feasibility of our method in several experiments.

1 Introduction

The number and variety of information resources containing spatial representations are increasing into wide areas of computer networks. Spatial computation technology can be used for information retrieval from those resources [9].

Location-based services for car or human navigation are applications of spatial computation technology, and such services are being popularized by the dissemination of such location-sensing technology as GPS. However, these services currently provide only limited information, since no mechanism has been established for the automatic extraction of geocodes from information resources. Here, a 'geocode' shows the latitude and longitude of a landmark located on Earth. 'Geocoding' indicates a process for extracting landmarks (and their geocodes) related to given text information.

The automatic extraction of geocodes from information resources would be very useful, because it would make it possible to apply spatial computation technology to very large information resources.

In this paper, we present a new geocoding method for information resources whose main features are summarized as follows.

(Feature-1) Geocoding with a context-recognition function

Our method performs context recognition to identify a single landmark indicated by a location name contained in an information resource. We define 'context' as a set of terms that fix the meaning (the single landmark) of the location name in the resource.

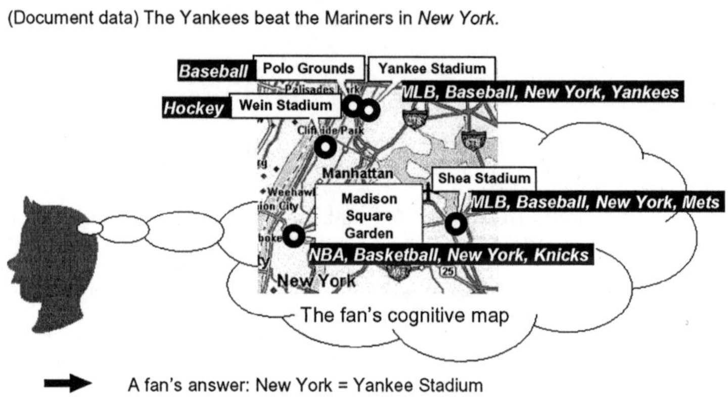

Figure 1: Our hypothesis

We illustrate the need for context recognition by using document data '*The Yankees beat the Mariners in New York*', which describes a Major League Baseball (MLB) game in the United States. The *Yankees* are a MLB team, and their home field 'Yankee Stadium' is located in New York. The *Mariners* are also a MLB team whose home field 'Safeco Field' is located in Seattle.

In the document data, *New York* is the target of geocoding. When we apply a conventional geocoding method to *New York*, the method retrieves many geocodes related to addresses that include the phrase *New York*. This is the crucial problem with a conventional geocoding method: it cannot retrieve the best landmark indicated by *New York*, unless the complete address of the landmark is given. Because the complete address is not included in the document data, it is impossible to translate *New York* into the best landmark 'Yankee Stadium' with a conventional method.

Note that a baseball fan solves this problem by recognizing the best landmark indicated by *New York* from the context of the document data. Then, we have the following hypothesis from the observation:

> *(Our hypothesis) A fan may select the best landmark suited to the context of the document data from landmarks on his cognitive map. He may memorize landmarks with features and retrieve from his cognitive map a landmark whose features are closely related to the context of the document data.*

Here, a user's cognitive map is the geographical memories that he has accumulated. For example, from information about the teams that appear in the document data, a fan can learn that *New York* indicates 'Yankee Stadium' (Figure 1). Our method selects the best landmark (and its geocode) by extracting the context of the document data and comparing it with such background knowledge as information about the teams. Thus, it is possible to translate part of a complete address into the best landmark.

(Feature-2) Geocoding without embedding additional information in landmarks into document data

Our method performs context-dependent geocoding by only using metadata that represent background knowledge about landmarks. The amount of metadata is almost equal to landmarks on a map. In other words, the amount of metadata does not change according to the amount of document data. Our method can extract geocodes from a huge amount of document data when the metadata of all landmarks are given.

On the other hand, another way of dealing with geocodes related to document data is to embed the geocodes into the related document data. See Dublin Core [4] and G-XML [7]. Although it is easy to implement a geocoding method with embedding, it is still necessary to embed geocodes into document data. But the embedding process for a huge amount of document data is very expensive. The creation costs of the metadata of our method are less than the costs of the embedding process because the number of landmarks is generally less than document data.

We conduct several experiments to clarify the feasibility of our method.

2 Related work

[17] developed an algorithm to extract addresses from Japanese document data and perform conventional geocoding for them. The system described in [2] was designed to explore the geocode of a web page by using the web server's IP address and the administrator's phone number. In contrast, our method translates a location name that appears in document data into the best landmark by recognizing the context of the document data.

[8] developed a thesaurus relating geographical names to geocodes that linked spatial terms to geographical names. The spatial terms indicate general ones with spatial meanings such as 'lake', 'bench', and so on. By using the thesaurus, it is possible to extract geocodes relating to the spatial context of document data. In contrast, our method relates non-spatial terms, such as names of teams and leagues to landmarks by recognizing the non-spatial context of document data. Thus, our method computes the best landmark (and its geocode) indicated by a location name contained in the document data.

[15] computed geographic locations related to a web page by analyzing the following information included in a web page: (1) address, (2) zip code, (3) phone number, (4) hyperlinks among web pages, and (5) geographic name. Our method can be applied to document data that do not include complete addresses, zip codes, phone numbers, or hyperlinks among web pages.

In terms of context recognition, the Mathematical Model of Meaning [12, 13] computes the semantic equivalence and similarity among words according to a given context. Our method is designed to translate a location name included in a document data into the best landmark by recognizing the context of the document data.

Geographic information retrieval systems for the WWW were described in [10, 11, 5, 6]. Those systems focused on a geographical context involved in information sources. The systems in [10, 11] were developed to measure the distance between geographical names from their ontology to perform query expansion for information retrieval. In [5, 6], spatial processing was based on the 'Semantic Web' concept. Our method deals with terms of document data as context to capture the meaning of a location name in the document data.

3 A context-dependent geocoding method for document databases

In this section we describe our method of implementing context-dependent geocoding for document databases. The method's main feature is that it translates a location name in given document data into a corresponding landmark by recognizing the context of the document data.

Our method can extract the best set of geocodes from document data that do not include such rich geographical information such as a complete address, zip code, phone number etc. It works according to the following procedure.

Preparation: Creating metadata for landmarks

To compute the correlation between input document data and landmarks, we relate each landmark to several words that represent its features. We define those words as 'landmark metadata' or simply 'metadata'. Metadata are designed as a fan's cognitive map. This procedure is performed by hand.

In the table of Figure 2, each of the five stadiums is referred to by three phrases that represent league name, franchise area, and team name. Attributes 'Geocode', 'City-and-State', and 'Landmark-Name' of the table are provided by the map's creator.

Step-1: Selecting landmarks suited to the context of input document data

This is performed as follows.

Step-1.1: Extraction of a context from input document data

A context is a set of terms in the input document data and is used to recognize a single landmark indicated by a location name. This is an essential procedure for context-dependent geocoding method because such context differs from the main topic of the input document data.

For the step, we implement the following context extraction mechanisms.

Mechanism-C1: Dealing with a set of terms in the first sentence and several terms that proceed and follow a location name as context

Context is frequently described in the first sentence of many newspapers. It also proceeds or follows location names in some Japanese newspapers. This is also one of the context extraction mechanisms for specific contents.

Mechanism-C2: Dealing with all the terms in the input document data as context

This is the simplest context extraction mechanism.

We compare the two mechanisms in Section 4.3.

Step-1.2: Computation of similarities between the extracted context and landmarks

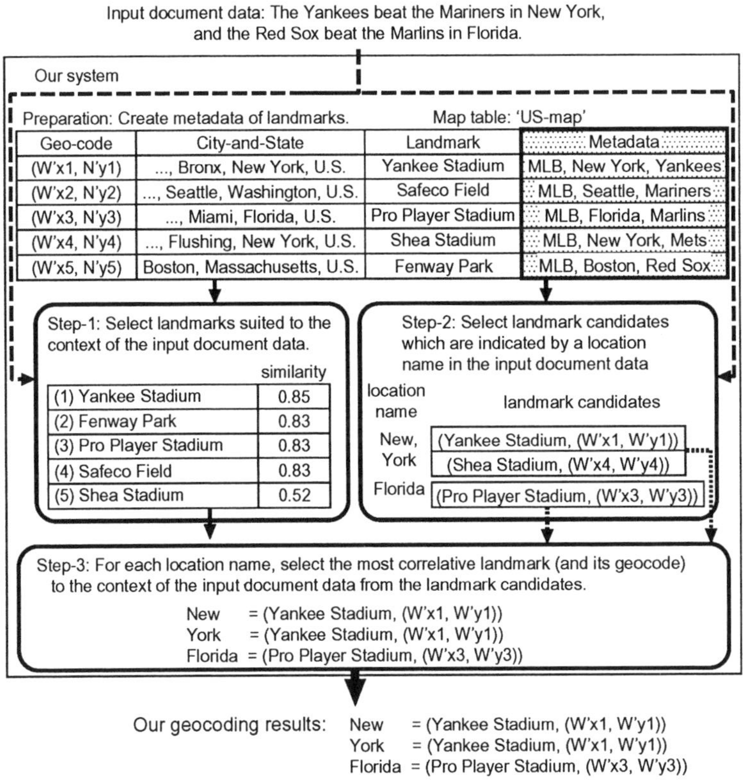

Figure 2: Geocoding procedures of our method

To select landmarks according to the extracted context, we use the similarity between such data and landmarks as their correlation. Our method recognizes the landmark with the highest similarity as the one best suited to the context. Similarity is computed by using a conventional document retrieval scheme.

For the step, we use a well-known vector space model based on the TF/IDF weighting scheme [18, 19], which is a way of weighting the relevance of a term to document data. The vector space model represents document data by an n dimensional vector composed of TF/IDF weightings of n terms and then computes the similarity between two sets of document data by calculating the inner product between the n dimensional vectors. In the vector space model, the inner product with the highest value indicates the most similar document data.

Our method selects landmarks suited to the context of the input document data according to the following procedure:

Step-1.2 (1): Creating vectors of landmarks by calculating TF/IDF weight-

ings from the metadata
We call these vectors 'landmark vectors'.

Step-1.2 (2): Creating a vector of the input document data by using the TF/IDF weighting scheme
We call this vector the 'document vector'.

Step-1.3 (3): Computing the inner products between the document vector and each landmark vector

In Figure 2, our method recognizes the home fields of four baseball teams ('the Yankees', 'the Mariners', 'the Red Sox', and 'the Marlins') as landmarks suited to the context of the input document data because the team names appear in the input document data.

Step-2: Selecting candidates of landmarks indicated by location names in the input document data

Step-2.1: Extracting nouns and undefined terms in the input document data

Our method extracts all nouns from the input document data because location names are generally indicated as nouns. This process is performed by using a part-of-speech tagger. Some part-of-speech taggers were designed to separate a document into terms and then return pairs of the term and its part of speech. For example, "Brill's Tagger" were developed for English documents [1], and "ChaSen" were developed for Japanese documents [14].

The left column of Figure 3 shows term and part-of-speech pairs captured by Brill's tagger. Here, 'NNP' and 'NNPS' indicate noun phrases. Eight nouns were extracted from the input document data of Figure 2.

Our method also extracts undefined terms from the input document data because they may indicate landmarks. Here, the undefined terms are ones not defined in the dictionary of the part-of-speech tagger. By dealing with undefined terms, our method can extract location names in which some unknown terms are included.

We call nouns and undefined terms extracted by this process 'candidate terms'.

Step-2.2: Selecting location names from the results of Step-2.1

Our method recognizes candidate terms as location names, if they appear in the address names of landmarks. Location names are selected by using the following SQL statement for each candidate term:

```
SELECT  Landmark, Geocode
FROM  [map table]
WHERE  City-and-State LIKE '%[candidate term]%',
```

where '[map table]' shows the map table name and '[candidate term]' shows one of the results of Step-2.1. If there are landmarks retrieved for a candidate term, our method recognizes the candidate term as a location name. For example, in Figure 2, landmarks indicated by the term 'New' are retrieved by the following SQL statement:

```
SELECT  Landmark, Geocode
FROM  US-map
WHERE  City-and-State LIKE '%New%'.
```

Linguistic information captured by Brill's tagger (Step-2.1)		Selection of nouns and unknown terms (Step-2.1)	Extraction of location names (landmarks) (Step-2.2)
The	DT		
Yankees	NNP	selected	
beat	VB		
the	DT		
Mariners	NNPS	selected	
in	IN		
New	NNP	selected	New (Yankee Stadium, Shea Stadium)
York	NNP	selected	York (Yankee Stadium, Shea Stadium)
,	,		
and	CC		
the	DT		
Red	NNP	selected	
Sox	NNP	selected	
beat	VB		
the	DT		
Marlins	NNP	selected	
in	IN		
Florida	NNP	selected	Florida (Pro Player Stadium)

Figure 3: Extraction procedure of location names from document data *'The Yankees beat the Mariners in New York, and the Red Sox beat the Marlins in Florida.'* in Fig. 2

'Yankee Stadium' and 'Shea Stadium' are obtained as a result. As shown in Figure 3, our method recognizes the candidate terms 'York' and 'Florida' as location names.

Step-3: Selecting the best landmark (and its geocode) indicated by each location name

When several landmark candidates are assigned to a location name in Step-2, our method extracts the landmark (and its geocode) that is the most correlative to the context of the input document data from the results of Step-2. This process is implemented to select the most correlative landmark from the results of Step-2 by referring to the results of Step-1.

Figure 2 shows that our method recognizes that the input document data is related to 'Yankee Stadium' and 'Pro Player Stadium'.

Our method cannot extract correct location names composed of such compound words as 'New' and 'York'. However, there are many syntactic analysis studies of natural language description [3]. The literature [17] also shows a method for extracting noun phrases with geographic information from document data. Using the study results, we may solve the problem. Moreover, since many location-based applications ignore the problem, we will not consider it here. For these applications, it is important to extract only the geocodes from document data to locate the document data on a map (Figure 4).

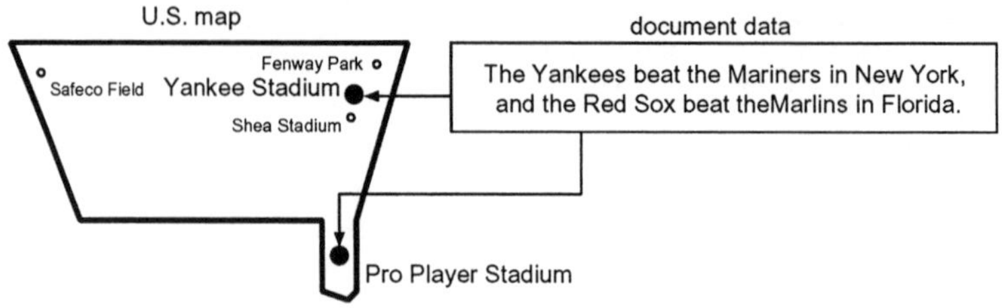

Figure 4: An application of our method

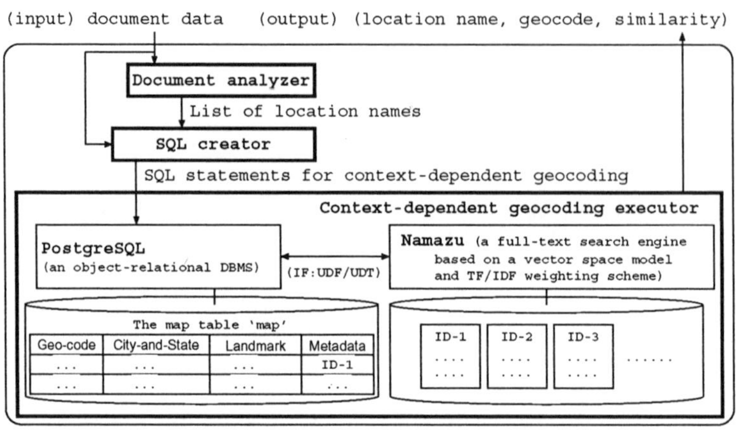

Figure 5: Our experimental system

4 Experimental evaluation

We have performed the following two experiments to clarify the feasibility of our geocoding method.

Experiment-1: Feasibility of our hypothesis and implementation method

Experiment-2: Feasibility of our context extraction mechanism

4.1 Experimental environments

4.1.1 Experimental system

We will describe the implementation of our experimental system that evaluated our method with Japanese document data.

Our experimental system was comprised of the following subsystems (Figure 5).

Document analyzer: This subsystem performs Step-2.1 described in Section 3 and returns a list of candidate terms. In the experiment, we used the 'ChaSen' part-of-speech tagger for Japanese document data.

SQL creator: This subsystem creates the following SQL statement for performing context-dependent geocoding,

```
SELECT  Landmark, Geocode, n0d_include(Metadata, [input document data])
FROM  map
WHERE  City-and-State LIKE '%[candidate term]%',
ORDER BY 2 DESC
```

where function 'n0d_include' computes the similarity between a landmark and input document data. The computation is performed by using 'Namazu' [16], which is a full-text search engine based on a vector space model and the TF/IDF weighting scheme.

The first column of the SQL statement results shows the geocode of the best landmark indicated by a location name.

We implemented function 'n0d_include' as a user defined function of 'PostgreSQL'.

Context-dependent geocoding executor: This subsystem executes an SQL statement created by the SQL creator and returns the execution results.

4.1.2 Experimental Data

To perform practical evaluations of our geocoding method, we used 1,000 Japanese articles from the Mainichi newspaper as experimental data. The Mainichi newspaper is a well-known Japanese newspaper. These articles described sporting events in 2002 and proved suitable for evaluation of our context-dependent geocoding method because different sporting events were often held in the same city.

We created metadata about stadiums, corporations, universities, and high schools. The amount of the metadata was 5,000.

4.2 Experimental-1

4.2.1 Experimental method

The goal of this experiment is to clarify the feasibility of our hypothesis and implementation method. We evaluated the precision of the geocoding results by using the following indicator:

$$Precision(x) = \frac{B(x)}{A},$$

where A is the number of terms recognized as location names by our method. $B(x)$ is the number of terms whose correct landmarks are within the top x of the ranking results of our method. In this experiment, A was 3952, so there were about 4 location names per article.

We measured the precision of our method in cases where x is 1, 3, and 10 respectively because our goal in the experiment is to demonstrate that our method can link location names to the correct landmarks.

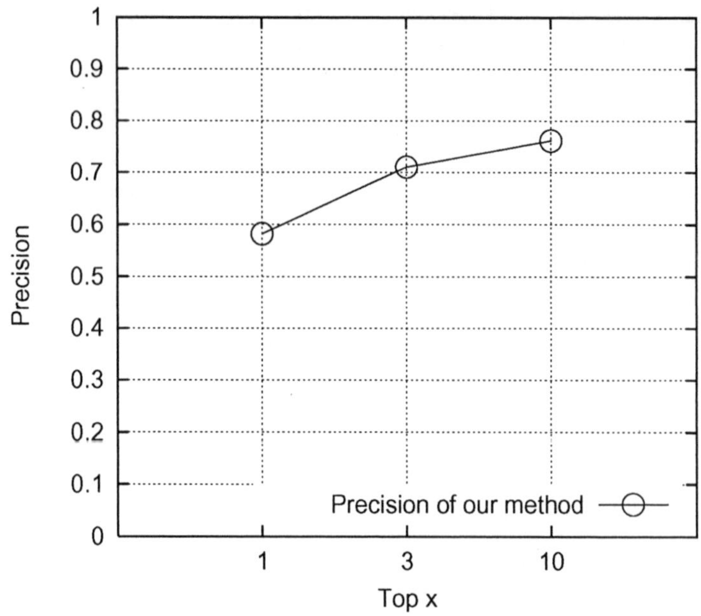

Figure 6: Precision of our method

4.2.2 Experimental results and considerations

Figure 6 summarizes the experimental results, which show that the precision of our method was over 70 % when a location name was translated to the top 3 landmarks. That is, our method gives high precision to the articles.

The small failures of our method are explained as follows.

Problem-1: Our method performed missing extraction of location names from input document data.

Figure 7 shows the average of the missing rates for articles with the same topics. The results shows that some terms where were not included in the geographical information, were recognized as location names. For example, in the articles on 'Sumo', which is traditional Japanese wrestling, our method extracted many terms indicating wrestlers' names because wrestlers often take names for places or their hometowns. The total average was over 10 %. By solving the problem, precision will rise to 80 %.

Solution: Today, there are many syntactic analysis studies of natural language description. Using such results, we hope to improve the location name extraction mechanism of our method (Step-2).

Problem-2: Our method could not exactly recognize contexts that appeared in a single document data.

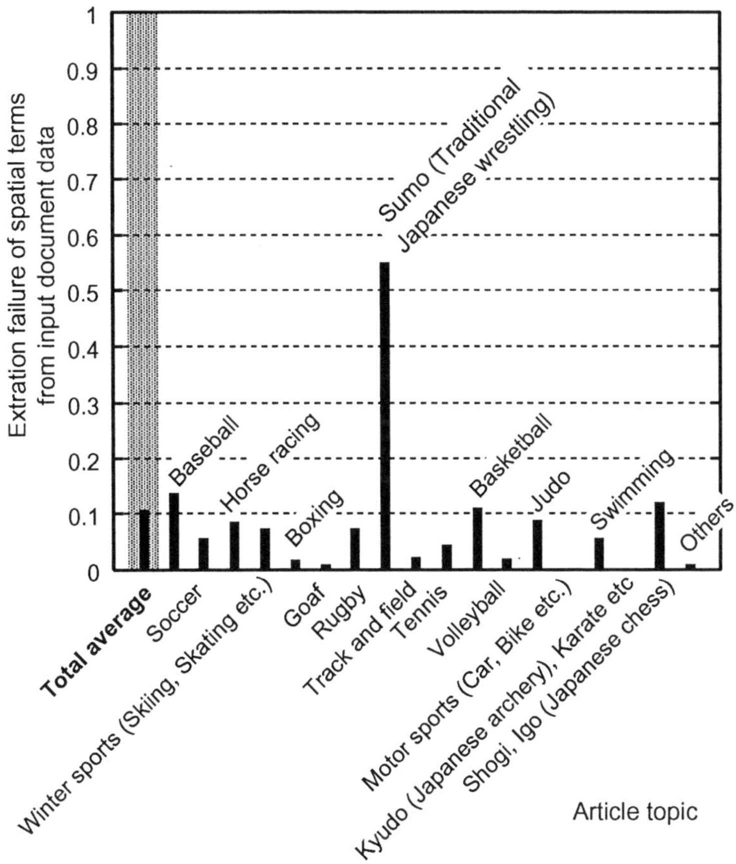

Figure 7: Extraction failure of location names from an input document data

Consider the example: '*Ichiro, who plays for the Seattle Mariners, attended an NBA game in Seattle*'. For such document data, our method could not distinguish between the two meanings of *Seattle* because our method was designed to capture the entire context of the document data.

Solution: The above considerations indicate the need to implement a mechanism that captures all contexts included in document data and recognizes their scopes. For the example document, such mechanism recognizes the difference between the two meanings of *Seattle* by capturing the two contexts of the MLB and the NBA and understanding their scopes.

Problem-3: Some document data did not include contexts for specifying the best landmark.

Consider the example document: '*The Yankees beat the Mets in New York*'. Both the MLB *Yankees* and *Mets* have their home stadiums in New York, but they are located in different places.

It is impossible to recognize the meaning of *New York* in the document data, because the context information of the document data is insufficient to determine a stadium in which the baseball game was held.

In such a situation, baseball fans can solve this problem. They can understand the meaning of *New York* when accompanied with such additional information as a baseball schedule.

Solution: The above considerations show that our method can solve this problem if other contexts for specifying the best landmark are given.

Problem-4: Home stadiums moved.

In the experiment, we created metadata that described features of recent home stadiums. However when landmarks move, their metadata, addresses and/or geocodes generally change. Consequently, our method did not recognize that past home stadiums were different from current home stadiums.

Solution: From the above considerations, our method needs a mechanism that captures temporal contexts of document data and manages landmarks' histories.

Problem-5: A few location names indicated places where the addresses did not include location names.

Consider the example: '*The New York Giants beat the Chicago Bears in New York*'. The home stadium of the *New York Giants*, which is an NFL team, is located in East Rutherford, New Jersey, not in New York. As a result, our method could not retrieve the home stadium of the New York Giants, because it selected landmark candidates from addresses. Since this problem affected such articles, the precision of our method for the articles was low.

Solution: The above considerations show the necessity of improving the mechanism for selecting landmark candidates indicated by landmark names.

The experimental results have demonstrated the feasibility of our context-dependent geocoding method. The discussion presented here have also clarified the scope of our method.

4.3 Experimental-2

4.3.1 Experimental method

The goal of this experiment is to clarify the feasibility of our context extraction mechanism. We compared the accuracy of the results of our geocoding method using Mechanism-C1 and Mechanism-C2 statistically. We implemented Mechanism-C1 for extracting the first sentence and the six terms that proceeded and followed a location name as context.

We evaluated the accuracy of the results of the two methods by using the same indicators used in Experiment-1.

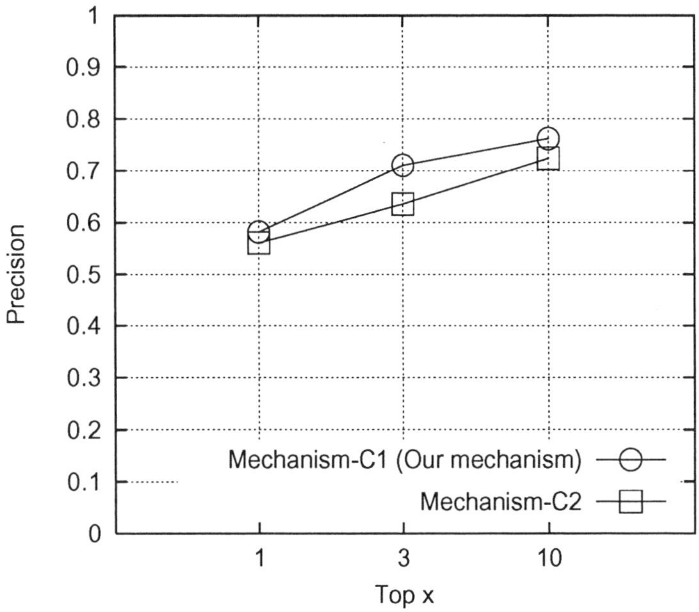

Figure 8: Comparison of precisions of proposed and comparative mechanisms (Mechanism-C1 is our mechanism for dealing with a set of terms in the first sentence and the six terms that proceeded and followed a location name as context. Mechanism-C2 deals with all terms of input document data as context.)

4.3.2 Experimental results and considerations

Figure 8 summarizes the experimental results, which show that our context extraction mechanism (Mechanism-C1) improved our geocoding method.

The main reason for the failures of our mechanism is that our method could not capture contexts from input document data and exactly recognize their scopes in it. However, there are many syntactic and semantic analysis studies of natural language description. Using such results, we hope to improve our context extraction mechanism. For such semi-structured document data as Web pages, our mechanism may be refined to recognize the scope of contexts from the structure of input document data.

We have demonstrated the feasibility of our context-dependent geocoding method.

5 Conclusions

Location-based services for car and human navigation are seen as the main applications of spatial computation technology. However, the information provided by these services has been limited, because no mechanism has been established for the automatic extraction of geocodes from the various information resources.

In this paper, we have presented a context-dependent geocoding method for document databases. Its main features are summarized as follows.

(Feature-1) Geocoding with a context-recognition function

Our method performs context recognition to identify the best landmark indicated by each location name given in information resources. Our method can be used to translate part of a complete address into the best landmark.

(Feature-2) Geocoding without embedding additional information in landmarks into document data

Our method performs context-dependent geocoding by only using the metadata representing background knowledge about the landmarks.

Our experimental evaluation has clarified that the precision of our method's results is high. Our method has also performed geocoding without embedding additional information about landmarks in the document data. That is, it is feasible to apply our method to a huge amount of document data.

In future work, we will solve the problems described in Section 4.2.2. We will implement an automatic creation mechanism for landmark metadata and evaluate the applicability of our method for document data in other fields besides sports and apply it to world wide location-based services.

Acknowledgements

This work is partially supported by The Information-Technology Promotion Agency, Japan (IPA), The Artificial Intelligence Research Promotion Foundation, Japan, and The Hori Information Science Promotion Foundation, Japan.

References

[1] Brill, E.: Some Advances In Rule-Based Part of Speech Tagging, *Proc. 12th National Conference on Artificial Intelligence (AAAI 94)*, pp.722–727 (1994)

[2] Buyukokkten, O., Cho, J., and Garcia-Molina, H.: Exploiting geographical location information of web pages, *Proc. Workshop on Web Databases (WebDB'99)*, pp.91–96 (1999)

[3] Chowdhury, G., G.: Introduction to Modern Information Retrieval, Facet publishing (2004)

[4] Dublin Core Metadata Initiative: http://dubli ncore.org/

[5] Egenhofer, M. J.: Toward the semantic geospatial web, *Proc. 10th ACM International Symposium on Advances in Geographic Information Systems (ACM-GIS 2002)*, pp.1–4 (2002)

[6] Farrugia, J., and Egenhofer, M. J.: Presentations and Bearers of Semantics on the Web, *Proc. 15th International Florida Artificial Intelligence Research Society Conference*, pp.408–412 (2002)

[7] G-XML : http://gisclh.dpc.pr.jp/gxml/contents

[8] Hill, L. L., Frew, J., and Zheng, Q.: Geographic Names. The implementation of a gazetteer in a georeferenced digital library, *Digital Library*, No. 5, Vol. 1, http://www.dlib.org/dlib/january99/hill/01hill.html (1999)

[9] Hosokawa, Y. and Kiyoki, Y.: A Computational Method of Spatial and Temporal Equivalence with Context Recognition Functions for Document Database, *to appear in Information Modelling and Knowledge Bases (IOS Press)*, Vol. 15, (2004)

[10] Jone, C. B., Alani, H., and Tudhope, D.: Geographical information retrieval with ontologies of place, *Spatial Information Theory, LNCS 2005*, pp.322–335 (2001)

[11] Jone, C. B., Purves, R., Ruas, A., Sanderson, M., Sester, M., van Kreveld, M. J., and Weibel, R.: Spatial information retrieval and geographical ontologies an overview of the SPIRIT project, *Proc. 25th ACM SIGIR 2002 Conf.*, pp.387–388 (2002)

[12] Kitagawa, T. and Kiyoki, Y.: The mathematical model of meaning and its application to multidatabase systems, *Proc. the 3rd IEEE Int. Workshop on Research Issues on Data Engineering: Interoperability in Multidatabase Systems*, pp.130–135 (1993)

[13] Kiyoki, Y., Kitagawa, T. and Hitomi, Y.: A fundamental framework for realizing semantic interoperability in a multidatabase environment, *Journal of Integrated Computer-Aided Engineering*, Vol.2, No.1, pp.3–20, John Wiley & Sons (1995)

[14] Matsumoto, Y., Kitauchi, A., Yamashita, T., Hirano, Y., Matsuda, H., Takaoka, K., and Asahara, M: Morphological Analysis System ChaSen version 2.2.1 Manual, http://chasen.aist-nara.ac.jp/index.html.en (2000)

[15] McCurley, K. S.: Geospatial Mapping and Navigation of the Web, WWW10 (2001)

[16] Namazu: a Full-Text Search Engine, http://www.namazu.org/index.html.en

[17] Sagara, T., Arikawa, M., and Sakauchi, M.: Spatial Document Management System Using Spatial Data Fusion, *IIWAS2001*, pp.399–409, (2001)

[18] Salton, G. and Buckley, C.: Term-weighting approaches in automatic text retrieval, *Information Processing and Management*, No. 24, Vol. 5, pp.513–523 (1988)

[19] Salton, G. and Buckley, C.: Improving retrieval performance by relevance feedback, *Journal of the American Society for Information Science*, No.41, pp.288–297 (1990)

Information Modelling and Knowledge Bases XVI
Y. Kiyoki et al. (Eds.)
IOS Press, 2005

A Description Model for Linking and Reconfiguration in Video Streams

Takahiro WADA[1] and Yuzuru TANAKA[1]
[1]*Meme Media Laboratory, Hokkaido University*
N13 W8, kita-ku, Sapporo, 060-8628, Japan
Email: {wada,tanaka}@meme.hokudai.ac.jp

Abstract. This paper proposes a new framework for video applications with dynamical linking, reconfiguration and reusing functions. We call this framework Hypermovie Framework. Separating a structure of a video stream from link references, our description model allows users to reuse their descriptions individually, and to assign multiple link references to a single anchor node. Using rule-based descriptions of reactive behaviors for link references, we can define multiple link actions for a single anchor node so that users may interact with various video objects either to seamlessly jump from one video clip to another or to invoke some application programs. This framework can dynamically reconfigure video applications as the result of evaluating each rule. Then, we represent a description method in our framework using XML and XQuery. It allows other document models like (X)HTML/XML to utilize each other and to reuse arbitrary parts of defined context in the video stream, even link conditions and link actions.Furthermore, using XQuery for describing rule-based behaviors, it enables the user to create or reconfigure video streams dynamically.

1 Introduction

Hypermedia documents including video media have grown rapidly on the Web, using evolving streaming technology for continuous media. Video media provide various visual aspects along with its timeline. Thus, video media have narrative structures and relationships among contents in it. Besides, VR(Virtual Reality) media, which render virtual reality space, have some structures specified by aspects based on 'angle' and 'angle + time' parameters. VR media denote either VR views (i.e. panorama view, object view) or VR movie (panorama view + timeline). With the use of such spatio-temporal aspects, VR media enable video streams to provide various contexts for the user, corresponding to when/where the user looks at. For example, e-learning systems, and digital video libraries are applications of this type.

Using these aspects in video streams, video integration models have been proposed. There are mainly two approaches: one is a metadata approach for describing inside contents as proposed by MPEG-7 [10], and the other is an approach for describing time synchronization and location coordination among various media data as proposed by SMIL [22]. Using these models, authors can provide video streams with contents information and contexts.

However, these models are not enough to provide multiple link references for one anchor corresponding to different user interactions and intentions. Because link references are treated as an attribute for a source media node, and an established link reference indicates a static destination node. In order to implement multiple link references to react different user interactions and intentions, we need a hyperlink model to dynamically react different user interactions. Furthermore, there is few reusability due to the definition of link references as attributes. To archive high reusability, we need to define link references independently from the structures of video streams.

We will consider the following two examples.

(i) The first example is a concentric panoramic view application. All the concentric panoramic views are taken from the same camera position. All the panoramic views are organized concentrically in the order from the lowest resolution view to the highest one. When the user presses and holds the mouse button on the movie, the movie changes the view to the next outer panoramic view at the same angle. If the user moves the mouse onto a particular anchor, the outer panoramic view pops up near the anchor. If he clicks the anchor, the linked Web page will pop up.

(ii) The next example is a news archive application. This application consists of movie archives for news programs and annotation data for news contents in each news program. These annotation data are stored in the external data storage and represent the details of each news content such as a news summary, a newsreel footage, and a time duration of each news content. When the user clicks a particular item in each news program, the system searches all the annotation data for all the news items with the topics related to the current news item. Then, the system creates a new news archive by concatenating all these related news items and presents it to the user.

From this point of view, we will propose a new description model for the video stream contextualization using dynamical linking and reconfiguration. We call this model Hypermovie Framework. First, we will introduce an idea of parameterized spatio-temporal media that organize video streams with respect to time and angle. Then, we will discuss rule-based reactive behaviors based on the Active-rule model [17]. Our new model enables the user to associate multiple link actions with a single anchor node. Each evaluation of such a rule associated with such an anchor dynamically creates or reconfigure video streams. Finally, we show a description method: the former part is described using XML to describe a structure of a video stream, and the latter part is expressed in XQuery [23] to describe link actions. It allows all kinds of (X)HTML/ XML documents to mutually utilize each other and to reuse arbitrary parts of defined contexts, link conditions and link actions.

This paper is organized as follows: in Section 2, we will explain an overview of Hypermovie framework and propose our new hyperlink model in this framework.In Section 3, we will discus rule-based link navigations in our hyperlink model. In Section 4, we will present a description method for our model using XML and XQuery. In Section 5, we will show some example descriptions. In Section 6, we will give some comments on related works. The last section concludes this paper.

2 Hypermovie

2.1 An Overview of Hypermovie Framework

Hypermovie framework provides a model for describing contextualization of video streams for dynamical linking and reconfiguration in response to various contexts. Separating a structure of a video stream from link references, our description model allows users to reuse their descriptions individually, and to assign multiple link references to a single anchor node. In order to achieve this goal, we will introduce rule-based reactive behaviors based on the Active-

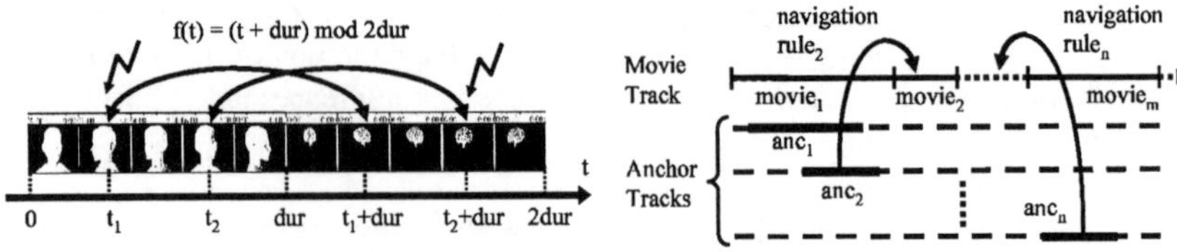

(a) Roy Stringer Framework (b) Hypermovie Time Model

Figure 1: Roy Stringer framework and Hypermovie time model

rule model [17] and a description method using XML and XQuery expressions. The details of a description method is described in Section 4.

Using the idea of rule-based reactive behaviors to assign link references, the system provides for the user to jump to the destination node dynamically in response to various types of events by evaluating each rule. Our hyperlink model has the following features: i) we can access time and angle information in each video stream; ii) we can associate various events with a single anchor; iii) we can set actions in response to various events; iv) we can set a condition to each link navigation; v) the system can create or reconfigure video streams through evaluating such rules.

2.2 *Roy Stringer Framework*

Roy Stringer [1] suggested a generic framework for video-centered applications. This is presented in Fig.1(a). In this figure, there are two movies; one shows a person who is rotating at a constant angular speed, and the other shows her internal structure rotating at the same angular speed. By clicking the former at some angle of the person, the user can jump to the latter movie showing the same angle of the internal view. In this framework, the application movie is constructed by putting one or more movies on a single timeline. Handling each frame in the movie as a node in hyperlink structure, it performs the link processing as the evaluation of a time shift function between frames. For the example in Fig.1(a), the time shift function is given as follows.

$$f(t) = (t + dur)\, mod\, (2 * dur), \quad (0 < t < 2dur, dur : duration)$$

2.3 *Hyperlink Model*

Fig.1(b) shows the temporal structure in our hyperlink model. This structure is an extension of the framework proposed by Roy Stringer. There are three parts: the first part is a *movie track* defining a single timeline. It is a cascade of one or more movies. The second part is a set of *anchor tracks*. These define anchors in each frame. The last part is a set of *navigation rules* that define link references from anchor tracks to movie track. We define each navigation rule as a reactive behavior for various events on each anchor. Each navigation rule, when invoked, returns a specific time position, i.e. a movie frame number, of the target movie track as the jump destination frame.

[1] Roy was an inspirational pioneer of the digital media industry, whom we had the good fortune to meet before his untimely death in 2001.

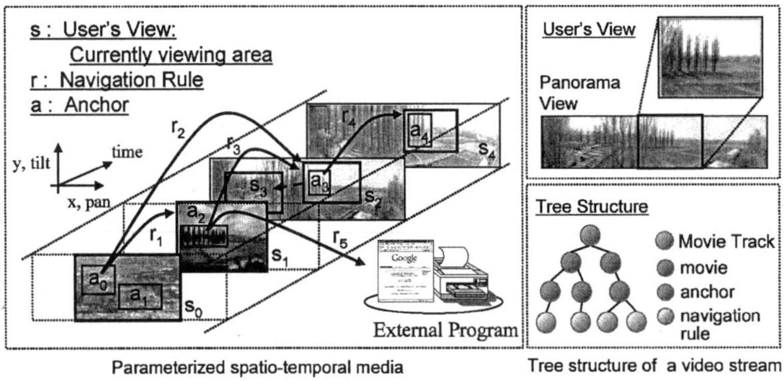

Figure 2: Parameterized spatio-temporal media and its tree structure

2.3.1 Parameterized Spatio-temporal Media

Now, we will propose a method for uniformly handling video streams. If a video stream has a VR view, we need to use its angle information (i.e. pan, tilt).We will define a notion of *parameterized spatio-temporal media* description that has a single timeline and some other parameters such as these angles indicating a specific view state at a certain time on the timeline. Fig.2 shows a structure of *parameterized spatio-temporal media* that consists of two movies and two panoramic views . Besides, a parameterized spatio-temporal media has a tree structure as the right part of this figure. The upper two levels in this tree structure may be used as destination nodes of hyperlinks. On the other hand, the lower two levels represent source nodes of hyperlinks (anchors) and link references (rules).

It is important for the modeling of interactive video-centered applications that their structures do not change. They only change their parameters even if the composed media may change. The *parameterized spatio-temporal media* object satisfies this property. This media structure enables the user to look at any location view at the specified time in the video stream. User's View ($s_0,\ldots,s_6$) in Fig.2 denote such views parameters.

2.3.2 Anchors

Anchors are sub tracks of a parameterized spatio-temporal media object. Each anchor has duration and region. In Fig.2, '$a_0,\ldots,a_4$' denote anchors. Anchors are classified into three types, corresponding to different spatio-temporal characteristics. They are fixed area anchors, moving area anchors, and timeline information anchors. Each fixed area anchor has a certain constant region during its appearance. For example, a rectangle region of each frame is relevant of this type. A moving area anchor has dynamically changing region during its appearance. The same person in consecutive frames may be associated with an anchor of this type. A timeline anchor has no area and is defined as a set of time intervals.

There are mainly two methods to specify anchor regions in a timeline movie. One is the automated anchor generation method [21], and the other uses the linear interpolation between two manually specified regions in two keyframes. On the other hand, in the case of VR contents (i.e. [6, 11]), the location is calculated by using current viewpoint, field-of-view and angle directions. We use QuickTimeVR [6] to view VR contents, and to specify anchors.

2.3.3 Link Navigation

We define link navigation to move to another view state from the current view state in parameterized spatio-temporal media as the repetitions of selecting and evaluating the designated rule for each event on each anchor. We call such a rule a *navigation rule*. In Fig.2, '$r_0, \ldots, r_5$' denote navigation rules.

Our link navigation method is based on the Active-rule model. Each navigation rule consists of three elements. It is expressed as an active-rule form ('event', 'condition', 'action'). The navigation rule is invoked, when the specified event is detected on the specified anchor. If checking process for the specified *condition* returns true, it executes the specified action. In the case of Fig.1, the navigation rule for anchors is defined as $mod2n$: $('click', 'true', '(t+movie.duration)\, mod\, (2*movie.duration)')$. Here, *mod2n* is a macro expression name of this rule. The details of our navigation model are described in the next section. In our model, we consider a special action pattern as an evaluation result of each navigation rule.It is an 'invoke message' action that invokes an external program (i.e. a web browser), or a notification to another anchor.

2.4 General Definition

The following three expressions denote generic definition of Hypermovie link navigations that consists of two parts; one is the user's view state (S_i at the first expression), and the other is the link navigation process. In these expressions, p indicates each status of the video stream (i.e. time, angle). R is a set of rules for the target anchor, and r is a rule. M in the third expression representing the invocation action.

$i)$	View State	$S_i : S(p_1, p_2, \ldots, p_n)$
$ii)$	Rule Selection	$event\ e \rightarrow targetAnchor.R(e)$
	check condition	$condition\ c_j \rightarrow r_k(e, c_j)$
$iii)$	Link Action	$r_k(e, c_j) \rightarrow S_{i+1}, M$

By evaluating navigation rules, the system guides the user to jump to another view state S_{i+1} from the current view state S_i. The link navigation processes consist of two parts: 'Rule Selections' (the second expression) and 'Link Action' (the third expression). In the 'Rule Selection' part, when an event e occurs on the anchor, the system collects a set of navigation rules R in response to the target event e. Then, the system selects a rule r_k in the collected navigation rules R if the evaluation result of the condition c_j returns 'true' in the 'check condition' phase. In the 'Link Action' part, the system evaluates the selected rule and executes the evaluation-result action.

Link navigation process of Hypermovie framework is represented as follows. When an event e_1 occurs on an anchor at S_0, the rule r_0 is selected in response to e_1 and executed, if the condition c_0 is true. As a result, the user is able to move to the next view state S_1.

$$\overset{\downarrow e_1}{S_0} \xrightarrow[r_0(e_1,c_0)\leftarrow R_0(c_0)]{R_0=select(e_1)} \overset{\downarrow e_2}{S_1} \xrightarrow[r_1(e_2,c_1)\leftarrow R_1(c_1)]{R_1=select(e_2)} S_2 \cdots$$

3 Rule-based Link Navigation

Link navigation is executed by repetitively selecting and evaluating navigation rules. This navigation mechanism is based on the active-rule model. It has the following features: i) we

Table 1: Events

	Type	Target	Event	Additional parameter
primitive	user	mouse	down,up,click,enter,leave,dblClick	key, time, position
			drag, drop	key, time, position, object
		keyboard	down, up	focus
	time	node	begin, end	node data
		real-time	wall-clock	clock data
	message	node	notification message	node data
		system	interrupt message	node data
composite	user	mouse	motion, etc	key, time, position
	time	interval timer	click+3s, nodeBegin+5s, etc	primitive event information
	other		other composite, user definition	each event information

may set various events to the anchor; ii) we may set arbitrary conditions to each link action; iii) the system creates or reconfigures a video stream dynamically by evaluating the rule.

3.1 Navigation Rule definition

In order to reveal each navigation process, we need to clearly articulate each factor of the navigation rule. The *event* means any events that occur on anchors. The *condition* specifies the condition in which link action process is executed for the event. The *action* denotes a hyperlink operation.

3.1.1 Events

Events are classified into two types: atomic events and composite events. A composite event is defined as a boolean conjunction of atomic events.

Table.1 shows an event taxonomy in our model. The attribute *node* means a movie track, an anchor track, or a movie in the movie track. The user event represents either a mouse or keyboard event, and the status of the mouse or the keyboard is used as the parameters of the event such as 'mouse position' and 'occurrence time'. There are three types of time events: node time events, real time events and interval timer events. A node time event denotes an event caused by video stream timeline, such as 'begin' and 'end'. The message event denotes an interruption event caused by system notice message or by an external program in another rule result. The node information is used as additional information for the event. A composite event consists of more than two primitive events. For example, 'mousedown + 5s' is a composite event that means 5 seconds passed after pushing a mouse button.

3.1.2 Conditions

Once an event is detected, the rule selection process selects one rule whose *condition* returns true. We determine two types of *conditions*. One type has incidental parameters of the occurred event, and the other type has additional data to the anchor. Incidental parameters are shown in Table.1. Additional data represent the content properties (name, color, and so on) for each anchor (person, building, and so on). These data are used as metadata for the anchor.

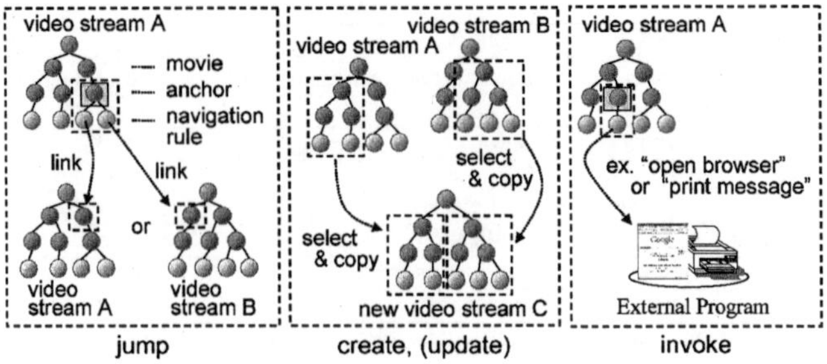

Figure 3: Action patterns

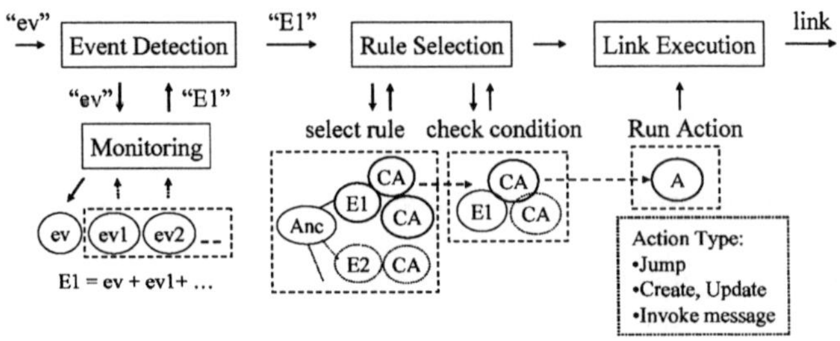

Figure 4: The process of the rule-based link navigation

3.1.3 Actions

The system evaluates the selected action as a link behavior if the condition phase returns true. Then the system executes the evaluation result action. We define four action patterns. They are '*jump*', '*update*', '*create*', and '*invoke*', as shown in Fig.3.

The user moves to the next view state in the same or other parameterized spatio-temporal media object obtained as the results of '*jump*'actions. Moreover, we prepare special kinds of jump actions, which are 'create' and 'update' actions. The '*create*' action enables users to create a new video stream as an evaluation result of the rule through navigation. The user is able to move to the newly created video stream as a '*jump*' action. The '*update*' action reconfigures a video stream as an evaluation result. Operations for parameterized spatio-temporal media objects, such as changes of a playing duration or movie contents, addition or deletion of an anchor information, are examples of '*update*' actions.

The '*invoke*' action consists of two types of actions: one invokes external programs such as a 'Web Browser'; the other sends a notification message to the video stream. In the second case, the system detects the notification message as an event to activate a particular anchor. In the first case, if the external program returns a value in response to an invocation message, the system detects it as an interrupt message event to activate the anchor.

3.2 Navigation Process

Our rule-based link navigation has three processes in each navigation. It is shown in Fig.4. The first process is the '*Event Detection*' process that detects each event which is associated

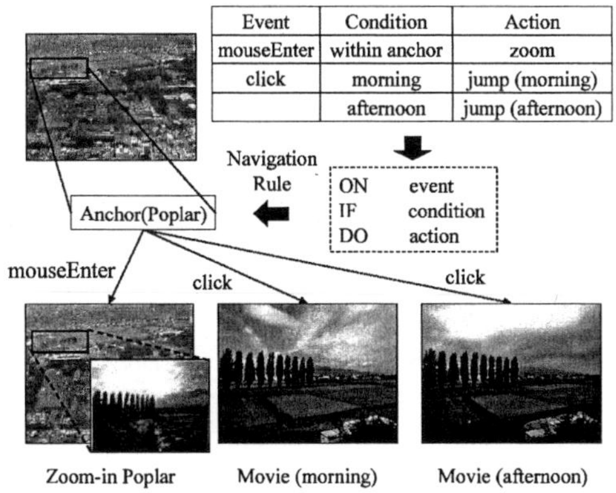

Figure 5: An example of the rule-based link navigation

with a certain anchor. The second process is the '*Rule Selection*' process. It selects a set of rules corresponding to the detected event on the target anchor, and checks the condition of each rule. As a result, it returns the first rule whose condition is true. The last process is the '*Link Action*' process. It evaluates the selected rule, and executes the action.

Fig.5 shows an example of a rule-based link navigation. The anchor, which is named 'poplar', reacts to two kinds of events ('click' and 'mouse enter'). When the user clicks this anchor, it navigates the user to the next movie defined as the next node, through checking the condition. In this case, the condition is whether it is now in the morning or in the afternoon. In addition, while the mouse cursor is over the anchor 'poplar', a preview of destination node appears.

4 Hypermovie Description

In this section, we propose a description method for our hyperlink model using XML for the structure description and XQuery [23] expression for describing the navigation rule. Separating the structure of video applications from the link references, our method allows us to reuse arbitrary portions of already defined contexts, link conditions and link actions for different applications or in different contexts. Consequently, by evaluating each XQuery expression representing a navigation rule, the system determines the next view state and create/update a video stream dynamically. Furthermore, any XML document can be regarded as a target for the condition of the navigation rule, because XQuery is a query language for XML documents. Table.2 shows eight elements and their basic attributes for representing a tree structure of each video stream in Fig.2.

4.1 Structure Expression

Our structure description is similar to SMIL [22] *seq* element, and extends the *anchor* element to describe a navigation rule represented in XQuery as its child node. A *seq* element is a root element of a video stream, and has some *media* elements as its child elements to represent each media object on the '*movie track*'. An *anchor* element is a child element of a *media*

Table 2: Elements and attributes of Hypermovie description

Element	Attributes	Roles of attributes
seq	id, dur	the duration of the 'Movie Track'
	time, pan, tilt, fov	the initial view state for playing
media	id, src,	URI for each media content
	clipBegin, clipEnd	begin and end time value in each media content
	pan, tilt, fov	the initial view state for each media content
anchor	id, begin, end	existence duration of each anchor in each media content
	z-index	overlap relationship between anchors
location	id, type	anchor type
rule	id	
on	id, trigger	event for a navigation rule execution
if	id, condition	macro expression of the condition for rule execution
do	id, action	macro expression of a hyperlink operation of each navigation rule

element, and represents an '*anchor track*'. A *location* element represents the location of the anchor at each time.Each '*navigation rule*' is described as a *rule* element, which is a child element of an *anchor* element, and has three elements as child elements to represent an active-rule format ('*on*', '*if*', '*do*').

The *dur* attribute in a *seq* element depicts the time duration of the video stream, and other attributes denotes attribute for initial view state in the video stream (start time, initial angle state). The *clipBegin* and the *clipEnd* attribute in the *media* element represent the 'start' and the 'end' time for playing each movie. The *angle* parameter represents the initial angle value of each VR content. In terms of the *anchor* element, it has two child elements that give a detail for its location (*location*) and navigation rules (*rule*) on it. The *begin* and the *end* attribute depend on the parent element timeline, and represent the existence time duration for the anchor. A *location* element has a set of position and angle values, which is used to specify anchor locations at different times.

An description example of the Hypermovie structure in Fig.1(a) can be described as follows. This example consists of two movies ('woman.mov', 'brain.mov'). Each movie has a single anchor track with a navigation rule. Here, we have removed unimportant attributes and details of *rule* elements.

```
<seq id="rotate_woman" dur="20s" time="0s">
 <media id="woman" src="woman.mov"
        clipBegin="0s" clipEnd="10s">
  <anchor begin="0s" end="10s">
   <location type="fix"> 0 0 100% 100% </location>
   <rule> here, rule description </rule>
  </anchor>
 </media>
 <media id="brain" src="brain.mov"
        clipBegin="0s" clipEnd="10s">
  <anchor begin="0s" end="10s"> ... </anchor>
 </media>
</seq>
```

4.2 Navigation Rule Expression

Each navigation rule is described in a *rule* element. An *on* element in a *rule* element has *trigger* attribute to specify some events as a trigger for rule execution. The '*condition*' part and the '*action*' part of each rule is represented as an *if* element and a *do* element respectively. Both elements contain XQuery expressions to specify their behaviors. The *if* element may be omitted if the condition part of the target rule is always 'true'. XQuery expression consists of three clauses: the first specifies variable bindings (*For*, *Let*), the second specifies condition handling (*Where*), and the last specifies creating returned values (*Return*). A *condition* attribute in the *if* element and an *action* attribute in the *do* element specify a rule template which is described in Section 4.3.

An example navigation rule for Fig.1(a) may be described as follows. Here, we omit the *if* element because the condition of this rule is always 'true'. In this example, when a 'click' event occurs on an anchor, the system evaluates an XQuery in the *do* element. The system obtains the time of the 'click' event in *Let* clause, and calculates the 'begin time' for the next node. As a result, the *Return* clause returns a *seq* element as a next node, whose initial time value is set to this calculated time value.

```
<rule id="mod2n">
 <on trigger="click"/>
 <do>{Let $vs  := doc("rotate_woman.xml")
      Let $dur := $vs/media[1]/@clipEnd
                      - $vs/media[1]/@clipBegin
      Let $et  := doc("event_history.xml")
                      //event[position()=last()]/@time
      Let $bt  := ($et + $dur) mod (2 * $dur)
      Return <seq id="{$vs/@id}" time="{$bt}">
                 {$vs/media}
             </seq>  }
 </do>
</rule>
```

To write the *do* element explicitly specifying what link action is executed, we may use macro expressions for four action patterns of navigation rules in a *return* clause. They are 'jump', 'update', 'create', and 'invoke' as described in Section 3.1.3. The evaluation result of a *return* clause is converted a *seq* element in the linking process. In the case of 'invoke' actions, there is no need for such a conversion because of the view state of current video stream regarded as the next view state.

```
jump:   <jump target="seq_id" time="" pan="" tilt="" fov=""/>
update: <update target="element_id">
          updated element data.
        </update>
create: <create>
          <seq id="new_seq_id"> ... </seq>
        </create>
invoke: <invoke target="program" param="value"/>
```

A *jump* element has attributes that indicates the next stream (*target*) and the viewing state (i.e. *time*, *pan*, *tilt*). An *update* element contains a changed value that is indicated by a *target* attribute. A *create* element contains a newly created video stream as a child element. An *invoke* element has two attributes to specify both the execution program (*target*) and parameter bindings.

Using the '*jump*' macro expression, the *Return*clause in the *do* element of the above-mentioned *rule* element is described as follows.

```
<rule id="mod2n">
 <on trigger="click"/>
 <do>{ Let ...
       Return <jump target="{$vs/@id}" time="{$bt}"/>  }
 </do>
</rule>
```

4.3 Rule Template

In preceding sections, we gave a fundamental method for describing navigation rules. However, it makes the whole description complicated, because the user needs to give detailed descriptions completely for all rules. The whole description is hard to understand for the user, and there is a little reusability of rule descriptions.

For these reasons, we introduce a notion of a '*rule template*'. It defines a macro for a rule expression. Users only need to assign the necessary parameters of such a macro. A template helps the user to understand the outline of a defined video stream context, and to use/reuse rules more easily. In the case of a rule 'mod2n', we need two parameters as input values: one is the duration of each movie and the other is the id of the target video stream. The template of 'mod2n' has the following using a declaration of an XQuery function.

```
declare function mod2n($dur as double, $tgt as string) as node() {
  Let  $et := local:last_event()/@time,
  Let  $bt := ($et + $dur) mod (2 * $dur)
  Return <jump target="{$tgt}" time="{$bt}" />
}
declare function local:last_event() as node() {
  Return doc("event_history.xml")//event[position()=last()]
}
```

By using such a rule template, the description of Fig.1(a) can be specified as follows.

```
<anchor begin="0s" end="10s">
 <location type="fix"> whole frame </location>
 <rule id="mod2n">
  <on trigger="click">
  <do action="mod2n('10', 'rotate_woman')"/>
 </rule>
</anchor>
```

4.4 Operation Flow of Link Navigation

Fig.6 shows an overview of an operation flow in a navigation process. These correspond to the three processes, which we discussed in Section 3.2. The '*Rule Selection*' process is composed of two phases: selection and evaluation. The selection phase takes one of the registered rules (XQuery), which is founded at the highest priority, which corresponds with the occurred event. The evaluation phase has a role of checking conditions and executing the action by evaluating the selected XQuery. In this process, we can use any XML documents such as Event History and Mpeg-7 annotation data for evaluating the rule. The '*Link Action*' process configures the result of rule evaluation to Hypermovie sequence and displays it to the user. Simultaneously, it registers *events* and *rules* for those anchors.

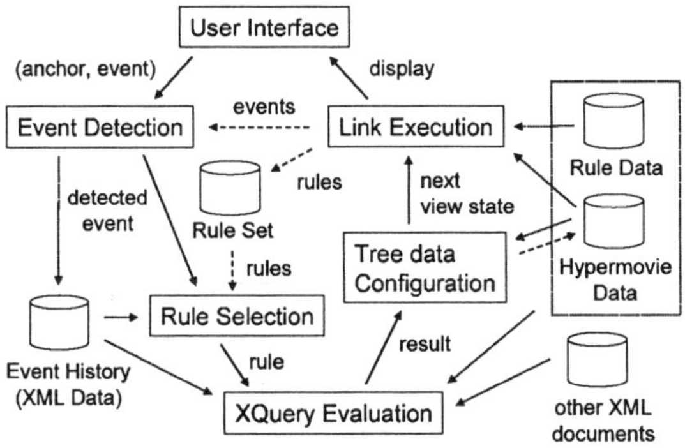

Figure 6: Operation flow of Hypermovie

In the case of Fig.1(a), the system stores all the *rule* elements into the '*Rule Set*', and registers a 'click' event, which is specified by the *trigger* attribute in the *on* element, into the '*Event Detection*' process. When a 'click' event is detected by the '*Event Detection*' process, the '*Rule Selection*' process selects and evaluates the rule 'mod2n'. The evaluation result is converted to a *seq* element that works as the next node in the '*Link Action*' process, and executes the result *jump* action. As a result, the user can jump into the next time position in the same video stream.

5 Example Scenario

5.1 Concentric Panoramic Views

We will show an example in which the user traverses through concentric panoramic views. The system consists of some panoramic views arranged concentrically with different resolutions. These panoramic views are taking from the same camera position. Outer views have higher resolutions. Fig.7 shows a part of this example. The user can move from an inner panoramic view to an outside panoramic view to zoom into same object.

All the panoramic views are organized in the order from the lowest resolution view to the highest resolution one. Firstly, the system stores three rules in the *seq* element. They are '*zoom_in*', '*open_url*' and '*show_preview*'. Then, the system registers corresponding events as rule execution triggers for the stored rules. The first one is the 'mouseDown+3s' event that moves to the next outer panoramic view. This event triggers a 'create' action to temporarily create the next view state. The second is the 'click' event that opens a related Web page. The last is the 'mouseEnter' event for previewing the next outer panoramic view. When the system detects a registered event on an anchor, it selects the corresponding *rule* element and evaluates its *do* element. Then, the system executes the evaluation result as a link action.

The following shows the rule template description corresponding to the registered events.

```
declare function zoom_in() as node() {
 Let  $c_m := local:current_movie(),
 Let  $n_id := $c_m/following-sibling/@id,
 Let  $a_p := $c_m/@pan, $a_t := $c_m/@tilt
 Return  <jump target="{$n_id}" pan="{$a_p}" tilt="{$a_t}"/>
```

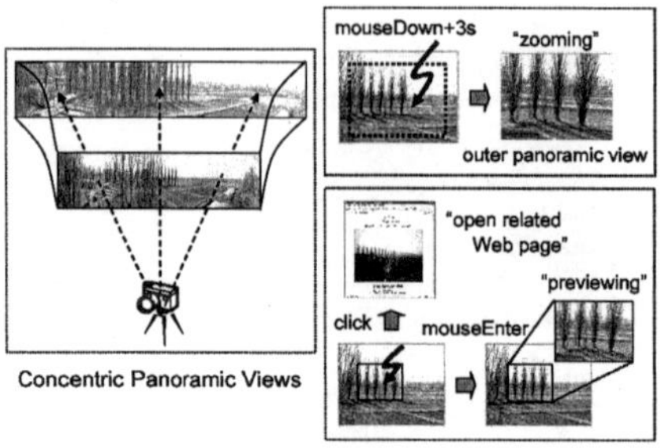


```
...
<media id="p1" src="poplar1.mov">
 <anchor id="frame_area" z-index="0">
  <rule id="zoom_in">
   <on trigger="mouseDown+3s"/>
   <do action="zoom_in()"/>
  </rule>
 </anchor>
 <anchor id="poplar" z-index="1">
  <rule id="open_url">
   <on trigger="click"/>
   <do action="open_url(url)"/>
  </rule>
  <rule id="show_preview">
   <on trigger="mouseEnter"/>
   <do action="show_preview(p2)"/>
  </rule>
 </anchor>
</media>
<media id="p2" src="poplar2.mov">
...
```

Figure 7: An example of concentric panoramic views

```
}
declare function open_url($url as string) as node() {
 Return  <invoke target="WebBrowser" param="{$url}"/>
}
declare function show_preview() as node() {
 Let  $c_m := local:current_movie(),
 Let  $n_m := $c_m/following-sibling
 Return
   <create>
     <seq id="temp_seq"  dur="1"
          time="0" pan="{$c_m/@pan}" tilt="{$c_m/@tilt}"/>
       <media id="{$n_m/@id}" src="{$n_m/@src}"/>
     </seq>
   </create>
}
```

The rule '*zoom_in()*' has no input parameters, and is evaluated in response to the event 'mouseDown+3s'. This rule returns a 'jump' action. The user can move to the next *media* element in the *seq* element. This is achieved by initiating the viewing angles to the same angles on the current panoramic view. The system gets this angle data from the 'Event History' that stores all the past events. The rule '*open_url()*' returns an 'invoke' action to open the Web browser with the specified url that is specified by the input parameter 'url'. The system executes the rule '*show_preview()*' while the mouse position is over the anchor 'poplar'. This rule returns a 'create' action. The system creates a temporary *seq* element to show a panoramic view with a higher resolution centered around the area of the anchor 'poplar'.

5.2 News Archive Application

We will show a description example of the news archive application. In this example, we specify each news item and character, such as an anchor person, as anchors. This application consists of movie archives for news archives and annotation data for news contents in each news program. These annotation data are stored into the external data storage and represent

Table 3: Elements and attributes for the news archive

Element	Attributes	Roles of attributes
news	id	particular id of each news archive
resource	mid	particular 'media' element id of corresponding 'seq' element
	src	media location url
time	date	broadcast date
	start,end	begin and end time of a news program
item	id	particular id of each news content
	begin,end	begin and end time of each news content
topic	id	topic id

```
<news id="news1">
 <resource mid="media1"
          src="news1.mov">
 <time date="2004-02-28"
       start="18:00:00"
       end="18:10:00"/>
 <item id="i1"
       begin="18:00:00"
       end="18:1:32">
  <topic>north korea</topic>
  <topic>nuclear</topic>
 </item>
 <item id="i2" .../>
 ...
</news>
```

(a)News annotation data

```
<seq id="archive1" time="0s">
 <media id="media1" src="news1.mov"
      clipBegin="0s" clipEnd="600s">
  <anchor id="telop1"
          begin="15s" end="30s">
   <rule>
    <on trigger="click"/>
    <do action="related_news
                ('news1','i1')"/>
   </rule>
  </anchor>
  ...
 </media>
 <media .../>
 ...
<seq>
```

(b)The structure of the news archive

Figure 8: An example of annotation data and the Hypermovie structure for the news archive

the details of each news content such as a news summary, a newsreel footage, and a time duration of each news content. Table.3 shows required elements and attributes of annotation data for each news contents. Fig.8(a) shows a description example of an annotation data for a news program stored into an external data storage.

For example, when the user clicks news items in each news program, the system searches all the annotation data for all the news item with the topics related to the current news item. Then, the system creates a new news archive by concatenating all the retrieved news items and presents it to the user.

Hypermovie description of a news archive is shown in Fig.8(b). A news archive has some news programs expressed as *media* elements. Each *media* element has some anchors corresponding to each news content in the news program. Each *anchor* element has a navigation rule to search all the annotation data for all news items with topics related to the current news item and to create a new news archive by concatenating all the retrieved news items. In this example, we set a navigation rule '*related_news*' to an anchor, which is named 'telop1', in response to a 'click' event. When the system detects a click event at the anchor 'telop1' in the '*Event Detection*' process, the '*Rule Selection*' process selects the rule '*related_news*'.The evaluation result of this rule is a 'create' action that creates a new news archive by concatenating all the news item with the topics related to the current news item.

The following shows the main part of rule template '*related_news*'.

```
declare function related_news($n_id as string,
                    $item_id as string) as node() {
 <create>
  <seq>
   {Let $tpcs := local:topics($n_id,$item_id)
    For $r_n in local:search_news($tpcs)/news
    Let $m := collection("news_seq")
               /seq/media[@id=$r_n/resource/@mid]
    Return
     <media id="{$m/@id}" src="{$m/@src}"
            clipBegin = "{local:begin_time($r_n)}"
            clipEnd = "{local:end_time($r_n)}"}>
               {$m/anchor}
     </media> }
  </seq>
 </create>
}
declare function topics($n_id as string,
                    $item_id as string) as node() {
 <topics>
  {Let $ns := collection("news_list")
              /news[@id = "$n_id"]
   For $tpc := $ns/item[@id="$item_id"]/topic
   Return $tpc }
 </topics>
}
declare function search_news($tpcs as node()) as node() {
 <result>
  {For $ns in collection("news_list")
   For $it in $ns/item
   Where every $wd in $tpcs/topic satisfies
           some $tp in $it/topic satisfies
             (contains($wd/text(),$tp/text()))
   Return <news id="{$ns/@id}">
            {$ns/resource} {$ns/time} {$it}
          </news> }
 </result>
}
```

The rule template '*related_news*' is evaluated in response to the 'click' event, and the evaluation result of this rule is a 'create' action to present a new news archive to the user. This rule has two input parameters: one specifies the *id* attribute of the *news* element for the annotation data; the other specifies the *id* attribute of the 'item' element for the annotation data. Both the '*search_news*' function and the '*topics*' function are used in the evaluation process of the rule '*related_news*'. The '*topics*' function returns all the *topic* elements included in the current *item* element where the user clicks. The '*search_news*' function returns all the *news* elements including the same *topic* elements as the given input parameter.

For example, when the user clicks the anchor 'telop1' in Fig.8(b), the system gets all the *topic* elements included in the current news item using the '*topics*' function. In this case, the '*topics*' function returns two *topic* elements shown in Fig.8(a). Then, the system searches all the annotation data for all the *news* elements including these *topic* elements using the '*search_news*' function. Finally, the system creates a *seq* element as a new news archive by concatenating all the *media* elements which include all the news items with topics related to the current news item. In this process, the system calculates the *clipBegin* attribute value and the *clipEnd* attribute value of all the retrieved *media* elements including required news items.

6 Related Works

6.1 Hypervideo Model

Early hypervideo models [2, 14, 16] provided restricted set of hypervideo functions. Aspen Movie Map [16] is a well known system that provides a walkthrough application for the central Aspen using four-way video media. The user is able to switch videos to the desired direction by his operation to walk through streets. Micon [2] demonstrates video-to-video linking in the Elastic Charles. This system has anchor node that the anchor is a miniaturized movie icon. This movie icon is called Micon (Movie icon). Looking at the Micon, the user can overview the details of the linked video and jump to the linked video by clicking Micon. In [14], video-to-video, video-to-text and text-to-video linking are provided to define more interactive and multiliner structures. Then, this system extends to adjust to the Web [15]. In these systems, video media are only used as terminal nodes (source node or destination node).

Hypercafe [19] provides narrative sequences to the user at a virtual cafe that consists of video clips with conversations. It represents a general hypervideo concept and framework for building narrative sequences and gives multiple link opportunities to video applications. MovieDNA [18], which is based on the Focus+Context visualization, is introduced to the navigation of video streams. The user browses the context of video stream by using a hierarchical summary. FlyAbout [11] is an application based on a VR movie. It enables the user to view different locations from arbitrary directions. For example, the travel application with a map-like interface allows the user to go through roads on the maps viewing the surroundings in arbitrary directions. Though these hypervideo models deal with narrativities and contexts for video, their structures are not open for other applications. Neither of them provided multiple link references for the same single anchor.

A Object-based Navigation [8] introduces dynamic linkages using characteristics of each object (i.e. color, shape, etc). This type of techniques for dynamic linkage among media objects is also used in Open Hypermedia System domain [1, 13] and in video-content retrieval application domain [7]. Recently, these techniques are applied to a meta level description language such as MPEG-7 [10].

In comparison with these video integration models, our framework can provide a uniform way of handling video streams and constructing link references dynamically by evaluating navigation rules triggered by events on anchors.

6.2 Video Integration to Hypermedia Document

In [12], MPEG-7 [10] is presented as a suitable front-end format for multimedia database system model. Its format is defined as an XML scheme. It is becoming a multimedia description standard for audio-visual streams. It provides the user/author with the search arbitrary of any objects, using their predefined metadata. Thus, it enables users or authors to use the context of video stream in other XML documents and applications of multimedia database system, and vice versa. However, because MPEG-7 does not define a context for each stream, it is only used as an exchange format. In MPEG-21 [9], MPEG group discuses the context definition method for the video stream like SMIL [22].

SMIL [22] is a standard for scripting synchronization and coordination of streams. It allows as to easily define a new context of multimedia documents by combining some media objects. In addition, Microsoft released an implementation for their Web browser(HTML+TIME [20]). It allows the user to integrate basic functionalities of SMIL and HTML, and to reuse other HTML tags. Celentano discussed a synchronization model of nested continuous media objects for user interactions in [3], which is similar to SMIL model. However, these models do not have enough description methods to provide different user interactions and multiple link references for each anchor.

In HTIMEL [4, 5], which is an extension of HTML, video objects are managed uniformly with other resources. It enables the user to integrate hypervideo aspects to HTML documents. HTIMEL is able to define multiple link destinations to realize different contexts and spatio-temporal link opportunities for source and destination nodes. On the other hand, link destinations are static references to target tags. Like SMIL, this model is also not enough to reuse hyperlink structures, because destination is represented as an attribute value in the source tag.

In comparison with these description models for video streams, our framework can provide reusability of descriptions by separating a structure of a video stream from link references. Furthermore, describing each navigation rule with XQuery expressions, our framework can establish link references dynamically corresponding to various events on anchors, and reconfigure video streams dynamically through the link navigation.

7 Conclusion

In this paper, we proposed a description model for the contextualization of video streams, which we call a Hypermovie framework. Our framework has mainly three features: (i) we can set various actions to a single anchor; (ii) the system can create or reconfigure video streams through evaluating such rules; (iii) separating a structure of a video stream from link references, our framework allows users to reuse their descriptions individually. We proposed a hyperlink model with a notion of parameterized spatio-temporal media and rule-based link navigation for uniformly handling video streams. We proposed a rule-based link navigation model to assign some navigation rules to a single anchor node as link references. By evaluating the designated rule for each event on each anchor, the user can move to the next node dynamically. Moreover, using XML and XQuery expressions to describe our hyperlink model, it allows users to reuse arbitrary parts of defined contexts in the video stream, and even link conditions and link actions. Our description method allows us to define rich navigation structures as a set of reactive behaviors rules associated with various types of events on anchors, and to assign multiple link references to a single anchor node. Using our framework, we can create a video stream by concatenating arbitrary parts of video media including both live-action video and computer graphics, and we can associate different user operations on video media with the execution of various application programs.

References

[1] K. M. Anderson, R. N. Taylor, and E. J. Whitehead, Jr. Chimera: hypermedia for heterogeneous software development enviroments. *ACM Trans. Inf. Syst.*, 18(3):211–245, 2000.

[2] H. Brøndmo and G. Davenport. Creating and Viewing the Elastic Charles – A Hypermedia Journal. In *HYPERTEXT II: State of the Art*, Navigation and Browsing, pages 43–51. 1989.

[3] A. Celentano and O. Gaggi. A synchronization model for hypermedia documents navigation. In *Proceedings of the 2000 ACM symposium on Applied computing*, pages 585–591, Como, Italy, 2000.

[4] T. Chambel, N. Correia, and N. Guimarães. Hypervideo on the Web: Models and Techniques for Video Integration. In *International Journal of Computers & Applications*, pages 90–98. Acta Press, Vol.23, #2.

[5] T. Chambel and N. Guimarães. Context Perception in Video-Based Hypermedia Spaces. In K. M. Anderson and S. Moulthtrop, editors, *Proceedings of the 13th ACM Conference on Hypertext and Hypermedia*, pages 85–94, New York, June 11–15 2002.

[6] S. E. Chen. QuickTime VR - An Image-Based Approach to Virtual Environment Navigation. In *Proceedings of the ACM SIGGRAPH '95*, pages 29–38, Los Angeles, 1995.

[7] N. Dimitrova, H.-J. Zhang, B. Shahraray, I. Sezan, T. Huang, and A. Zakhor. Applications of Video-Content Analysis ad Retrieval. *IEEE MultiMedia*, 9(3):42–55, 2002.

[8] K. Hirata, S. Mukherjea, Y. Okamura, W.-S. Li, and Y. Hara. Object-Based Navigation: An Intuitive Navigation Style for Content-Oriented Integration Environment. In *Proceedings of the Eighth ACM Conference on Hypertext*, Information Retrieval, pages 75–86, 1997.

[9] ISO/IEC JTC1/SC29/WG11 N5231. MPEG-21 Overview. http://www.chiariglione.org/mpeg/.

[10] ISO/IEC JTC1/SC29/WG11 N5525. MPEG-7 Overview.

[11] D. Kimber, J. Foote, and S. Lertsithichai. FlyAbout: Spatially Indexed Panoramic Video. In *Proceedings of the ACM Multimedia 2001*, pages 339–347, Ottawa, 2001.

[12] H. Kosch. MPEG-7 and Multimedia Database Systems. *ACM SIGMOD Record*, 31(2):34–39, June 2002.

[13] P. H. Lewis, H. C. Davis, S. R. Griffiths, W. Hall, and R. J. Wilkins. Media-based navigation with generic links. In *Proc of the ACM Hypertext'96*, pages 215–223, 1996.

[14] G. Liestøl. Aesthetic and Rhetorical Aspects of Linking Video in Hypermedia. In *Proceedings of the ECHT'94 European Conference on Hypermedia Technologies*, pages 217–223, 1994.

[15] G. Liestøl. Video on the Web: Principles of Continuity Linking and Narrative Flow. In *WebNet'99*, Honolulu, Hawaii, USA, Oct. 1999.

[16] J. Nielsen. *HYPERText&HYPERMedia: The Internet and Beyond.* Academic Press, 1995.

[17] N. W. Paton and O. Diaz. Introduction. In *Active Rules in Database Systems*. Springer Verlag, 1999.

[18] D. Ponceleon and A. Dieberger. Hierarchical Brushing in a Collection of Video Data. In *Proc of HICSS'34 (Hawaii International Conference on Systems Science)*, 2001.

[19] N. Sawhney, D. Balcom, and I. Smith. HyperCafe: Narrative and Aesthetic Properties of Hypervideo. In *Proceedings of the 7th ACM Conference on Hypertext*, pages 1–10. ACM Press, Mar. 16–20 1996.

[20] P. Schmitz, J. Yu, and P. Santangeli. Timed Interactive Multimedia Extensions for HTML (HTML+TIME) Extending SMIL into the Web Browser.

[21] J. M. Smith, D. Stotts, and S.-U. Kum. An Orthogonal Taxonomy for Hyperlink Anchor Generation in Video Streams using OvalTine. In *Proceedings of the Eleventh ACM Conference on Hypertext*, pages 11–18, San Antonio, Texas, United States, 2000.

[22] W3C. Syncronized Multimedia Integration Language 2.0 Specification.

[23] W3C. XQuery 1.0: An XML Query Language.

Information Modelling and Knowledge Bases XVI
Y. Kiyoki et al. (Eds.)
IOS Press, 2005

Concepts are structured meanings

Marie DUŽÍ, Daniela ĎURÁKOVÁ, Marek MENŠÍK
Department of Computer Science,
VSB – Technical University of Ostrava
17. listopadu 15,
708 33 Ostrava, Czech Republic
e-mail: marie.duzi@vsb.cz
daniela.durakova@vsb.cz
mensikm@centrum.cz

Motto: "*...Ich verstehe ... unter dem Worte* ***Erkenntniss*** *ein jedes Urtheil, das einen wahren Satz enthält, oder (was eben so viel heisst) der Wahrheit gemäss oder richtig ist.*" (Bolzano, *Wissenschaftslehre*)

There is nothing better for Practice but a good Theory (SOFSEM'86)

Abstract. Theoretical results presented in previous papers are illustrated by practical application of building up an intelligent AI-like system. We provide a case study on business personal management implemented in Prolog, which makes use both of mathematical results of the classical set-theoretical Formal Concept Analysis, and of the hyper-intensional approach of non-traditional procedural Theory of Concepts. Due to the rich ontology of Transparent Intensional Logic the well-known paradox of logical / mathematical omniscience common to epistemic logics with intensional semantics is avoided. Explicit knowledge is represented by TIL logical constructions structured from the algorithmic point of view, and the inference machine implemented in the Prolog programming language makes it possible to perform deductive reasoning on structured knowledge base, deriving thus implicit knowledge from the explicit one.

Key words: Formal concept analysis, ontology, structured knowledge, hyper-intension, inference machine, paradox of omniscience

Introduction

The problem of *structured meanings* is a hot topic of semantic analysis since Frege's paper on *Sinn und Bedeutung* appeared [Frege 1982]. In [Duží 2003] and [Duží, Materna 2002] we advocated for the new non-traditional theory of concepts assigned to expressions as their meanings: concepts are conceived as 'instructions for arriving at the denoted entity', procedures that are structured in an 'algorithmic way'. We claimed that our theoretical hyper-intensional framework provides an extension of the traditional extensional / intensional set-theoretical approach. In this paper we briefly describe a case study of an application that makes use both of the classical Formal Concept Analysis and of the hyper-intensional procedural approach based on the Transparent Intensional Logic.

The paper is organised as follows: in Chapter 1 we recapitulate the Formal Concept Analysis (FCA) as presented by R. Wille and B. Ganter in [Ganter,Wille 1999]. In Chapter 2 we characterise knowledge and briefly summarise basic axioms of epistemic logics. Transparent Intensional Logic is presented in Chapter 3 where the hyperintensional theory of concepts as algorithmic instructions is described. In Chapter 4 we examine partial ordering on (equivalent classes of) procedurally structured concepts. FCA is examined here from the logical point of view, its qualities as well as shortcomings are formulated and possible extensions proposed. Chapter 5 describes an application – the case study on a business personal policy management. A thorough specification in the TIL 'language of constructions' is presented together with its implementation in the Prolog programming language.

1. Formal Concept Analysis

The 'Formal Concept Analysis' (FCA) project originated around 1980 in Darmstadt, Germany. Research group led by R. Wille began to systematically develop a framework for knowledge representation using lattice theory applications. The first publication by R. Wille was presented in 1982 [Wille 1982], and in 1996 Ganter and Wille published a textbook on the mathematical foundations of FCA, which has been translated into English [Ganter, Wille 1999]. The method has been originally proposed as a mathematical model of data analysis. Much of the mathematics required for the application has been borrowed from lattice theory [Birkhoff 1948]. However, it achieved results that are of much a broader interest than originally intended, and the method is useful in particular as a tool for *structured knowledge representation*.

Formal concept analysis starts with the primitive notion of a *(formal) context*.

Definition:

A *formal context* is a triple (G, M, I), where G and M are non-empty sets and I is a binary relation called *incidence relation* between G and M, $I \subseteq G \times M$. The elements of G are called *objects* and the elements of M are *attributes* of the context.

We write *gIm* or $(g,m) \in I$ and say "the object g has an attribute m". ♦

Fundamental operators defined at the formal context characterise particular associations (connections) between objects and attributes given by the incidence relation. They are defined as follows:

Definition:

Given a formal context (G, M, I) and a subset $A \subseteq G$ of objects, we define

$A^{\uparrow} = \{m \in M \mid gIm \text{ for all } g \in A\}$ and similarly for a subset $B \subseteq M$ of attributes we define

$B^{\downarrow} = \{g \in G \mid gIm \text{ for all } m \in B\}$. ♦

Operators ↑, ↓ form the so-called *Galois connection* between power sets of G and M, P(G) and P(M), respectively, characterized by conditions indicating a natural "duality" between objects and attributes[1]. We can see that $A^{\uparrow}$ is the set of all the attributes common to objects of A, and $B^{\downarrow}$ is the set of all the objects possessing attributes from B. Hence two operators were defined:

$\uparrow : P(G) \rightarrow P(M)$ ("from objects up to their attributes")

$\downarrow : P(M) \rightarrow P(G)$ ("from attributes down to the objects")

Definition:

A *formal concept* of the context (G,M,I) is a pair (A,B) such that $A \subseteq G$, $B \subseteq M$, and $A^{\uparrow} = B$, $B^{\downarrow} = A$. We say that A is the *extent* and B is the *content* of the concept (A,B). The set of all concepts is denoted by B(G,M,I). ♦

Note: R. Wille uses the terms 'extent' and 'intent' of a concept, traditionally 'extension' and 'intension' of a concept have been used as well. We will use the terms as defined above, not to cause confusion with possible-world semantics terminology (PWS-intension, PWS-extension).

Traditionally, it is taken for granted that the *law of inverse proportion* between the content and extent holds: increasing the content causes decreasing the extent and *vice versa;* (extending the 'demands' decreases the number of objects meeting them – the more attributes the less objects falling under them, and *vice versa*). This intuition makes it possible to define a partial ordering on the set of formal concepts following the "information content" of a formal concept:

---

[1] See [Birkhoff 1948]

Definition:
Let (A_1, B_1), (A_2, B_2) be formal concepts of B(G,M,I). We say that the concept (A_1, B_1) is a *subconcept* of the concept (A_2, B_2) iff $A_1 \subseteq A_2$ (which is equivalent to $B_2 \subseteq B_1$), and we write $(A_1, B_1) \leq (A_2, B_2)$. The concept (A_2, B_2) is a *superconcept* of (A_1, B_1). ♦

Relation ≤ is obviously reflexive, antisymmetric and transitive, it is a partial ordering on the set B(G,M,I) of all formal concepts of the given context. Due to the fixed context it is easy to show that each two-elements subset of B(G,M,I) has a supremum and infimum. Hence the structure ⟨B(G,M,I), ≤ ⟩ is a lattice where meet and join are defined as follows:

$$(A_1, B_1) \wedge (A_2, B_2) = (A_1 \cap A_2, (B_1 \cup B_2)^{\downarrow\uparrow}),$$

$$(A_1, B_1) \vee (A_2, B_2) = ((A_1 \cup A_2)^{\uparrow\downarrow}, B_1 \cap B_2).$$

Moreover, it is a complete lattice, according to the Fundamental theorem of FCA:
Let (G, M, I) be a context. Then ⟨B(G,M,I), ≤ ⟩ is a complete lattice, called the concept lattice of (G,M,I), for which infimum and supremum of any subset can be described as follows:

$$\bigvee_{t\in T}(A_t, B_t) = \left(\bigcap_{t\in T} A_t, \left(\bigcup_{t\in T} B_t\right)^{\downarrow\uparrow}\right)$$

$$\bigwedge_{t\in T}(A_t, B_t) = \left(\left(\bigcup_{t\in T} A_t\right)^{\uparrow\downarrow}, \bigcap_{t\in T} B_t\right)$$

The proof and further details can be found in [Ganter, Wille 1999].

A formal context can be viewed as an elementary model of formal representation of knowledge, which yields even a more structural representation of conceptual knowledge by its concept lattice. Formal contexts are usually described by a *cross-table* (or incidence matrix) while concept lattices are effectively visualized by *labelled line diagrams (Hasse diagrams)*.

Simple as this theory is, it has many useful applications in the area of conceptual modelling, when performing "conceptualisation and/or classification", i.e., building ontologies of the area of interest. It is used not only in a narrow sense, modelling objects and their database-like attributes, but also in a broader sense, modelling (*'de dicto'*) associations of higher-order objects like properties, individual concepts, etc. In the area of information retrieval FCA has been applied to reduce the 'dimension of the problem' of the search in the huge amount of data spread in web files. An incidence matrix can thus model not only the relation of 'having an attribute', but also the relation (between higher-order objects) of being a requisite or being a typical property[2]. Or, it can be the relation of 'occurring in' between particular documents and words. The latter is in particular interesting, leading to surprising results in building an 'intelligent thesaurus'. We can also use 'finer-grained' methods of representing dependencies between attributes of a conceptually scaled many-valued context, or even a fuzzy context, where a degree of affiliation is examined. Besides, efficient algorithms of concept lattice derivation from a given context have been proposed.

The procedure of building up a formal concept lattice usually consists of the following steps:

- Collecting data on the objects of interest.

---

[2] See [Duží 2003]

- Selecting relevant data organised in an incidence matrix that serves as an input for the algorithm of the concept lattice derivation.
- Based on the partial ordering of formal concepts defined by the incidence matrix a concept lattice is derived.
- Graphic representation of the lattice using Hasse diagram serves to interpret the result.

Before presenting our case study on the 'business personnel policy management', we first characterise the notions of implicit/explicit knowledge and the problems connected with the 'Epistemic Closure'. Using Transparent Intensional Logic we further explicate the Formal Concept Analysis from the logical point of view, in order to deeper understand the problem and possibly to propose an extension of the theory.

2. Characteristics of knowledge, epistemic logic

The fundamental notion of 'knowledge' is commonly used in practice, but the question 'What kind of entity knowledge is?' has been discussed for more than two hundred years as a central epistemological problem, and the discussion is far from being exhausted. Traditionally, knowledge is characterised as a *true justified belief*. But distinct people have distinct beliefs, and we can ask "whose knowledge?", "whose beliefs?". From the point of view of an epistemic logic, the following like argument is of a central interest:

Allen *knows* that Bert is *convinced* that Cynthia is lying.
Allen *believes* that Bert is infallible.
Hence Allen *believes* that Cynthia is lying.

Thus another possible characteristics might be

Knowledge is a relation between an agent[3] *A and a statement P.*

However, not every relation to a proposition could be called knowledge. For instance, 'Allen *wishes* that Bert were infallible' could hardly be characterised as Allen's knowledge. Perhaps the most plausible definition is:

Knowledge is a justified belief of an agent A that the statement P is true.

Logical systems of epistemic logic make it possible to represent knowledge, and according to their expressive power to realise less or more powerful inference machines, which enable us to infer logical consequences (*implicit knowledge*) from the set of explicitly represented propositions (*explicit knowledge*). A necessary condition of an adequate correct inferring (possibly) all the pieces of implicit knowledge is thus a fine-grained analysis of statements like

Agent A knows that P.

Following the above characteristics, almost all the epistemic systems accept a rule of knowing being a *factivum*:

Agent A knows that P $\vdash$ *P is true.*

Systems of epistemic logics are logics of knowledge from a modal point of view stressing the role of inference. They axiomatise "behaviour" of epistemic operators **K** (is known) and **B** (is believed), which are not truth-value operators but concern *meanings of statements*. There is a similarity between logic of knowledge and belief on the one hand,

[3] By an agent we mean here an agent in the broadest sense: an individual, a computer system, a group of human beings (common knowledge), etc. We will address an agent by 'he'.

and logic of necessity and possibility on the other hand: if an agent *knows* that P, then it is (from his/her point of view) *necessary* that P. The modal notion of a possible world is here understood as a world compatible with that what the agent knows. Basic axioms are (φ,ψ are formulae):

- All the axioms of 1st order predicate logic
- The axiom of logical racionality ('Epistemic closure'):
  (K) $[K\varphi \wedge K(\varphi \Rightarrow \psi)] \Rightarrow K\psi$
- The rule of necessitation: $\varphi \mid\!- K\varphi$

Other accepted axioms (characterising some features of knowledge):

- (T) $K\varphi \Rightarrow \varphi$ knowledge implies truth
- (D) $K\varphi \Rightarrow B\varphi$ knowledge implies belief
- (S4) $K\varphi \Rightarrow KK\varphi$ positive introspection
- (S5) $\neg K\varphi \Rightarrow K\neg K\varphi$ negative introspection

Epistemic theories are usually assigned *intensional semantics* based on Kripke's frames. Kripke's model M is a tuple ⟨W, R, I⟩, where W is the set of possible worlds, R is a binary relation of accessibility on W, and I is a function assigning to each formula φ a subset of W. Formula φ is true in a world *w* iff $w \in I(\varphi)$, denoted $(M,w) \models \varphi$. A formula $K\varphi$ is understood as φ being *true in all the worlds accessible* to an agent according to his/her knowledge: $(M,w) \models K\varphi$ iff $(M,w') \models \varphi$ for all *w*' such that *w*' R *w*. Axioms (T), (S4) are accepted, if the relation R is reflexive and transitive, theorems of (T), (S4), (S5) are valid if the relation R is an equivalence.

Applicability of intensional systems

Systems based on intensional semantics are applicable without any problems if the statement *Agent A knows that P* is understood as

Agent A has enough information so that he ***might*** *safely* ***act*** *as if P were true.*

This is a model of an *implicit knowledge* assigned to the agent "from the outside". A more natural notion is the notion of *explicit knowledge* (represented, for instance, by agent's memory – knowledge base state). In such a case we not only want that the agent *can* safely act as if P were true, but also that the agent is ***actually able to act.*** Typically the agent is demanded to be able to answer whether P is true. Intensional epistemic systems are applicable if:

a) we neglect *time* (and other factors) necessary for agent's deriving the logical consequences of his explicit knowledge,
b) the proof calculus (the agent's inference machine) is *complete* and *decidable.*

In other words, this is the case of an idealised *logically/mathematically omniscient agent.* Due to famous Gödel's results, it is not a realistic case even when considering the computability power of a powerful computer system inference machine.

Explicit knowledge relates an agent to the *meaning* of the embedded clause P. In an ideal case the semantic representation should perfectly reflect *synonymy of* sentences, because the meaning is just what really synonymous sentences (even in distinct languages) have in common. Intensional semantics that represents the meaning as an intension (proposition = a subset of possible worlds) is too coarse-grained: it conflates synonymy with equivalency, which leads to the *paradox of omniscience.*

3. Hyperintensional approach of TIL

To be able to model *explicit knowledge* in a realistic way, we need a fine-grained logical analysis, which assigns distinct meanings to distinct sentences A and B, whenever A and B are not strictly speaking synonymous, i.e., whenever it is possible to believe one of them without believing the other. Tichý and his followers[4] show in a persuading way that the set-theoretical approach is always just an approximation to real synonymy, for meanings are structured in an *algorithmic* way. An essential improvement of the set-theoretical approach is provided by the two-level semantics of Transparent Intensional Logic (TIL): between an expression E and the denoted entity D (intension in case of empirical expressions) there is a *logical construction,* i.e. the *concept C* identifying (or failing to identify in case of empty concepts) the denoted entity D. This semantic shift of Frege-Church schema is illustrated as follows:

$$\text{Expression E} \xrightarrow[\text{expresses}]{} \text{Concept C = meaning of E} \xrightarrow[\textit{identifies}]{} \text{Denoted entity D}$$

Basic notions of TIL have been presented in [Duží 2003], [Duží, Materna 2002]. Here we briefly recapitulate. *Concept* is a closed logical *construction,* i.e. an abstract procedure, structured from the *algorithmic* point of view, unlike 'mathematical structure' structured from the mereological point of view. The sense of an expression consists in specifying the *way (procedure, instruction)* of identifying the denoted entity, the expression talks *about.* TIL constructions are specified in an adjusted formal version of Church's typed λ-calculus, where terms (interpreted in a fixed way – the intended interpretation) denote not the entity that is constructed but just the construction itself.

The TIL language is fully transparent; all the semantically salient features are inherently present. In particular, we need to know the *type* of the denoted entity and the *way* of its identifying, i.e. the concept of the denoted entity. Moreover, we need not only to *use concepts* within the theory, but also to *mention* them (knowledge is a relation to the meaning = construction!). In such a case the denoted entity is a (less-structured) construction, and it is identified by a more structured construction of a higher order.

Thus the ontology of TIL is organised into a two-dimensional infinite hierarchy, which at the lowest level has to stem from a certain base. TIL is an open system and the choice of the base is arbitrary. For instance analysing a language of arithmetic, the base consisting of truth-values and of the set of natural numbers would suffice. For the purpose of natural language analysis we use the so-called *epistemic base* {ο, ι, τ, ω}, where:

ο – the set of truth values
ι – universe of discourse (the set of individuals)
τ – the set of time points
ω – the set of possible worlds

On a "horizontal line" mereological complexity is increased by the rule of forming partial functions: if α, β_1, ..., β_n are types then the collection of partial functions $\beta_1 \times \ldots \times \beta_n \to \alpha$ is a type, denoted $(\alpha\, \beta_1 \ldots \beta_n)$.

In the "vertical line" algorithmic complexity (the order of a construction) is increased: Entities of the ***1st order*** type are not structured from an algorithmic point of view, they are not procedures – concepts, nor do they involve concepts in their domains.
Collection of constructions identifying 1st order entities forms a type of ***order 2*** – $*_1$.

---

[4] See [Tichy 1988], [Materna 1998], [Duží 2003a]. Similar ideas are presented in [Moschovakis 1994]. Theory of concepts based on the algorithmic conception has been first presented in [Materna 1998].

Constructions identifying entities of order 1 and 2 form a type of ***order 3*** – $*_2$.
And so on … $*_3$, $*_4$, …
Pavel Tichý introduced in his book [Tichy 1988] six kinds of constructions, and later on for the needs of conceptual modelling constructions of tuples and projections have been introduced. In most cases four basic kinds are used:
(Notation: An object O of a type α – O/α. A construction X identifying an entity of type α: X $\rightarrow \alpha$.)

- *Variables x, y, z, …, p, q, …* (for each type countably many variables) identify objects of the respective type dependently on a valuation v, they v-construct.
- *Trivialisation* of any entity X identifies simply X without "a help" of any other concepts, denoted ^{0}**X**.
- *Closure* $[\lambda x_1 \ldots x_n\, X]$ constructs a partial function F/($\alpha\ \beta_1 \ldots \beta_n$), where $x_i \rightarrow \beta_i$, X$\rightarrow \alpha$, by abstracting X from $x_1, \ldots, x_n$.
- *Composition* $[X\ X_1 \ldots X_n]$ constructs the value (if any) of the function F constructed by X at the argument v-constructed by $x_1, \ldots, x_n$. (Composition can fail to construct, because the partial function F does not have to return a value at an argument, there may be value gaps.)

Intensions are 1st order entities; β-intension is modelled as a function from possible worlds ω to a type α, which is most frequently a chronology of β-objects: $((\beta\tau)\omega)$, abbreviated $\beta_{\tau\omega}$. Examples (sets are modelled by characteristic functions):
Proposition – $((o\tau)\omega)$; abbreviated $\mathbf{o}_{\tau\omega}$ – 'Charles knows that the Pope is in danger'
Individual office – $\iota_{\tau\omega}$ – 'the Pope'
Property of individuals – $(\mathbf{o\iota})_{\tau\omega}$ – 'being in danger'
An attribute (empirical function) – $(\alpha\beta)_{\tau\omega}$ – 'father of …'

Higher-order entities are *hyperintensional* – involving constructions: $*_n$, $(o*_n)$, $(o\iota*_n)_{\tau\omega}$,…

To adduce a simple example, consider knowledge of some mathematical fact T (theorem). Since all the mathematical facts are analytically necessary, they are equivalent (but not synonymous!), true in all the possible worlds. When somebody knows T, he is not related to the truth value True (nor to the proposition TRUE), but to the *way* of obtaining the truth – the respected concept. (Note that our theory of concepts assigns constructions, meanings to any meaningful expressions, not only to *universals*.) Hence ***Know*** / $(o\ \iota *_n)_{\tau\omega}$.
The simplest way of identifying an entity X is trivialisation ^{0}X – the simple concept of X. Intensions are functions; using variables $w \rightarrow \omega$, $t \rightarrow \tau$, intensions are constructed by concepts of a form $\lambda w \lambda t$ Y. Composition defining the intensional descent of an intension constructed by Y $\rightarrow \alpha_{\tau\omega}$ is of a form [[Y w] t], abbreviated by Y_{wt}.

Examples of analysis { $x \rightarrow \tau$, Bijection / (o $(\tau\tau)$) }:
Function Successor (adding 1) returns 4 at 3:

$[^0{=}\ [\lambda x\ [^0{+}\ x\ ^01]\ ^03]\ ^04]$, or using infix notation $[[\lambda x\ [x + {}^01]\ ^03] = {}^04] \rightarrow$ o (True).

Function Successor is a bijection:

$[^0\text{Bijection}\ \lambda x\ [^0{+}\ x\ ^01]\ ] \rightarrow$ o (True).

Agent A knows that the function Successor is a bijection:

$\lambda w \lambda t\ [\ ^0\text{Know}_{wt}\ ^0\text{A}\ ^0[^0\text{Bijection}\ \lambda x\ [^0{+}\ x\ ^01]]\ ] \rightarrow o_{\tau\omega}$ (proposition True/False in w,t).

This fine-grained analysis is an adequate modelling of explicit knowledge. It does not presuppose *any inference abilities* of an agent (as an example of such a logical idiot imagine a knowledge base without any software procedure that would make it possible to infer and/or compute other facts entailed by the knowledge base, except of storing and updating data). All the inference abilities of particular agents have to be explicitly stated in a form of postulates. For instance, if the agent A knows that Successor is a bijection and *if he is able to use the rule* of existential generalisation, then he also knows that there are bijective mappings:

$\lambda w \lambda t\ [\ {}^0\mathrm{Know}_{wt}\ {}^0\mathrm{A}\ {}^0[{}^0\mathrm{Bijection}\ \lambda x\ [{}^0{+}\ x\ {}^0 1]]\ ]$

$\lambda w \lambda t\ [\ {}^0\mathrm{Know}_{wt}\ {}^0\mathrm{A}\ {}^0[\ [{}^0\mathrm{Bijection}\ \lambda x\ [{}^0{+}\ x\ {}^0 1]] \Rightarrow \exists f\,[{}^0\mathrm{Bijection}\, f]\ ]$ $\quad f \to (\tau\tau)$

$\lambda w \lambda t\ [\ {}^0\mathrm{Know}_{wt}\ {}^0\mathrm{A}\ {}^0[\ \exists f\,[{}^0\mathrm{Bijection}\, f]\ ]$

We obtained an instance of the axiom (K) of agent's logical rationality.

Logic specifying the behaviour of a multi-agent system, where particular agents are autonomous and intelligent, but nevertheless resource-bounded, has to relate agent's epistemic closure to particular rules known to the agent. TIL language of constructions enables us to represent explicit knowledge, and thus it makes it possible to integrate syntactic formal methods with the rigorous semantics[5].

4. Ordering structured concepts

In Chapter 1 we presented Ganter & Wille mathematical theory of formal concepts viewed as couples ⟨Extent, Content⟩ defined in a fixed context, with nice mathematical properties, in particular forming a complete lattice. Such a structure is not a hyperintensional entity we advocated for. However, to be able to make use of its nice mathematical features, we have to explicate the theory from the viewpoint of TIL, and to propose an extension of the approach. First we have to define the extent and intent of a TIL concept:

Definition:
A concept C_1 is *contained in a concept* C_2, iff C_1 is a sub-construction of C_2.
Content of a concept C is the set of concepts that are contained in C.
Extent of a concept C is the object E, which is constructed by C.
An empirical concept is such a concept C_E, the extent of which is an α-intension I (/ $\alpha_{\tau\omega}$).
The *extent of an empirical concept C_E in a world/time w,t* is the value of its extent I in w,t, i.e. I_{wt} (/α).
An empty concept is a concept the extent of which is an empty set.
A strictly empty concept is a concept that does not have an extent (does not construct anything, the construction is v-improper, like, e.g., the greatest prime).
A simple concept (primitive with respect to a given conceptual system) of an object X is ^{0}X (it cannot be strictly empty). ♦

Relation of intensional containment, as defined above, is the relation of partial ordering on the set of concepts; it is evidently reflexive, anti-symmetric and transitive. Still, in general, it cannot be used for a correct definition of a semantic conceptual lattice, because it does not have to be in accordance with semantic entailment. Just an enumeration of the set of contained concepts *does not suffice*. We also have to specify *the way* in which these contained concepts are composed together to form a complex. To do that, we need the notion of a requisite of a property (/of an office) and of a property typical for a property (/for an office):

[5] Axiomatisation of the logic of explicit knowledge in TIL has been a subject of further research.

Definition:
Let P, Q, G be constructions of any properties and U of an office. We define:
$[\mathrm{Req}^{pr}\ \mathrm{P\ Q}] = \forall wt\ \forall x\ [[\mathrm{Q}_{wt}\ x] \Rightarrow [\mathrm{P}_{wt}\ x]]$ (*P is a requisite of Q*)
$[\mathrm{Req}^{of}\ \mathrm{P\ U}] = \forall wt\ [[{}^0\mathrm{E}_{wt}\ \mathrm{U}] \Rightarrow \forall x\ [[\mathrm{U}_{wt} = x] \Rightarrow [\mathrm{P}_{wt}\ x]]]$ (P *is a requisite of* U)
$[\mathrm{TP}^{pr}\ \mathrm{P\ Q\ G}] = \forall wt\ \forall x\ [\neg[\mathrm{G}_{wt}\ x] \Rightarrow [[\mathrm{Q}_{wt}\ x] \Rightarrow [\mathrm{P}_{wt}\ x]]]$ (*P is typical for Q, unless G*)
$[\mathrm{TP}^{of}\ \mathrm{P\ U\ G}] = \forall wt\ [[{}^0\mathrm{E}_{wt}\ \mathrm{U}] \Rightarrow \forall x\ [\neg[\mathrm{G}_{wt}\ x] \Rightarrow [[\mathrm{U}_{wt} = x] \Rightarrow [\mathrm{P}_{wt}\ x]]]\ ]$
(*P is typical for U, unless G*) ♦

Note: In artificial intelligence the condition G is called *the guard* of a rule. The existential condition E / $(o\ \iota_{\tau\omega})_{\tau\omega}$ is a property of an office, returning True (in *w,t*) when the office is occupied (in this *w,t*). Generally, existential presupposition concerning the 'office rules' can also be conceived as a guard of the rule. Such an existential guard is indispensable in case of offices, because if the office U is vacant in some world/time *w,t*, then the composition U_{wt}, and thus the whole composition $[[\mathrm{U}_{wt} = x] \Rightarrow [\mathrm{P}_{wt}\ x]]$, are *v*-improper (partiality[6] is propagated up).
Examples:
A typical property of a bird is flying, unless it is a penguin or an ostrich.
A typical property of a swan is being white, unless it comes from Australia or New Zealand.
A requisite of the King of France is being a ruler.
A requisite of a whale is being mammal.

Finally, we define a semantic partial ordering on the set of (equivalence classes of) concepts that can induce a *conceptual lattice*:

Definition:
Let C and D be empirical concepts such that C constructs a requisite R of the extent I constructed by D. Then *C is weaker than or equivalent to D,* denoted $\mathrm{C} \leq \mathrm{D}$. ♦

Claim: Let properties EC, ED be extents of concepts C, D, respectively, such that $\mathrm{C} \leq \mathrm{D}$. Then necessarily, i.e., in all world/times *w,t* it holds that $\mathrm{ED}_{wt} \subseteq \mathrm{EC}_{wt}$.
Proof: Follows from the definition of requisites.

As a special case, if the number of *all* the requisites R^i of an intension I is *finite*, a concept C can construct I by means of "conjuncting" R^i:

$$\lambda w \lambda t\ \lambda x\ ([\mathrm{R}^1{}_{wt}\ x] \wedge \ldots \wedge [\mathrm{R}^n{}_{wt}\ x]).$$

Then, obviously, for any C_j such that
$\mathrm{C}_j = \lambda w \lambda t\ \lambda x\ ([\mathrm{R}^{i1}{}_{wt}\ x] \wedge \ldots \wedge [\mathrm{R}^{ik}{}_{wt}\ x])$, where $\{i_1,\ldots,i_k\} \subseteq \{1,\ldots,n\}$ it holds that $\mathrm{C}_j \leq \mathrm{C}$ and in any world/time the extent of C is a subset of the extent of C_j ($\mathrm{EC}_{wt} \subseteq \mathrm{EC}_{jwt}$), i.e., the *law of inverse proportion* is valid.

We have obtained the classical Ganter-Wille *conjunctive conception*, which is just a special case of our general definition.

Another *weaker semantic ordering* on the set of concepts can be defined on the basis of typical properties: Let C and D be empirical concepts such that C constructs a property T typical for the extent I constructed by D. Then $C \leq_G D$; Now the inverse relation between the extents in world/times holds only on the condition that the respective guards are satisfied (which can be formulated as some non-intrinsic integrity constraints, and their validity will be checked by a program):

---

[6] For details on quantification logic of partial functions, see [Duží 2003a]

Claim: Let intensions EC, ED be extents of concepts C, D, respectively, such that C $\leq_G$ D. Then a *guarded rule* is valid on the assumption that the respective guard G is satisfied (in world/time *w,t*): $ED_{wt} \subseteq EC_{wt}$ (in case ED is an office and the office is not vacant, then $ED_{wt} \in EC_{wt}$).
Proof: Follows from the definition of typical properties.

Now a *semantic content* of a concept C might be defined in two ways: Strictly speaking, it is the set of those concepts C_i that identify the *requisites* of the entity constructed by C, i.e., $C_i \leq C$. In a weaker sense, it is the set of concepts C_j that identify *typical properties* of the entity constructed by C, i.e., $C_i \leq_G C$. To obtain members of the semantic content of the concept C we use inference rules of TIL (natural deduction rules).

We obtained an extension of the classical Ganter-Wille conception, where particular attributes are just monadic predicates connected in the conjunctive way, interpreted in a fixed context. In TIL specification concepts of any complex structure can be used. A typical example of the need of such an extension is the *transitive closure* problem: the relation of an ancestor (in its broadest sense) cannot be defined by conjuncting a finite number of unary predicates; a recursive definition is necessary.

5. Case study: managing business personnel policy

First, collecting business data we create particular incidence matrixes. Two principles are followed: a) matrix of *contingent* assigning attributes (properties and roles) to people, and b) matrix of *necessary* relations between properties and roles, i.e., skills required for a given role and/or property. Afterwards, particular conceptual lattices are created, which illustrates the personnel structure of a company in a clear way providing a lot of useful information for the management. These data serve as an input to fill the knowledge base of an intelligent information system. Enriching the system with recursive rules whenever needed, we obtained some interesting results. The powerful query / specification language[7] of TIL constructions enables us to raise complex queries, i.e. concepts of intensions. Answering a query means obtaining *implicit knowledge* – the value of the intension in a world/time couple *w,t,* which is rendered by a knowledge base state. The inference machine has been implemented in the Prolog programming language, which provided a sufficient implementation background for a fragment of query specification language formulated in the TIL language of constructions. In particular, the '*de dicto / de re* interplay', i.e. distinguishing using / mentioning entities of any order proved to be a very fruitful method.

5.1. Application of the Formal Concept Analysis

a) business *roles* (members of the G set of 'objects'):

Roles	Manager	Secretary	Technologist	Designer	Accounter	Worker	Keeper	Foreman
Abbr.	Ma	Se	Te	De	Ac	Wo	Ke	Fo

b) required *skills* (requisities and typical properties candidates, members of the M set of 'attributes'):

Required features	Common Sense	Work disposal	Construct. technique	Techno-logy	Invoicing	Ordering	Product-ion	Evidence
Abbr.	co	wd	ct	te	in	or	pr	ev

[7] For details see [Duží, Materna 2002]

Analysing business know-how, required skills have been assigned to particular business roles. This knowledge is given by the cross-table. The table 1 can be understood as a description of formal context: its objects are business roles heading the rows of the table, and their requisites and/or typical properties are represented by the columns.

abbr	co	wd	ct	in	or	pr	ev	te
Ma	2	2	0	1	1	0	1	0
Se	2	1	0	2	2	0	2	0
Te	1	0	0	0	0	2	0	2
De	1	0	2	0	0	2	0	0
Ac	1	0	0	2	2	0	2	0
Wo	0	0	0	0	0	2	0	0
Ke	1	0	0	2	2	0	2	0
Fo	2	2	1	1	1	2	1	1

Table 1: Cross-table: Features of Roles (requisities / typical properties of roles)

Explanatory notes:
0 without – a skill is not required
1 typical – a skill is typically required, or is typical for a role
2 necessary – a skill is a requisite of a role

To use the FCA algorithm of creating the conceptual lattice it is necessary to adjust input values to Yes/No values (1/0). For this purpose we use the scaling technique described in [Ganter,Wille 1999]. To scale the basic table we used the method of assigning a nominal scale to each many-valued attribute. The adjusted cross-table in the needed form of context is recorded by the table Scaling values:

abbr	co1	co2	wd1	wd2	ct1	ct2	te1	te2	in1	in2	or1	or2	pr2	ev1	ev2
Ma	0	1	0	1	0	0	0	0	1	0	1	0	0	1	0
Se	0	1	1	0	0	0	0	0	0	1	0	1	0	0	1
Te	1	0	0	0	0	0	0	1	0	0	0	0	1	0	0
De	1	0	0	0	0	1	0	0	0	0	0	0	1	0	0
Ac	1	0	0	0	0	0	0	0	0	1	0	1	0	0	1
Wo	0	0	0	0	0	0	0	0	0	0	0	0	1	0	0
Ke	1	0	0	0	0	0	0	0	0	1	0	1	0	0	1
Fo	0	1	0	1	1	0	1	0	1	0	1	0	1	1	0

Table 2: Scaling values: nominal scale for each many-valued attribute from original cross- table.

The FCA algorithm produced a conceptual lattice. Hasse diagram of the lattice is presented in the Appendix (Figure 1). Note that each sublattice of the diagram represents a group of business activities that have something in common (accounting, production, technology, ...).

Other tables have been created by collecting (contingent) data on particular employees, applicants and collaborators. Their roles, skills, education, etc. have been recorded. An example of the result of such a data collection is presented in the following table 3.
The resulting lattice is presented in the Appendix (Figure 2).

People:
Vanessa (Va), Jane (Ja), Alberto (Al), Bill (Bi), George (Ge), David (Da), Peter (Pe)

abbr	co	wd	ct	te	in	or	pr	ev
Va	1	1	0	0	1	1	0	1
Ja	1	1	0	0	1	1	0	1
Al	0	0	1	1	0	0	1	0
Bi	1	0	1	0	1	1	1	1
Ge	1	0	1	1	1	0	0	1
Da	1	1	0	1	1	1	0	1
Pe	0	0	1	0	0	0	1	0

Table 3: Knowledge and skills (persons and their skills)

5.2. TIL query specification and Prolog implementation

The inference machine of a fragment of the powerful TIL framework has been implemented in Prolog. Particular queries are precisely specified using the TIL language of constructions. First, using incidence matrixes, Prolog knowledge base has been created. Besides, an 'intelligent thesaurus' assigning types to simple expressions as well as to business idioms has been built up and recorded in the Prolog knowledge base. This feature enables us to reject non-reasonable questions without even 'looking at the database'. A fine-grained TIL analysis makes it possible to handle ambivalent questions by asking for additional information. Making use of the lattice structure and requisites / typical properties associations resulted in an 'intelligent behaviour' of the system: thus, for instance, in case of obtaining a notice the system proposes further actions according to the structured conceptual knowledge, in order to promote a smooth recovery.

The scaled matrix of requisites / typical properties has been easily recorded (due to the linear resolution strategy of Prolog, roles are assigned effectively):

```
Table3:
data(va,  [co,wd,in,or,ev]).
data(ja,  [co,wd,in,or,ev]).
data(al,  [ct,pr,te]).
data(bi,  [co,ct,in,or,pr,ev]).
data(ge,  [co,ct,in,ev,te]).
data(da,  [co,wd,in,or,ev,te]).
data(pe,  [ct,pr]).
```

Table2: *requisites and typical properties are recorded in the 1st line, the 2nd line contains only requisites*

```
req([co,wd,in,or,ev],      ma).
req([co,wd],               ma).

req([co,wd,in,or,ev],      se).
req([co,in,or,ev],         se).

req([co,pr,te],            te).
req([pr,te],               te).

req([co,ct,pr],            de).
req([ct,pr],               de).

req([co,in,or,ev],         ac).
req([in,or,ev],            ac).
```

```
req([co,in,or,ev],         ke).
req([in,or,ev],            ke).

req([co,wd,ct,in,or,pr,ev,te], fo).
req([co,wd,pr],                fo).

req([pr],                  wo).
req([],                    wo).
```

Note: Assigning roles to particular people can be done in two ways: either using the table 'makes', which contains data on roles actually performed by the people, or assignment of roles to people is executed by the program. The program offers for each X those roles that X is able to perform. When confirmed by the user, the program can even assign particular roles effectively, minimising human resources.

Ambivalent queries

In case of raising an ambivalent query the system asks for additional information.

Q1: ***Which properties does Mr. X have as an accountant?***

There are two readings of the query. Either we simply want to know properties of Mr. X who happens to be an accountant, but his profession is actually not of particular interest. It serves just for testing whether we found the right person. Or, we actually ask for the properties of Mr. X that he has and that are requisites of the accountant. In either case the answer will provide a set of properties / $(o\,(o\iota)_{\tau\omega})$.

Typing:

Has_prop	/	$(o\,\iota\,(o\iota)_{\tau\omega})_{\tau\omega}$
Performs	/	$(o\,\iota\,(o\iota)_{\tau\omega})_{\tau\omega}$
X	/	ι
accountant	/	$(o\iota)_{\tau\omega}$

The analysis of the former case (*de re*):

QR: $\lambda w\, \lambda t\, \lambda prop\, [[{}^0\text{Has\_prop}_{wt}\, {}^0X\, prop] \wedge [{}^0\text{Performs}_{wt}\, {}^0X\, {}^0\text{Accountant}]]$

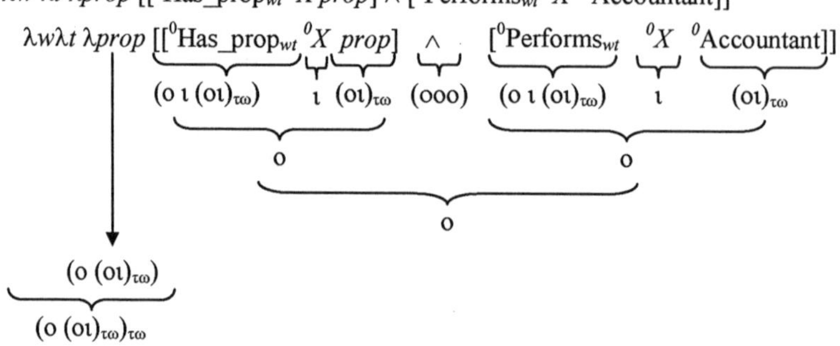

The analysis of the latter *de dicto* case:

QD: $\lambda w\, \lambda t\, \lambda prop\, [[{}^0\text{Req}\, prop\, {}^0\text{Accountant}] \wedge$
$[{}^0\text{Performs}_{wt}\, {}^0X\, {}^0\text{Accountant}] \wedge [{}^0\text{Has\_prop}_{wt}\, {}^0X\, prop]]$

In the program the query Q1 is written as follows:

```
has_properties([ac,[x]],Which).
```

/* ac – accountant, Which – variable that will contain properties */

The respective part of the program:

```
has_properties([Role,[Who]],Which)        :-
                type_checking([Role,[Who]]),
                ambivalence(Role,Who,X),
                has_pr([Role,[Who]],Which,X).

      /* Answering QD: */
      has_pr([Role,[Who]],Which,req)      :-
                makes(Who,Role),
                data(Who,Which1),
                req(Requisity,Role),
                select_properties(Requisity,Which1,Which),
                !.

has_properties([Role,[Who]],Which)        :-
                not (type_checking([Role,[Who]])),
                Which='Failed Analysis of query!',!.
has_properties([Role,[Who]],Which)        :-
                not (type_checking([Role,[Who]])),
                Which='Wrong query no typing',!.

has_properties([Role,[Who]],Which)        :-
                type_checking([Role,[Who]]),
                ambivalence(Role,Who,X),
                has_pr([Role,[Who]],Which,X).
      /* if X is not an accountant */
      has_pr([Role,[Who]],Which,req)      :-
                not (makes(Who,Role)),
                Which=Who-'not performing role '-Role,!.

has_properties([Role,[Who]],Which)        :-
                type_checking([Role,[Who]]),
                ambivalence(Role,Who,X),
                has_pr([Role,[Who]],Which,X).
      has_pr([Role,[Who]],Which,who)      :-
      /* Answering QR */
                not (makes(Who,Role)),
                Which=Who-'not performing role '-Role,!.

has_properties([Role,[Who]],Which)        :-
                type_checking([Role,[Who]]),
                ambivalence(Role,Who,X),
                has_pr([Role,[Who]],Which,X).
      has_pr([Role,[Who]],Which,who)      :-
      /* Answering QR */
                data(Who,Which).

ambivalence(Role,Who,X)                   :-
```

/* This procedure is to decide whether Q is a query of type QR or QD */

```
          nl, write(Ambivalent query!'),nl,
          write('--------------------'),nl,
          write('If you want the requisites of role - '),
          write(Role), write(' - / req'), nl,
          write('If you want the properties of - '),
          write(Who), write(' - /prop'), nl,
                read(X),nl.
```

Emergency situation: losing a key man.

Q2: ***In which roles is Mr. XY indispensable?***

Indispens / $(o\ \iota)_{\tau\omega}$
Performs / $(o\ \iota\ (o\iota)_{\tau\omega})_{\tau\omega}$
role / $(o\ \iota)_{\tau\omega}$

$\lambda w\ \lambda t\ \lambda role\ [[{}^0\text{Indispens}_{wt}\ {}^0XY] \wedge [{}^0\text{Performs}_{wt}\ {}^0XY\ role]]$

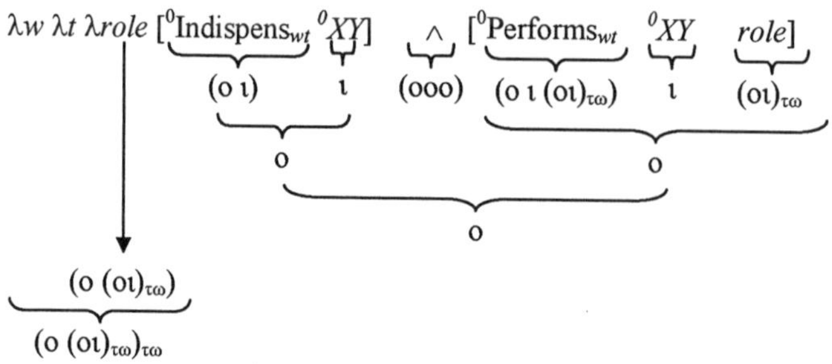

Q3: ***What happens if we lose XY ?***

a) XY is indispensable

– In case the man is indispensable, the program looks for those who have some requisites of the role performed by XY, so that they can be easily trained to surrogate XY

```
losing(Whom,What_happens)                          :-

                     makes(Whom,Role),
                     indispensable([Whom],Role),

                     What_happens='Function '-Role-' CANNOT be
                     performed by anybody else !'.
```

– We have to train somebody so that particular people can perform the work.

```
losing(Whom,What_happens)                          :-

                     makes(Whom,Role),
                     indispensable([Whom],Role),
                     findall(Cl,najdi_cl1(Cl,Role),People1),
                     sort(People1,People),

                     What_happens='After training, the work can
                     be performed by '-People.
```

b) XY is not indispensable

– In case the man is dispensable, other people can perform the work immediately without any problems:

```
losing(Whom,What_happens)                          :-

                     makes(Whom,Role),
                     not (indispensable([Whom],Role)),
                     people_req(Role,People),

                     What_happens='Function '-Role-' can be
                     performed by '-People.
```

– And after training even by others:

```
losing(Whom,What_happens):-
        makes(Whom,Role),
        not (indispensable([Whom],Role)),
        findall(Cl,najdi_cl1(Cl,Role),People1),
        sort(People1,People),
        What_happens='Function '-Role-' can be
        performed after traning by'-People.
```

Non-reasonable question

Q4: ***Which roles are performed by number 5?***

Performs	/	$(o\,\iota\,(o\iota)_{\tau\omega})_{\tau\omega}$
5	/	τ
role	→	$(o\iota)_{\tau\omega}$

$$\lambda w\, \lambda t\, \lambda role\, [{}^0\mathrm{Perfroms}_{wt}\ \ {}^0 5\ \ role]$$

$$(o\,\iota\,(o\iota)_{\tau\omega})\quad \tau\quad (o\iota)_{\tau\omega}$$

Failed !

In the program the query Q4 is written as follows:

```
makes_role(5,Role).
```

The query is rejected due to the type checking:

```
makes_role(Who,Role)                    :-
        type(Who,_),
        type(Role,_),
        type_checking([Role,[Who]]),
        makes(Who,Role).
makes_role(Who,Role)    :-
        type(Who,_),
        not (type_checking([Role,[Who]])),
        nl, write('Failed Analysis of
        query!'),nl,!,fail.
type_checking([X,[Y]])  :-
      typ(X,[Typ,[TypY]]), typ(Y,TypY),
      test(Typ,[[X,[Typ,[TypY]]],[[Y,[TypY]]]]),!.
```

6. Concluding Remarks

In February 2004 a group of logicians in the Department of Computer Science, Technical University of Ostrava founded the Research Laboratory of Intelligent Systems (see, http://www.cs.vsb.cz/labis/index_eng.html) in order to integrate research activities in the area of *Logic and Artificial Intelligence.*

Our research covers the following areas of Artificial intelligence: knowledge representation and management, communication and process management. The main goal can be characterised as the management of *multi-agent systems* of autonomous intelligent agents that are resource-bounded. The system has to be robust, able to handle emergency and critical situations. Particular agents have to communicate in a (pseudo-)natural language, in order to provide relevant information whenever and where-ever needed. In our opinion, meeting such an ambitious goal is not possible without a powerful logical background. We decided to make use of the Transparent intensional logic, the system that provides a framework for a correct fine-grained logical semantics. The presented study can be viewed as the first step to fulfil the goal.

Next we are going to specify, design and realise a model that would simulate behaviour of autonomous intelligent agents. Each agent will have its own intelligence and recourses, which he/she/it has to use to meet particular demands. This model framework will serve the purpose of theoretical research as well as practical experiments, and it will be designed in such a way that the results could be used in real applications.

Theoretical research will be further concentrated on building an inference machine deriving implicit knowledge from the explicit knowledge-base. After an appropriate axiomatisation we have to either adjust the logic programming language (Prolog) or to build up an interpret of a new one, so that the TIL rigorous specification could be implemented in full. The precise logical specification will also make a formal verification of the system feasible.

This work has been promoted by the grant project AS CR No 401/03/1403 – Principles of Logical Analysis, and the grant project AS CR No 401/03/H047 – Logical Foundations of Semantics and Knowledge Representation.

References

[Birkhoff 1948] G. Birkhoff: *Lattice theory.* Revised edition, New York.

[Duží, Materna 2002] M. Duží, P. Materna: Intensional Logic as a Medium of Knowledge Representation and Acquisition in the HIT Conceptual Model. In *Information Modelling and Knowledge Bases XIV.* Ed. Hannu Kangassalo, Eiji Kawaguchi, Amsterodam, Netherlands. IOS Press, 2003, Vol. 94, pp 51-65, ISBN 1 58603 318 2

[Duží 2003] M. Duží: *Concepts, Language and Ontologies.* In *Information Modelling and Knowledge Bases XV.* Ed. by Hannu Kangassalo, Eiji Kawaguchi, IOS Press 2004, Amsterdam, Netherlands, pp. 193-209.

[Duží 2003a] M. Duží: Do we have to deal with partiality? In *Miscellania Logica*, 2003, Tom V, 45-76

[Frege 1982] G. Frege: Sinn und Bedeutung. *Zeitschrift fur Philosophie und philosophische Kritik 100,* pp.25-50

[Ganter, Wille 1999] B. Ganter, R. Wille: *Formal Concept Analysis.* Springer – Verlag, 1999. ISBN 3-540-62771-5

[Ganter, Wille 1996] B. Ganter, R. Wille: *Applied Lattice Theory: Formal Concept Analysis*, report in Darmstadt, 1996

[Materna, 1998] P. Materna: *Concepts and objects.* Acta Philosophica Phennica, Vol.63, Helsinki 1998

[Moschovakis 1994] Moschovakis, Y.: Sense and denotation as algorithm and value. In *Lecture Notes in Logic, #2 (1994),* Springer, J. Oikkonen and J. Vaananen, eds., pp. 210-249.

[Tichy 1988] P. Tichy: *The Foundations of Frege's Logic.* De Gruyter 1988.

[Wille 1982] R. Wille: Restructuring lattice theory: an approach based on hierarchies of concepts. In: I. Rival, editor, *Ordered sets,* Reidel, Dordrecht-Boston, 1982, 445-470

Appendix

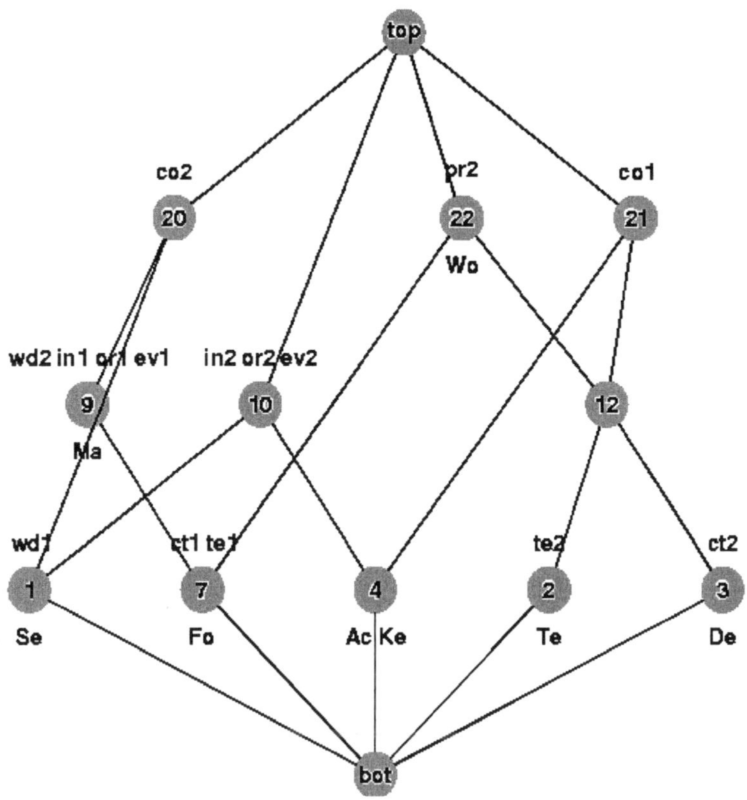

Figure 1: The concept lattice of scaling context

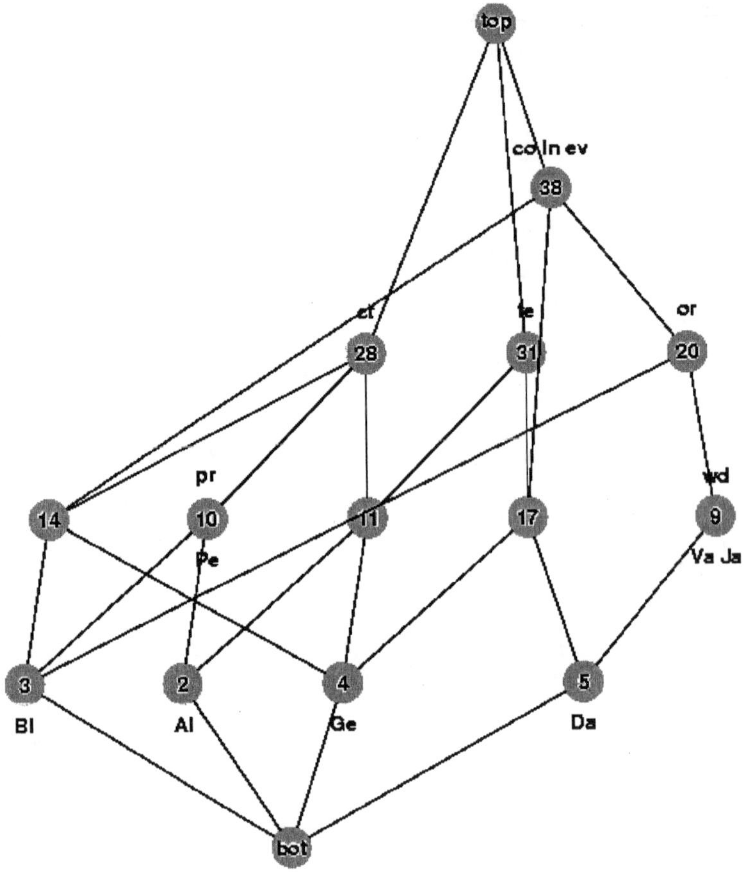

Figure 2: The concept lattice of context of people with their declared features.

Information Modelling and Knowledge Bases XVI
Y. Kiyoki et al. (Eds.)
IOS Press, 2005

Emergence of Lexicon and Names

Jaak HENNO
Tallinn Technical University, Tallinn 19086, Estonia

Abstract. Herein we study emergence of common lexicon with names in a multi-agent society in a set-up, where objects can be composed of other objects and agents do not have to use full descriptions of communication objects (or some words are lost in communication channel). Computational modelling shows, that after initial state of exchanging grounded messages (messages with indicating the communication object) agents can continue learning new words and names using pure (without indicating the object) messages only. The process has several features similar with children's word-learning process. The results help to understand emergence of human language and can also be used to develop robots which are able to learn to communicate.

Introduction

In the Introduction to recent book "Language Evolution" [1] M. H. Christensen and S. Kirby noticed [2] that:

- language evolution may be the hardest problem in science (p.1);
- we still know relatively little about the origins and evolution of this peculiar human trait (p.1-2); currently many different (and often controversial) perspectives are expressed (p 12);
- the strongest point of consensus is the notion that to fully understand language evolution, it must be approached simultaneously from many disciplines (p.12-13);
- (another) area of consensus is the growing interest in using mathematical and computational modeling to explore issues relevant to origin and evolution of language; given the number of different factors that may potentially influence language evolution, our intuitions about their complex interactions are often limited; when multiple processes have to be considered together, modeling becomes a useful and necessary tool (p. 13).

Here is studied emergence of common lexicon and names in society of agents. The used here setup differs from other publications on this topic (J. Batali [3], Hurford [4], Zuidema [5], Steels [6], [7], Kirby [8], Christiansen [9] etc) in the following:

- Vertical transmission (generations, natural selection) is not used – agents learn what they need rather quickly without dying and selection;

- Agents give names to meaning space objects and objects can be composed of other objects; object's hierarchical structure helps to understand problems connected with compositionality (see below);

- while at the beginning agents exchange only grounded messages (messages with indication of message's object), they soon become able to learn new words from ungrounded (without object) messages; this is similar to humans language-learning: most of the words that we know and use are not taught to us, we have learned them from context - picked them up from conversations, from read texts etc [10], [11].

The proposed message-exchange algorithm was tested using a computational model (program). The model shows several features reported in child language learning studies (understanding develops quicker than actual lexicon; there is some kind of vocabulary spurt). The used model is functional (a Prolog program), not based on neural nets; neural net models can show, that something is achievable, but do not show the "inner" mechanism of the process (as complained in [12]: "we would like to inspect the ...grammar directly. Unfortunately, the grammar acquired by our system is implicit in the learned feature weights").

The considered here algorithms help to understand better human language, but they can also be used for developing robots or software agents, who are able to learn to communicate.

1. The setup

1.1 Compositionality and semantics

Language is a discrete mapping between signals (words, phrases, sentences) and their meanings (objects of the semantic space). The most important property of any natural (human) language is its compositionality: the meaning of a signal (phrase, sentence) is composed of meanings of its parts (words, morphemes), and composition complexity (e.g. depth) is in principle not restricted; we restrict our message's (e.g. sentences) complexity only to ease understanding. Human language is (potentially) an infinite discrete mapping, able to cope with (potentially) infinite set of meanings (real-word objects). All communication systems (languages) used by animals are either analogue (mimics, pointing) or can express only a finite set of meanings, the main difference between discrete human language and all animal languages is infinite expressive power achieved by making the system discrete and using the Frege's Principle of compositionality [13].

The principle of compositionality is very natural. However, the closer examination reveals that (as many narrative definitions) it allows many very unnatural models, e.g. such as Gödel numbering discussed in [14] - intuitively we expect compositionality to be something continuous. Even when this mapping is computationally modelled [15] (computational implementation usually helps to clear problems with narrative definitions), treatment of compositionality evokes questions: "A compositional language is a mapping which preserves neighbourhood relationships – neighbouring meanings will share structure, and that shared structure in meaning space will map to shared structure in signal space" (p. 6). This definition assumes topology in the meanings and signal spaces: "We use the Hamming and Levenstein (string edit) distance measures to quantify inter-meaning and inter-signal distances respectively" (p. 11). Meanings are vectors of features (attributes) and signals – strings of characters. This definition seems to indicate that semantically similar objects (having many features in common) should be denoted by words or phrases, having many common characters (words) in same position, but this obviously does not hold in human languages. Nothing is said about ordering of features in meanings – changing the order of features in the feature vector changes totally the Hamming distance, but *red*, *big*, *round* denotes still the same object as *round*, *red*, *big*. If the above definition is considered on level of words (i.e. similar objects should be described by phrases which use (mostly) same words in same position, then there will be problem with languages, where word order (in sentence or phrase) is not very strongly determined; such are e.g. Estonian and Finnish: "Juhan armastab Maret", "Maret armastab Juhan", "armastab Maret Juhan" etc (John loves Mary) are all understandable phrases.

In linguistic analysis compositionality is considered as correspondence between syntactic and semantic trees [16], the Chomsky's "deep structure"[17]. The later is similar in all languages: "...in some important sense, all languages are equal"[2] and is usually expressed as a hierarchical structure, e.g. *verb*(*object*, *subject*), but sentence constituents verb (V), object (O) and subject(S) can be substituted with sub trees making the sentence semantic tree more and more complex. Sentence itself is the labelling of leaves of this tree, so compositionality means that however complex the sentence tree is, it is possible to reconstruct the tree having only the labelling of its leaves. The semantic tree *verb(object, subject)* expresses rather complex structure of dependencies – two sub trees with different roles (not interchangeable). Adding attributes (adjectives) to object, adverbs to verb, replacing subject with noun phrase etc makes the tree more and mode complex and there appears mode and more interdependencies. Human languages have invented clever ways to code labels at the tree leaves (words, morphemes) in such a way that we are still able from listing of labels at leaves (the sentence) to restore the whole tree, however deep and complex it is. Different languages use for this different tricks: some use word order (and all possible orderings: SVO, SOV, VSO, VOS, OVS, OSV occur in world languages), some use pro- and postpositions, suffixes and prefixes etc, i.e. the mapping: "syntax" $\leftrightarrow$ "semantics" is in different languages very different. This may be (one of) the reason(s) why linguists are still "slow to join the resurgence of interest in the evolution of language" – they still "are not in agreement about how to characterize what evolved"[2], p.4. The "deep structure" is the same, but different languages have invented very different syntaxes (i.e. tricks) and it seems to be very difficult to find some unifying principles behind different encodings (syntaxes). Thus it may be good to study some simpler (but still unbounded) hierarchical structures. Such hierarchical structures are formed by objects containment and naming – here all sub trees have the same role. When we consider an object, we always add one (un-perceivable) attribute which makes the whole object structure hierarchic – its name. Instead of vector of features (*round*, *big*, *red*) we consider hierarchical structure *ball*(*round*, *big*, *red*); the word (noun) *ball* is something, which we add to object with features *round*, *red*, *big* and which serves as a database key – it determines the object. When an object is composed from other objects names are labels at tops of corresponding sub trees. Compositionality for hierarchical structure of the semantic (object) space means that the mapping from object space to signal (language) space should preserve this hierarchical structure. We should know where "to put brackets" and for this the following conditions seem to be sufficient [12]:

(i) constituents (i.e. labelling at sub-trees) do not cross each other and

(ii) constituents occur in constituents contexts
i.e. the language's mapping *structure_tree* $\rightarrow$ *phrase* should be a homomorphism (similar definition of compositionality is used in [18]).

In the following computational models are considered semantic objects with hierarchical structure induced by naming; the emerging language satisfies (i), (ii); order of labels at leaves inside a constituent (where to put the name – first or last) can be set either to pre- or postorder.

1.2 The semantic and symbol spaces

Let $\mathcal{F} = \{f_1,...,f_n\}$ be the set of (semantic) features (semantic attributes). All features are binary, i.e. they are either present or not (all features can be coded using binary features only). Let $\mathcal{W} = \{w_1, w_2,..., w_i,...\}$ be the (potentially infinite) symbol space; $w_1, w_2,...w_i,...$ are called *words*.

The meaning space $\mathcal{M}$ (the space of *objects* O and their depth $d(O)$) is defined inductively; in the following word "list" denotes a repetition-free set:

1. finite lists $[f_{i1},..., f_{il}]$, $2 \le l \le l_{\max}$ are objects;
2. if $a_1,...a_l$, $2 \le l \le l_{\max}$ are features or already defined objects, i.e. $a_i \in \mathcal{F} \cup \mathcal{M}$, then a list $[a_1,...,a_l]$ is also an object; $d([a_1,...,a_l]) = \max_i(d(a_i))+1$; ($d(f)=0, f \in \mathcal{F}$).

The idea behind the above condition $l \ge 2$ is, that in real word an object can not consist of only one feature or sub-object (such objects were simply renaming a feature or object), e.g. "red" – every object which possesses the attribute "red" possesses also some other attributes (e.g. "round", "big"). Although the size of objects is bounded (agents are not able to perceive however many features, this restriction is an analogue to the limits on humans short-term memory: 5..7 items simultaneously), there still may be an infinite number of objects – objects are compositional, i.e. may contain other objects (in simulations depth of objects was bounded).

Language is a mapping $\lambda : \mathcal{M} \cup \mathcal{F} \rightarrow \mathcal{W}$ with the following properties:

1. $\lambda(f_i) \ne \lambda(f_j)$, if $i \ne j$;
2. $\lambda([f_{i1},..., f_{il}]) = p(\lambda(f_{i1}),...,\lambda(f_{il}),\lambda([f_{i1},...f_{il}]))$,

   $\lambda([O_1,...,O_l]) = p(\lambda(O_1),...,\lambda(O_l),\lambda([O_1,...,O_l]))$,

Here p is any permutation of $l+1$ elements and $\lambda : \mathcal{M} \rightarrow \mathcal{W}$ is the one-to-one *naming* function satisfying:

3. $\lambda([O]) \ne \lambda(f)$ for any $O \in \mathcal{M}, f \in \mathcal{F}$
4. $\lambda(O_i) = \lambda(O_j)$, $O_i, O_j \in \mathcal{M}$ iff objects O_i, O_j are the same (they differ only in ordering of their components).

1.3 Communication act

Above where described properties of the "language at the limit", i.e. language, which emerges in a computational experiment. In the following will be shown how a society of agents $\mathcal{A} = \{A_1,...,A_k\}$, $k \ge 2$ can develop a language with the described above properties. Every agent $A_i \in \mathcal{A}$ maintains its own lexicon $\mathcal{W}^i$ and language $\lambda_i : \mathcal{M} \cup \mathcal{F} \to \mathcal{W}^i \subset \mathcal{W}$ and doesn't know anything about languages of others; in the process they modify their lexicon and languages, so that they become more and more similar and their properties closer and closer to the above described "ideal" properties. At the beginning agents are "tabula rasa", i.e. $\mathcal{W}^i = \varnothing,\ i = 1,...,k$.

The mapping $\lambda : \mathcal{M} \to \mathcal{W}^i$ consists of two sub-mappings. At first agents create names for objects. Names are "ideal" attributes, created and belonging to agent, other agents perceive only the word, which agent created for the name. The naming mapping $\eta : \mathcal{M} \to \mathcal{N}^i$ creates for every object $O \in \mathcal{M}$ name $\eta(\mathcal{O})$ and adds this name to list $\mathcal{N}^i$ of objects and names (for every name is indicated the corresponding object as list of its attributes - features or sub-objects). Thus $\lambda(O) = \lambda(\eta(O))$ for every $O \in \mathcal{M}$. At the beginning agents do not know anything about the environment, i.e. $\mathcal{N}^i = \varnothing,\ i = 1,...,m$. Agent creates name attribute and adds corresponding object to its list of names (agent's model of external word) only when he needs it, i.e. when it is composing a message about a previously unknown object or when receiving a message about some previously unknown object.

Agent's lexicon has a simple structure. Since names are mostly handled the same way as features, in the following the word "attribute" will denote both features and names. An agent A_i maintains for every attribute a list of words W_a^i (when agent's identity is clear from context, notation W_a is used). This list contains all known to him denotations (words) for this attribute. With every word is stored also its use count, i.e. number of previous uses. When agent needs a word for an attribute or name (in step 3), he selects from this list the word with highest use count; if there are yet no words for some attribute, agent "invents" a new word. In [19] were considered also some other tactics of word selection, e.g. agents can "stick" to their "own" word (what they invented earlier) etc.

The language is developed in a sequence of communication *acts* (iterated learning [20]). A communication act consists of following steps:

1. $A_S, A_R \in \mathcal{A}$ - the speaker (sender) and receiver are selected; usually they are selected randomly, but "teaching" (when sender and receiver are fixed) and some hierarchical communication systems were also considered;
2. a communication object $O \in \mathcal{M}$ is selected (randomly); at the beginning are used objects with depth 1 and only gradually are used objects with greater and greater depth (in simulations was depth ≤ 3);
3. speaker A_S creates message, describing the name and attributes of the communication object $O = [a_1,...,a_k]$ using words, which already are in his lexicon (or creates a new word if necessary); if O contains sub-objects, he creates first corresponding to them messages using the same procedure; speaker is allowed not use/describe all the features of the communication object O; (humans also do not consider all the features significant, instead of *big red round ball* we use *big ball*, *red ball* or simply *ball*; some words may be also lost in communication channel); attributes described in the message are selected randomly;

4. receiver A_R gets the created by speaker description of the communication object (list of words) and the communication object is "pointed", i.e. he also receives the communication object, so that a message is a meaning-signal pair $m = <O,[w_1,...,w_n]>$ (this setup follows the gestural theory[21],[22] which sees language as originating from a system of manual gestures); when agents already have learned some words, also "pure" messages $m = <[w_1,...,w_n]>$ without indication of the object are used; "pointing" is an analogue language, thus the process can also be considered as the first step in transformation from analogue language to a discrete one; the two-part messages are mix of analogue and discrete communication needed to create meanings for discrete signals;
5. receiver uses message to improve its lexicon (learn); formally, if $m = <O,[w_1,...,w_n]>$, O=$[a_1,...,a_k]$, $k \geq n$, then
   1. receiver checks, if he already knows the object O, i.e. $O \in \mathcal{N}^R$; if $O \notin \mathcal{N}^R$, he creates a new name $\eta(O)$, adds pair $[\eta(O), O]$ to his lists of names $\mathcal{N}^R$ and creates a new empty wordlist $W^R_{\eta(O)} = [\ ]$ for the new attribute $\eta(O)$;
   2. he finds known words and attributes:
      $\mathcal{W}_1(m) = \{w_i, w_i \in \{w_1,...,w_n\}, \exists a_j \in \{a_1,...,a_k\}, w_i \in W_j^R\}$ (known words);
      $\mathcal{W}_2(m) = \{w_1,...,w_n\} \backslash \mathcal{W}_1(m)$ (unknown words);
      $A_1(m) = \{a_j \in \{a_1,...,a_k\}, \exists w_i \in \{w_1,...,w_k\}, w_i \in W_j^R\}$ (known attributes);
      $A_2(m) = \{a_1,...,a_k\} \backslash A_1(m)$ (unknown attributes);
   3. modifies its lists of words:
      1. for every $a \in A_1(m)$, $w_i \in \mathcal{W}_1(m)$, $w_i \in W_a$ he increases the use count of w_i in W_a by 1; after that he removes from W_a all words, which have very low use count, i.e. words with use count $\leq \max_a - p$, where $\max_a$ is the biggest use count in W_a (at the moment) and p is a parameter (for p values $5..m/4$ were used, where m is the number of agents)
      2. for every $a \in A_2(m)$, $w \in \mathcal{W}_2(m)$ he adds pair $[w,1]$ to W_a (1 indicates, that the word w has been used once)
      3. for all $a \notin \{a_1,...,a_k\}$ he removes all words $w_1,...,w_n$ from all lists W_a

Lemma. The upper bound to agent memory is

$$k * \log_2(n + |\mathcal{M}|) + |\mathcal{M}| * l_{\max} * \log_2(n)$$

i.e. linear in number of objects $|\mathcal{M}|$; here $k = |\mathcal{A}|$ is the number of agents, $n = |\mathcal{F}|$ is the number of features, $|\mathcal{M}|$ is the number of objects and $l_{\max}$ - the maximal number of object attributes.

Upper bound for learning time after receiving a message $m = <O,[w_1,...,w_n]>$ is $c*|O|*n$; $|O|$ is the number of features of the object O.

Proof. Agent's data structures are the lists W_a and the list of names $\mathcal{N}$. Since $a \in \mathcal{F}$ or a is a name of an object, there are $n + |\mathcal{M}|$ attributes, i.e. lists W_a; their length is $\leq k$. In the list of names $\mathcal{N}$ there are $\leq |\mathcal{M}| * l_{\max}$ elements (descriptions of objects); length of description of an object with $\leq l_{\max}$ components is $\frac{n!}{(n-l_{\max}+1)!} \leq n^{l_{\max}}$. The time bound follows directly from the above description of actions in step 5.

2. An Example

1. This example shows first steps in the development of a common lexicon. In the example $\mathcal{F} = \{1,2,3,4\}$, $\mathcal{M} = \{[2,4],[2,3],[3,[2,4]]\}$, $\mathcal{A} = \{A_1, A_2\}$. At the beginning agent's lexicons and lists of objects are all empty.

1. Agent 1 creates a message about the first object:
   $m = <[2,4],[a4\_1, a2\_1, a101\_1]>$
   (word $a4\_1$ denotes attribute 4, word $a2\_1$ - attribute 2 and $a10\_1$ is the name of the object [2,4] in agent 1 lexicon; agent 1 "invented" all those words).
2. Agent 2 receives the message *m*, creates for the object [2,4] new name 101 and modifies its lexicon:
   $W_2 = [2, [[a101\_1, 1], [a2\_1, 1], [a4\_1, 1]]]$
   $W_4 = [4, [[a101\_1, 1], [a2\_1, 1], [a4\_1, 1]]]$
   $W_{101} = [101, [[a101\_1, 1], [a2\_1, 1], [a4\_1, 1]]]$
   i.e. he added to his lexicon all received words as denotations for both attributes 2,4 and also as possible denotations for the name 101 (which he created; that string "101" occurs also in the name a101_1 is purely incidental, agents do not analyse words and handle them as indivisible symbols).
3. Suppose next agent 2 receives from agent 1 message about object [2,3]:
   $m = <[2,3],[a3\_1, a2\_1, a102\_1]>$
   This message already simplifies the lexicon of agent 2, which now becomes:
   $W_{102} = [102, [[a102\_1, 1], [a3\_1, 1]]]$
   $W_2 = [2, [[a4\_1, 1], [a2\_1, 2], [a101\_1, 1]]]$
   $W_3 = [3, [[a102\_1, 1], [a3\_1, 1]]]$
   $W_{101} = [101, [[a4\_1, 1], [a101\_1, 1]]]$
   $W_4 = [4, [[a4\_1, 1], [a101\_1, 1]]]$
   - agent 2 could remove the word $a2\_1$ from the lists W_{101}, W_4
4. Suppose next agent 1 sends message to agent 2 about object [3,[2,4]]:
   $m = <[3,[2,4]],[a3\_1, a103\_1, a4\_1, a2\_1]>$
   After receiving this lexicon of agent 2 contains:
   $W_3 = [3, [[a3\_1, 2], [a102\_1, 1]]]$
   $W_2 = [2, [[a101\_1, 1], [a2\_1, 2], [a4\_1, 1]]]$
   $W_4 = [4, [[a101\_1, 1], [a4\_1, 2]]]$
   $W_{103} = [[103, [[a2\_1, 1], [a4\_1, 1], [a103\_1, 1]]]$

If agent 2 sends now message to agent 1 about whatever object from $\mathcal{M}$ or about a "next level" object (object, obtained composing objects from $\mathcal{M}$), the message will be understood correctly.

3. Emergence of common lexicon and understanding

3.1 Parameters

In computational experiments were followed several parameters of the process:

- how quickly lexicon develops, i.e. for many attributes and objects agents already have a word, how many words are in their lists, how are their use counts distributed - after a while use count of one word (the "right" one) will be much higher;
- for how many attributes and names agents already use the same word;
- how many different most frequent words for every attribute and object are there in the whole society; when speaking, agent uses his most frequent words, thus number of different most frequent words for an attribute or for an object indicates number of synonyms for this attribute or object.

3.2 Emergence of understanding

Understand a word means to pick correctly the corresponding meaning (a feature or an object). To find the corresponding attribute receiver searches his lists of denotations to find the list where the word has highest use count and selects the corresponding attribute as the most probable meaning of the word.

Understanding was measured sending messages without indicating the object (so called "pure" messages). Receiver had to return the object, which he considered most probable. For instance, if agent 2 after the previous example is sent the message [*a*3_1, *a*2_1, *a*102_1] he can correctly calculate the communication object $O = [3, 2]$.

After some number of communication acts agents could understand even messages consisting of object's name only, i.e. messages which did not contain indication to the communication object ("pure" messages); even more, they could understand "pure" messages consisting of only name of the object (without any description, i.e. words referring to attributes of the communication object). The following graph shows percentages of correctly understood pure messages and pure messages consisting of only a name:

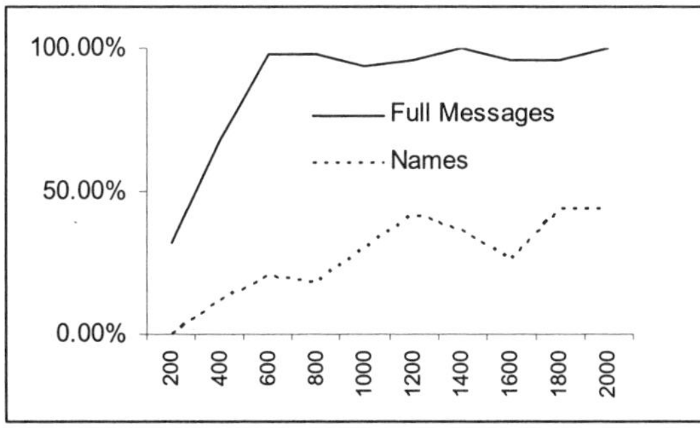

Figure 1. Development of understanding – percentages of correctly understood (full) messages and messages consisting of only object's name; 50 objects, 6 agents, control messages (50 messages) sent after every 200 communication acts.

3.3 How environment's structure influences learning speed

Let us call a meaning space $\mathcal{M}$ *learnable* if there exists society of agents $\mathcal{A} = \{A_1, ..., A_m\}$, $m \geq 2$ for which percentage of wrongly understood messages (without object) can be made however small with increase of number of communication acts; *strongly learnable*, if percentage of wrongly understood words (which may also describe a feature, not an object) can be made however small with increase of number of communication acts. Since our lexicon is grounded, the negative results[23] obtained in formal language learning theory[24] are not relevant here.

Meaning of something (a feature or name) can be found only by comparing this feature or object to other attributes or objects. For this there should be differences in the world, e.g. to distinguish a feature A from a feature B there should be objects, which separate these features, i.e. where in one object occurs A and not B and in other - B and not A. The more there are such separating pairs of objects the quicker the difference of attributes A, B can be learned.

The *indiscernibility relation* [25] ι is defined on the set of features $\mathcal{F}$ by: $(f_1, f_2) \in \iota$ iff $\forall O \in \mathcal{M}\ f_1 \in O \Leftrightarrow f_2 \in O$

Lemma 1. A meaning space is strongly learnable iff the indiscernibility relation ι is trivial (separates all features).

In one direction proof is trivial. If two features are always present together then it is not possible to separate corresponding words. In other direction the mathematical proof seems to be rather complex because of many random moments in the setup (especially the step 5.3.3); however, all computational experiments show that the statement holds.

An example of learnable, but not strongly learnable object space would be e.g. a space consisting of only one object $\mathcal{M} = \{[1, 2]\}$.

Somewhat surprisingly computational experiments show that also the following holds:

Claim. Every meaning space is learnable.

Since "learnable" is in this set-up the limit of a process including many random events, this claim is very difficult to prove with mathematical methods. However, in experiments with many different meaning spaces we were not able to develop a counter-example.

Computational experiments show, that this somewhat surprising claim follows from the ability of agents to produce (potentially) unlimited number of different signals (words). At the beginning of the experiment the difference of signals for different objects and from different agents was introduced to allow following how words created by different agents are "sinking" into the emerging common vocabulary. Later it was understood, that just this was the (or at least one of the) main features why everything seemed to work so fine (i.e. common lexicon emerged, agents become able to understand "pure" messages etc). Suppose agents were able to produce only a limited number of different words. In this case the "semantic load" (number of meanings) for every word will be so big, that the emergence of singular meanings will be impossible or at least very slow. It is obvious, that in the extreme case when agents are able to produce only one (or a finite number) of signals they can not separate (potentially unlimited) number of real-word meanings. Necessity of sufficiently developed communication (speech) organ (i.e. able to produce unlimited number of signals) for language emergence has been stressed also in human language emergence studies, see e.g. [26], also the Chomsky's "property of discrete infinity"[27].

In order to investigate, how structure of objects influences speed of emergence of common lexicon, a *separation coefficient* of objects was introduced (something opposite to indiscernibilty and tolerance relations, studied in rough set theory, see e.g. [28]):

$$s(O_1,O_2) = |(O_1 \backslash O_2) \cup (O_2 \backslash O_1)|$$

i.e. the number of features which occur only in one of them. Separation coefficient of an object:

$$s(O) = \sum s(O,O_i),\ O_i \in \mathcal{M}$$

The separation coefficient characterises rather well the "difficult" attributes, for which the common word emerges much slower than for other attributes; in the following table is shown number of different words with the highest use count (10 objects, 6 agents) in one of experiments:

Table 1. Object separation coefficient and number of words per object

O	1	2	3	4	5	6	7	8	9	10
s(O)	528	590	568	346	346	612	582	628	346	558
# of words	1	1	1	6	6	1	1	1	4	1

4. Speed of emergence of common lexicon

New information is added to agent knowledge base only after receiving a message, when agent is trying to find correct attributes for words from received message. Therefore it is natural to assume, that the speed of learning depends only from the total number of word guesses and does not depend e.g. on whether an agent was taught (supervised learning, roles of listening/speaking fixed) or language was emerging in a series of communication acts, where roles of speaker/listener were selected randomly (unsupervised learning). To make the number of word guesses equal (for an agent), the number of tests should also be correspondingly changed. For instance, if an agent is taught using *N* series of tests, where every test consists of *K* communication acts, then for 2 agents the unsupervised learning experiment should also consist of *N* series, but already with $2*K$ communication acts in every series and if instead of two are used *m* agents, $m > 2$, then the number of tests in every series should be $2*K*m$.

Experiments show, that with these corrections to number of tests in series the speed of developing understanding is approximately equal and does not depend whether the unsupervised or supervised learning was used and how big was the group of agents; in bigger group the distribution is also bigger, but with more tests the speeds become more and more similar.

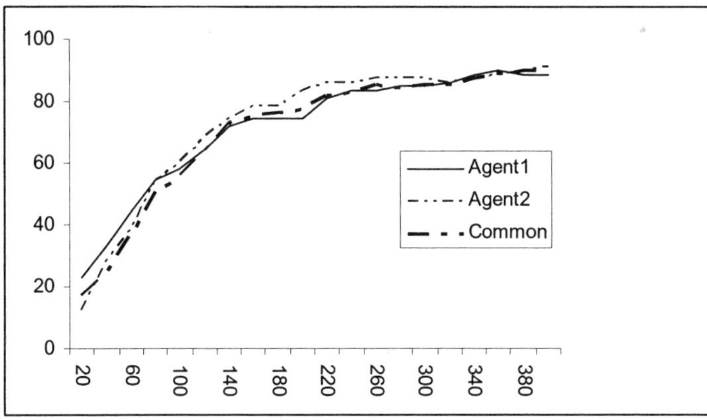

Figure 2. Emergence of common words in series of communication acts between two agents (roles of speaker/listener selected randomly), altogether 400 communication acts (10 attributes, 80 objects), percentage of common words was measured after every 20 acts.

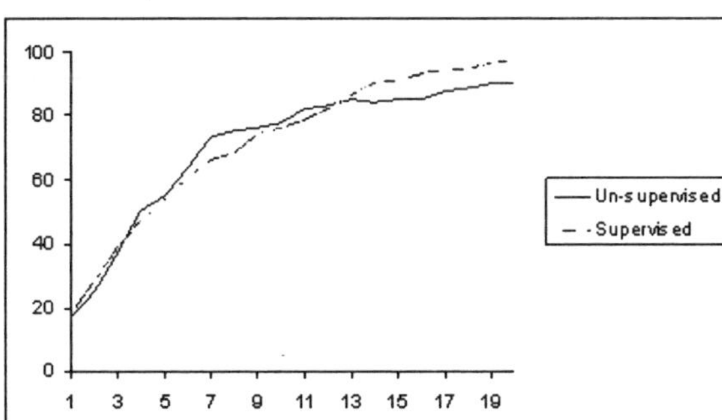

Figure 3. Emergence of common words 1) in unsupervised learning (roles of speaker/listener were selected randomly) and 2) in supervised learning (roles of speaker/listener fixed). To make the number of world guesses approximately equal, the number of tests (between measurements) was in the second test reduced to half (20 in the first test, 10 in the second). In both tests was used the same set of 79 objects (over 10 attributes). The speed of developing understanding is approximately the same.

Somewhat unexpected was, that understanding of "pure" messages (also messages containing only a name) developed much quicker than common vocabulary:

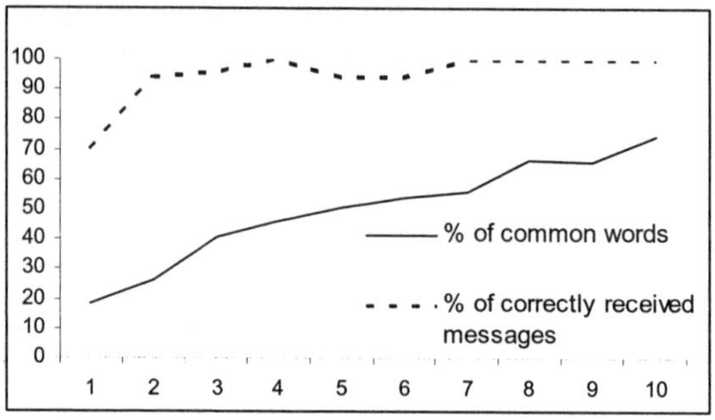

Figure 4. Development of percentage of common words (i.e. both agents used the same word for an attribute or object) and percentage of correctly understood messages from a series of 50 test messages after 10000 communication acts (10 agents, 20 attributes).

This phenomenon has been noticed also in studies concerning language learning of children: "what children *say* ...is a quite imperfect measure of what they *know*"[29].

Self-learning

Most of the words what we know and use are not taught to us. We have learned them from context - picked them up from conversations, from texts etc [30], [31].

Since agents are after some training able to understand pure messages (messages without indication of communication object), they can also learn from pure messages (messages without object).

Pure messages were used only after some initial period of exchanging normal full messages. Most of them unify vocabularies, i.e. increase the number of attributes, for which all agents use the same word. For instance, when 3 agents learned four objects (over four attributes) in 5 test series of unsupervised learning, 10 communication acts (with rich messages) in every series, then after 50 communication acts they had four attributes (two proper attributes and two names) for which they used more than one word. After that was arranged 50 self-learning acts (i.e. sender and receiver selected randomly, but messages were sent without communication act). At the end of this second series there was only one attribute and one object, which had synonyms.

Below are results from a bigger test.

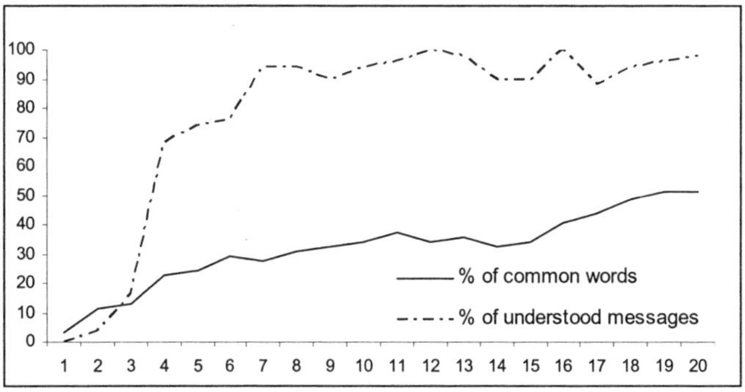

Figure 5. Emergence of understanding (% of correctly understood words): 1) supervised learning, 20 series of 10 tests; 2) unsupervised learning, 20 series of 20 tests; 2 agents; 3) unsupervised learning, 20 series of 120 tests, 6 agents. In all tests was used the same set of 70 objects (10 features).

In the above graph are results of test, where 10 agents learned 50 objects (over 12 attributes). The first 10 series contained 100 full messages in every series, the next 10 series were self-learning - 100 pure messages in every series. At the end of every act was for a pair of agents measured % of common words (i.e. attributes and names, for which they used the same word when speaking) and % of correctly understood messages (from series of 50 test messages). The percentage of common words grows with pure messages nearly the same way as with full messages.

5. Vocabulary spurt

Many specialists in child learning claim, that at about 16 months of age, or when a child learns about 50 words children word learning sharply accelerates [32]; this is called word burst, word spurt, vocabulary burst, naming explosion, word explosion, and so on (some researchers are skeptical[29]). It is confusing, what actually means "learn a word", since children know much more words than they actually produce in messages. In this experiment the main parameters followed were the number of words in common and percentage of correctly understood messages. While the number of words in common changed all the time more or less the same way, the percentage of correctly understood messages really increased sharply after some initial period of learning. Below are plotted percentages of correctly understood messages in series of communication acts with different number of objects (12 agents):

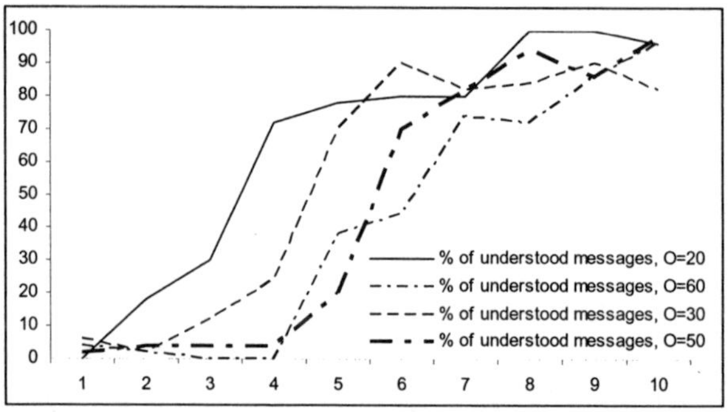

Figure 6. Acceleration of learning (vocabulary spurt) in different experiments.

Similar acceleration in percentage of correctly understood messages can be seen also on several earlier graphs, so agents did have some kind of "vocabulary burst".

6. Conclusions

Here was considered the process of emergence of discrete language from analogy-based language (pointing, mimics, "body language", facial expressions" etc) on assumption that agents are able to produce sufficiently large number of different signals. Only the very first step in the development of discrete language was considered – emergence of vocabulary for denoting single objects and invention of names for objects. Agents followed a very simple algorithm based on noticing differences both in environment (the meaning space) and in received signals. They developed rather quickly a common vocabulary and the process shows several features similar to children's word learning: understanding develops quicker than vocabulary; there is some kind of "vocabulary spurt" etc. After some preliminary training period with mixed (analogue and discrete, i.e. language messages) agents were able to continue learning using discrete (language) messages only. Naming objects seems to induce also creation of "inner model" of outside word – when agents do not create names, "inner model" is not needed.

References

[1] Language Evolution. Morten H. Christensen and Simon Kirby, eds. Oxford University Press, 2003

[2] M. H. Christiansen, S. Kirby. Language Evolution: The Hardest Problem in Science?. In: Language Evolution, M.H.Christiansen, S. Kirby (Eds.), Oxford University Press 2003, pp 1-16

[3] J. Batali. Computational simulations of the emergence of grammar. In: J. Hutford, C. Knight, M. Studdert-Kennedy (Eds), Approaches to the Evolution of Language: Social and Cognitive bases, Cambridge University Press, 1998, 405-426

[4] J. Hurford. Social transmission favours linguistic generalisation. C. Knight, J. Hurford,, M. Studdert-Kennedy (Eds.), The emergence of language, 1998

[5] W.H. Zuidema, Emergent syntax: the unremitting value of computational modelling for understanding the origins of complex language.
http://citeseer.nj.nec.com/cache/papers/cs/22259/http:zSzzSzarti.vub.ac.bezSz~jellezSzresearchzSzecal_zuidema.pdf/zuidema01emergent.pdf

[6] L. Steels. The synthetic modeling of language origins. Evolution of Communication, 1:1 (1997), 1-34.

[7] L. Steels, F. Kaplan. Bootstrapping grounded word semantics. In: T. Briscoe (Ed), Linguistic Evolution through Language Acquisition: Formal and Computational Models. Cambridge University Press, 2002
[8] Kirby, S. Natural Language from Artificial Life. Artificial Life, 8:2 (2002), 185-215.
[9] N.L. Komarova, M. A. Nowak. Language, Learning and Evolution. In: Language Evolution, M.H. Christiansen, S. Kirby (Eds.), Oxford University Press 2003, 317-337
[10] R.J. Sternberg. Most vocabulary is learned from context. In: M.G. McKeown & M.E. Curtis (Eds.) The nature of vocabulary acquisition. Hillsdale, NJ: Erlbaum, 1987
[11] Nagy, W.E. & Herman, P.A. (1987). Breadth and depth of vocabulary knowledge: implications for acquisition and instruction. In M.G. McKeown & M.E. Curtis (Eds.) The nature of vocabulary acquisition. Hillsdale, NJ: Erlbaum
[12] D. Klein, C. D. Manning. Natural Language Grammar Induction using a Constituent-Context Model. In: Thomas G. Dietterich, Suzanna Becker, Zoubin Ghahramani (Eds). Advances in Neural Information Processing Systems 14 (NIPS 2001), vol. 1, 35-42
[13] G. Frege. Die Grundlagen der Arithmetik. Eine logisch-mathematische Untersuchung über den Begriff der Zahl. Breslau: W. Koebner. Reprint published by Georg Olms, Hildesheim, 1961.
[14] Gelder, T. van (1990). Compositionality: A connectionist variation on a classical theme. Cognitive Science, 14, 355-384.
[15] K. Smith, H. Brighton, S. Kirby. Language Evolution in a Multi-agent Model: the cultural emergence of compositional structure. http://citeseer.nj.nec.com/533946.html
[16] T. M. V. Janssen. Compositionality. In: Handbook of Logic and Language, Elsevier, Amsterdam; MIT Press, Cambridge, J. Van Benthem & A. ter Meulen (Eds), 1997, 417-473
[17] Noam A. Chomsky. Deep structure, surface structure, and semantic interpretation. In: Danny D. Steinberg and Leon A. Jacobovits, editors, Semantics: An interdisciplinary Reader in Philosophy, Linguistics, and Psychology, pages 183-216. Cambridge University Press, Cambridge, 1970.
[18] Edward P. Stabler, Travis C. Collier, Gregory M. Kobele, Yoosook Lee, Ying Lin, Jason Riggle, Yuan Yao, Charles E. Taylor. The learning and emergence of mildly context sensitive languages. Proceedings of the European Conference of Artificial Life (ECAL) 2003; http://ludoviko.linguistics.ucla.edu/stabler/loop.pdf
[19] J. Henno. Emergence of communication and creation of common lexicon in multi-agent environment. Proceedings of the 12th European-Japanese Conference on Information Modelling and Knowledge Bases. Krippen, Swiss Saxony, Germany. May 27-30, 2002, pp 229-233
[20] Kenny Smith, Simon Kirby, Henry Brighton. Iterated Learning: a framework for the emergence of language. Artificial Life, 9: 4 (September 2003); citeseer.nj.nec.com/565681.html
[21] Michael A. Arbib. The Evolving Mirror System: A Neural Basis for Language Readiness. In: Language Evolution, Morten H. Christiansen, Simon Kirby, eds. Oxford University press 2003, pp 182-200
[22] Michael C. Corbballis. From hand to Mouth: The Gestural Origins of Language. In: Language Evolution, Morten H. Christiansen, Simon Kirby, eds. Oxford University press 2003, pp 201-218
[23] L. G. Valiant. A theory of the learnable. Communication of the ACM (1984), 1134-1142
[24] E. M. Gold. Language Identification in the Limit. Information and Control 10 (1967), 447-474
[25] E. Orlowska. Algebraic approach to database constraints. Fundamenta Informatica 10 (1987) 1, 57-68
[26] Studdert-Kennedy, M., Goldstein, L. Launching Language: The Gestural Origin of Discrete Infinity. In: Language Evolution, M.H.Christiansen, S. Kirby (Eds.), Oxford University Press 2003, pp236-254
[27] Chomsky, N.A., New horizons in the Study of Language and Mind. Cambridge, Cambridge University perss, 2000
[28] Komorowski, J., Pawlak, Z.,Polkowski, L., Skowron, A. Rough Sets: A tutorial. http://citeseer.nj.nec.com/komorowski98rough.html
[29] Bloom, P. Myths of word learning. In: D.G. Hall & S.R. Waxman (Eds.) Weaving a lexicon, Cambridge, MA; MIT Press
[30] Sternberg, R.J. (1987). Most vocabulary is learned from context. In M.G. McKeown & M.E. Curtis (Eds.) The nature of vocabulary acquisition. Hillsdale, NJ: Erlbaum.
[31] Nagy, W.E. & Herman, P.A. (1987). Breadth and depth of vocabulary knowledge: implications for acquisition and instruction. In M.G. McKeown & M.E. Curtis (Eds.) The nature of vocabulary acquisition. Hillsdale, NJ: Erlbaum
[32] Dromi, E. Early lexical development. Cambridge University Press, 1987.

Entity-Relationship Models recast as Grammars

Hans Bruun & Jørgen Fischer Nilsson

Informatics and Mathematical Modelling
Technical University of Denmark

Abstract: Entity-relationship models have achieved widespread use with ER-diagrams serving as informal or semi-formal tools for conceptual modelling. As a supplement to the standard set theoretical understanding of the diagrams in the paper ER-diagrams are reconstructed and analysed as a restricted form of BNF-grammars, where entity-classes form non-terminals and the relationships of the diagram become production rules. Emphasis is put on the ISA inclusion relationships which are to be handled as inverse of the syntactical derivation relation. The generative abilities of grammars then account for the inheritance in ISA-hierarchies, which is crucial in taxonomic specifications such as in ontologies.

Keywords: Entity-relationship models, grammars, ontologies.

CR categories: H.2.1: Data models; I.2.4:Representation languages; F.4.3: Formal languages.

1 Introduction

The entity-relationship model (ER-model) introduced in [Chen 1976], see also [Chen 1997], is a graph-oriented data model presentation, which has proven useful in the contexts of conceptual modelling and in database design. In this paper we have in mind the model also from the perspective of ontology engineering, where emphasis is put on taxonomies. For a contemporary informal presentation of ER-diagrams see e.g. [Garcia-Molina *et al.*]. A formal exposition is found in [Thalheim]. Similarly the conceptual graph model proposed independently [Sowa] has gained much interest in the area of formal knowledge representation and reasoning. These formalisms owe much of their appeal to their diagrammatic forms, which makes them handy for practitioners.

We are concerned with the relationship between these diagrammatic formalisms and a well-known theoretical framework in computer science, namely context-free formal languages. The purpose of this study is firstly to elucidate the formal underpinnings of the entity-relationship modelling tool from other viewpoints than the prevailing set and logic comprehension. The present study serves to transfer theoretical insights and results to practitioners applying the model.

The paper is structured as follows: After a review of the basics of ER-diagrams, sect. 3 presents a production rule view on ER-diagrams. Finally, section 4 mentions an application of these views to ontologies.

2 Entity-Relationship Models - *apercu*

ER-Models take form of ER-diagrams as described below. An ER-model comprise entity types or classes ***C*** and relationships ***R***.

2.1 ER-triplets

The building blocks of ER-diagrams may be conceived as triplets depicted e.g. as

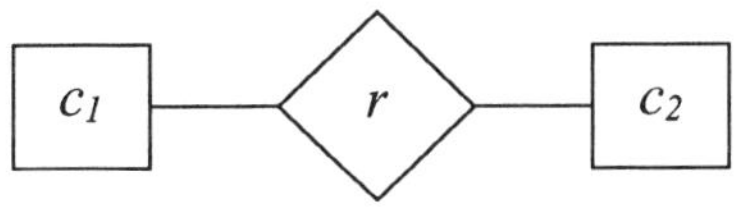

or

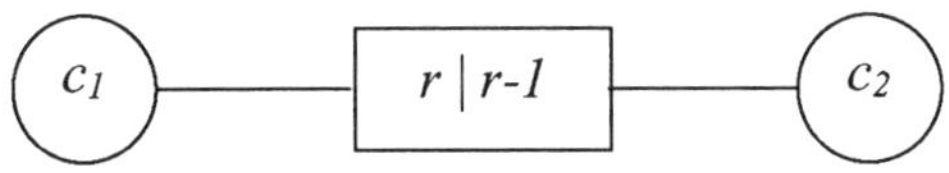

where c_i are entity-classes denoting sets of entities E_i, for short just classes. The relator r is to indicate the existence of a relationship between the relata being entity classes c_1 and c_2.

We distinguish entity classes c_i from their contained set (their extension) E_i since two classes may contain the same set of entities, yet being considered distinct. Thus diagrams are intensional, implying that co-extensionality does not imply identity of classes.

The triplet specifies that a relation (in the mathematical sense of a set of ordered pairs) exists which is relevant to the model. Thus if E_1 and E_2 are respectively the set of entities in c_1 and c_2 in an extensional view the (possibly time dependent) relation r belongs to the power-set

$$2^{E_1 \times E_2} \text{ that is } r \subseteq E_1 \times E_2$$

This means generally speaking that r, strictly, is a relation between <u>entities</u> rather than a relation between entity <u>classes</u>. More is to come concerning this point.

The second diagram-form of the triplets above also leaves room for stating the inverse relationship bound to exist mathematically, which as default is termed *r-1* here for the relation r. It may, however, be assigned a distinguished name.

Chiefly for the sake of simplicity in the paper we choose to consider binary relationships, only. However, it should be born in mind that an n-ary relationship between n (not necessarily distinct) classes may be replaced by binary relations at the expense of an extra entity class becoming related to the said classes through n new binary relationships, cf. e.g. [Garcia-Molina *et al.*].

2.2 ER-diagrams

ER-diagrams are constituted by triplets composed to form a finite, directed graph. The classes c_i appear as nodes and relationships r_i are labelled arcs. The example consists of 2 triplets sharing an entity class:

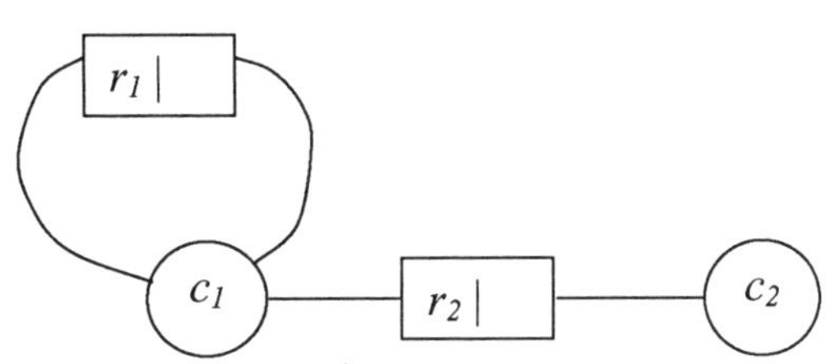

As indicated the diagram may contain cycles. Also, there may be multiple arcs between any pair of classes provided that they are labelled with distinct relations.

Relationships may possess (be imposed) functional properties (understood as constraints on the relations rather than an accidental occurrence of functionality). The functional properties may be indicated with numbers as in:

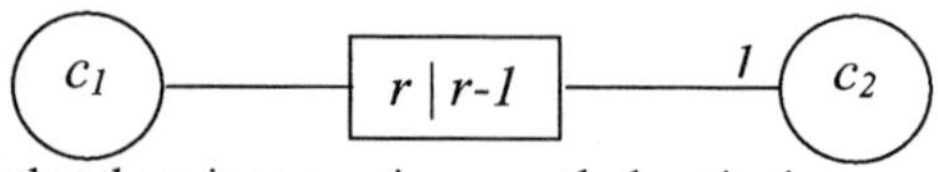

This triplet declares that there is at any time exactly 1 entity in c_2 associated with any entity in c_1. Thus r is a total function from c_1 to c_2. Further the interval mark 0..1 might be used to tell that r is a partial function.

2.3 Class Inclusion (ISA relationships)

There is a distinguished entity-class relationship, the ISA relationship, depicted as shown to the right, and thus devoid of explicit labelling. This class inclusion relationship states that the entities of c' (the sub-class) forms a subset of the entities of c'' (the super-class):

$$E' \subseteq E''$$

The ISA relationships form a (weak) partial order on all the entity classes in the diagram, since ISA is antisymmetric and transitive:

- if c_1 ISA c_2 and c_2 ISA c_1 then $c_1 = c_2$
- if c_1 ISA c_2 and c_2 ISA c_3 then c_1 ISA c_3

Arcs following from transitivity are usually not drawn:

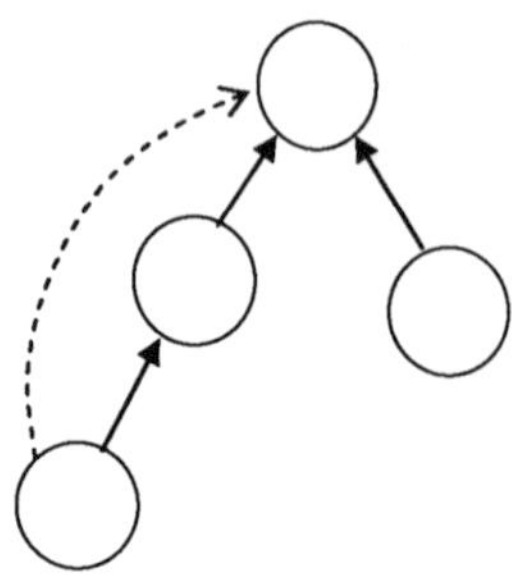

Since there are no cycles present in partial orders there is a preference for consistent upwards orientation of the ISA arrows as customary in classifications, cf. ontologies, as shown to the right.

The partial order may preferably be completed with introduction of a special universal class $\top$. Extensionally this class comprises the totality of entities. All entity classes are born sub-classes of $\top$.

Conventionally, ISA besides class inclusion also serve to express membership of an individual in a class. This membership may be handled as class inclusion by considering individuals as a singular class, i.e. a class whose extension is a singleton set.

One may observe that in contrast to the original exposition of ER-diagrams in [Chen 1976] in the present view the ISA-relationships methodologically form the backbone of the diagrams, which are then extended with application-specific relationships. This disposition becomes apparent in sect. 3 and 4. In particular in ontologies the ISA is the *sine qua non* relationship, whereas in traditional data base modelling it may be insignificant.

One should also observe that the ISA relationship is a relation between <u>classes</u> i.e. a subset of $\boldsymbol{C} \times \boldsymbol{C}$. By contrast, as observed above, an ordinary relationship r between say c_1 and c_2 is a relation between the <u>entities</u> of these classes.

2.4 Example ER-diagram: Kinship model

The above discussed notions in ER-diagrams are exemplified with the below diagram in which parents appear as a sub-class of persons. The diagram tells that there is a subclass of persons being parents. That is to say that they stand in a father or mother relationship to one or more persons. Thus the diagram has a recursive (cyclic) structure through the parent class. A person is recorded as having 0 or 1 maternal parent and 0 or 1 paternal parent -- since the parents may not be known.

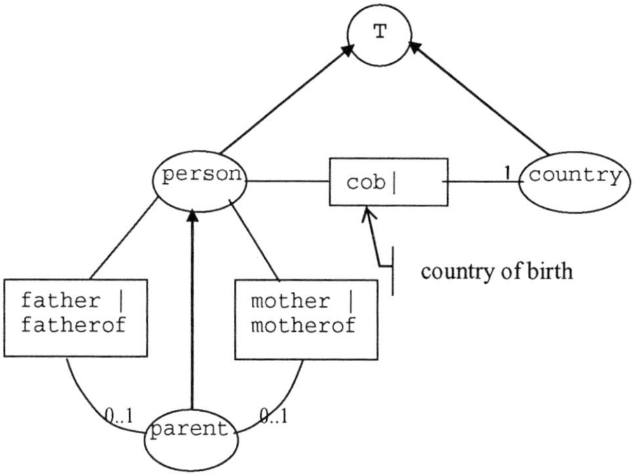

2.5 Non-Hierarchical Inclusion

Since the ISA relation is a partial order the inclusion structure need not be hierarchical. The below leftmost figure introduces a cross-class c_{12} being extensionally the overlap of c_1 and c_2.

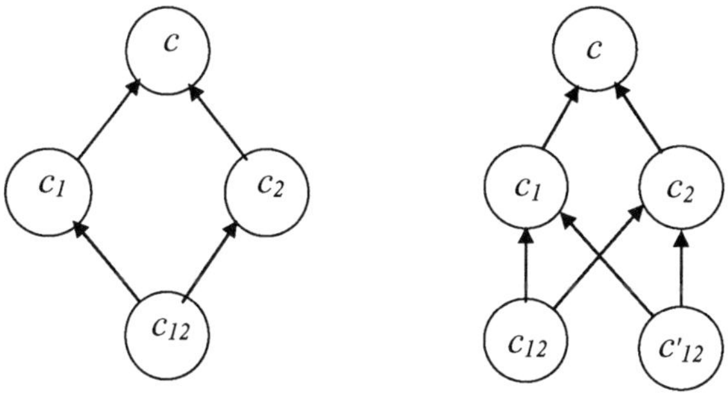

As examples of such cross-classes one may model for instance the class boy as the cross classes of male and child, these in turn being subclasses of person.

There may be more than one cross-category as shown in the rightmost figure: The categories c_{12} and c'_{12} are both cross-categories of c_1 and c_2. Absence of any cross-class for two classes by default means disjointness of the classes.

2.6 Attributes in ER-models

In ER-models entity classes may be adorned with attributes. In the figure the entities in the class c are now equipped with an attribute A.

Informally this is taken to mean that each entity in c comprise or is adorned with an A labelled value. In general there may be a bundle of attributes associated with the (members of) a class. However in the present context attributes are reconceived as functional relationships *cum* attribute-valued entity classes:

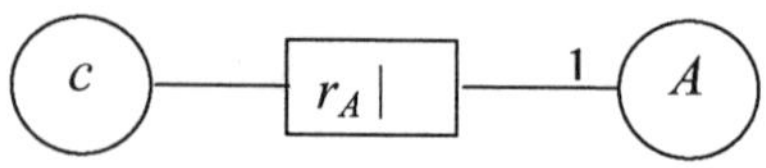

Thus adornment of attributes is conceived as triplets. The attribute class A often comprises concrete data entities such as identifiers or numbers.

This reduction of attribute ascription significantly simplifies the following formalisation without essential loss of generality. The key point is that the reduction supports an understanding of entity classes as composed of non-decomposable abstract entities (individuals possessing attribute-values by way of distinguished attribute-relationships). This is in contrast to the formalisation in [Thalheim] in which entities are modelled as compound objects with attributes.

3 ER-diagrams Formalised as Grammars

This section presents a formalisation of ER-diagrams as BNF grammars, serving the following purposes:

- Grammars facilitate a textual (i.e. 1-dimensional) specification of ER-diagrams.
- The syntactic conception of diagrams opens for ER-diagrams as a generative formalism for production of compound concept descriptions as to be explained.
- A rule-based understanding of the ER-model prepares for a subsequent logico-algebraic analysis of the ER-model, e.g. using description logic.

The production rule view on diagrams is of value in particular in ontologies as discussed in [Andreasen&Nilsson]; see also [Spyns et al.] for a general discussion of data models vs. ontologies.

3.1 Classes and Relationships as Production Rules

In our proposal classes become non-terminals in the grammar (with the universal class $\top$ becoming the start symbol). An immediate ISA relationship gives rise to a production rule. Accordingly the first diagram from sect. 2.3 becomes

$$c'' ::= c'$$

The non-hierarchical diagram from 2.5 similarly becomes

$$c ::= c_1 \mid c_2$$
$$c_1 ::= c_{12}$$
$$c_2 ::= c_{12}$$

Thus c_{12} can optionally be derived from c via c_1 *or* c_2, cf. also sect. 3.4 below.

The more general principle for production rules appear from the following representative diagram pattern (where c_1' and c_2' in particular may be the ever present universal class $\top$):

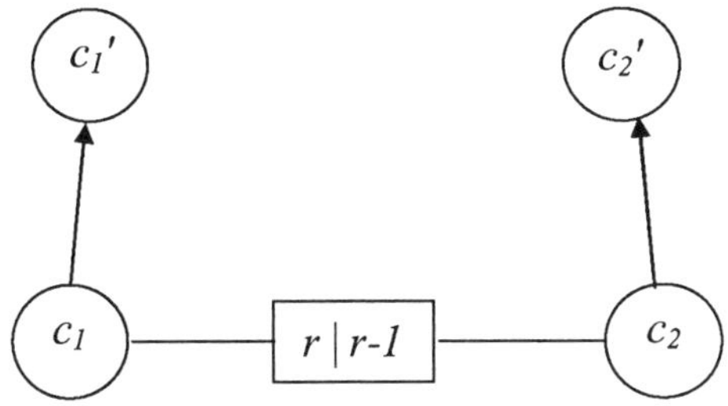

This diagram (fragment) gives rise to the following pair of complementary rules:

$$c_1' ::= c_1 \{ (r: c_2) \}$$

$$c_2' ::= c_2 \{ (r\text{-}1: c_1) \}$$

where the metasymbols *{...}* represent 0, 1 or more occurrences of enclosed symbols, so that the former rule admits production of c_1 from c_1' with as well as without attached relational connection $(r: c_2)$.

Relationships which are total functions, viz.: force mandatory presence of attached relationship:

$$c_1' ::= c_1 (r: c_2)$$

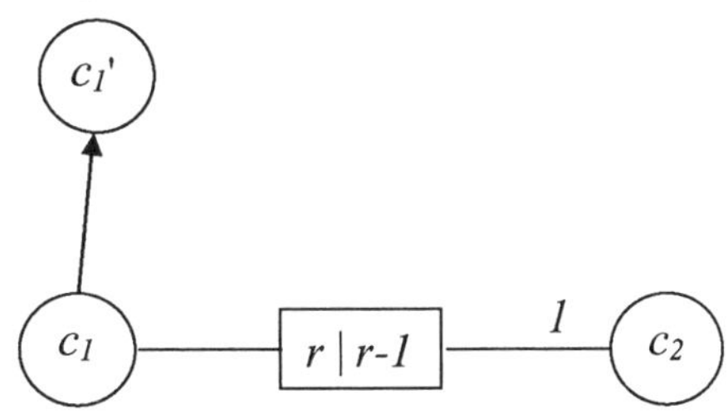

Similarly the metasymbols *[...]* representing optional occurrence of enclosed symbols can be used for indicating 0..1 adorned relations.

3.2 Multiple Relationships

More generally the bundle of relationships shown on the figure to the right yields (seen from *c*) the production rule

$$c' ::= c \begin{Bmatrix} (r_1 : c_1) \\ (r_2 : c_2) \\ \ldots \\ (r_n : c_n) \end{Bmatrix}.$$

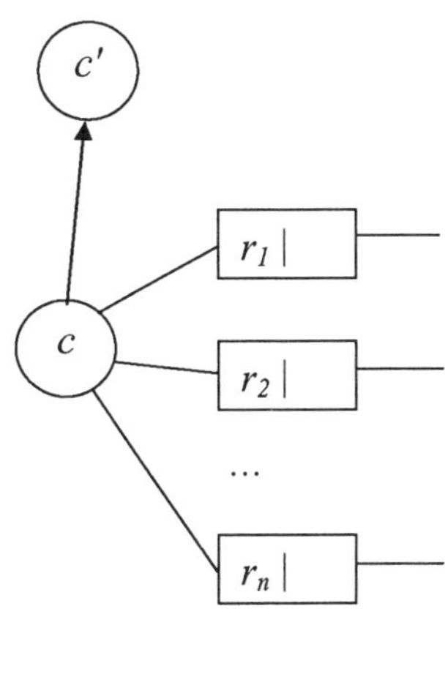

Thus the grammar serves as a link between the ER-graph representation and a feature structure representation offering optional as well as mandatory entries.

3.3 Grammatical Derivation and the ISA Relationship

According to the above principles the grammatical derivation relation becomes the inverse of the ISA relationship. The syntactical (multistep) derivation

$$A \rightarrow + \; B$$

conforms with the inclusion relationship *B* ISA *A,* recalling the transitivity property. Thus syntactic derivations express concept specialisations into sub-classes.

The ISA relation being a partial order there are no self-derivations:

$$c \rightarrow + \; c$$

However, the symbols of ER-classes may still be recursive via relations as in the production

$$c' ::= c \; [(r : c)]$$

as illustrated in the figure to the right:

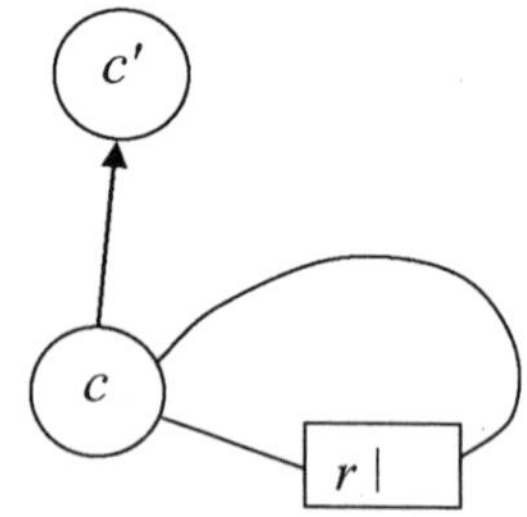

Presence of recursive symbols (corresponding to cycles in the ER-diagram) renders the ER-grammar generative in the sense that infinitely many sentential forms are producible. The producible sentential forms represent (compound) concept descriptions inherent in the ER-diagram as exemplified in sect. 3.6.

3.4 Inheritance of Relationships

It is a crucial feature that relationships for a class *c* are inherited to all sub-classes of *c* by the syntactic derivation mechanism as indicated in the diagram:

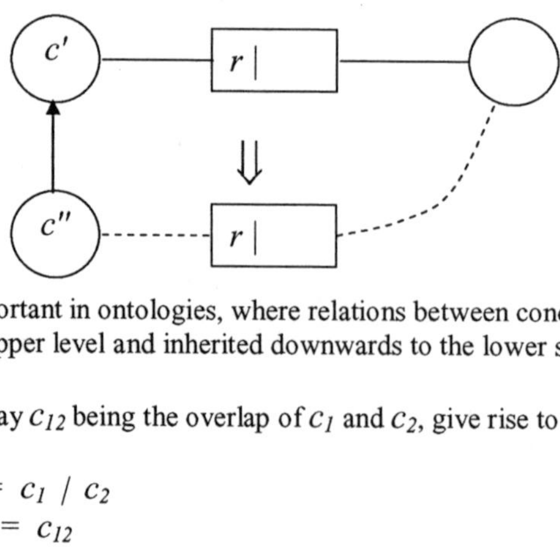

This feature is important in ontologies, where relations between concepts may be introduced at the upper level and inherited downwards to the lower strata in the ontology.

Cross-categories, say c_{12} being the overlap of c_1 and c_2, give rise to ambiguous grammars as in

$$c ::= c_1 \; / \; c_2$$
$$c_1 ::= c_{12}$$
$$c_2 ::= c_{12}$$

The relationships inherited to c_{12} depends on the derivation being via c_1 or c_2.

3.5 *Kinship* Example Revisited

The ER-diagram of sect. 2.5 gives rise to the following production rules

$\top$ ::= *person (cob : country) [(father : parent)] [(mother : parent)]*

$\top$ ::= *country [(cob-1 : person)]*

person ::= parent [(fatherof : person)] [(motherof : person)]

where *fatherof* as it appears is the inverse relation of *father* and analogously for mother.

The constraint that a human can not be both a father and a mother might be expressed by replacement of the third production rule with the rules

person ::= parent [(fatherof : person)]

person ::= parent [(motherof : person)]

3.6 Producing Sentential Forms

The sentential forms producible by the grammar represent concepts supported by the ER-diagram. Assume that *country* contains individuals such as Denmark, Sweden etc. through

country ::= Denmark / Sweden / ...

and that *person* is similarly made to comprise individuals Peter, John etc. Then the grammar specification produces concept descriptions in the form of sentential forms such as

person (cob : Denmark) (father : parent (cob : Sweden))

in this case specifying persons born in Denmark whose father is born in Sweden. The derivation process in the grammar provides limited reasoning (subsumption) with such descriptions. Observe that the producible set of such descriptions is unlimited once recursion is admitted.

The derivation is as follows

$\top$ → *person (cob : country) (father : parent)*
→ *person (cob : Denmark) (father : parent)*
→ *person (cob : Denmark) (father : parent (cob : country))*
→ *person (cob : Denmark) (father : parent (cob : Sweden))*

As it appears we assume a distinguished derivation relation in which a non-terminal *N* (viz., *parent*) is attached all the relationships which are specified on a derivation path from $\top$ to *N*. In other words, if a derivation exists from $\top$ to a sentential form having *N* as a prefix:

$$\top \rightarrow^{+} N\,(r_1 : c_1)\ \ldots\ (r_j : c_j)$$

then we have an implicit production rule

$$(r : N\,) ::= (\,r : N\,(r_1 : c_1)\ \ldots\ (r_j : c_j)\,)$$

3.7 Relationships between Relations

The grammatical framework may readily be extended to account for relationships between relations in the diagram (in addition to the inverse relationships of sect. 2.1) by means of production rules for relations. This can be exemplified in the context of the above example with the rule *parentof ::= fatherof | motherof*

which introduces a novel binary relation *parentof* as the "union" as it were (cf. super class) of two given ones.

4 Application to Ontologies

We are interested in applying ER models for specification of formal ontologies – general purpose top ontologies (cf. e.g. [Sowa 2000]) as well as application-specific ontologies.

In our approach to formal ontologies [Andreasen *et al.*,Andreasen&Nilsson] the ISA structure forms the backbone of an ontology extended with partitive, causative, locative, temporal etc. relationships. The ontology is specified as a grammar (cf. sect. 3) giving rise to a novel notion of *generative ontologies* [Nilsson 2002].

In this view concept descriptors come about in a syntactical derivation process starting with the most general top categories and stepwise specialising sentential forms, as in

$\top \rightarrow$ *state* $\rightarrow$ *state (causedby : state)* $\rightarrow^{+}$

symptom (causedby : state) $\rightarrow^{+}$

symptom (causedby : lack (wrt :stuff)) $\rightarrow$

symptom (causedby : lack (wrt : vitamin)) $\rightarrow$

symptom (causedby : lack (wrt :vitaminC))

representing a path down through the ontological ER-diagram.

The formal language framework and in particular the relation-algebraic frameworks of sect. 3 and 4 facilitate logical analysis of the semi-formal ER diagrams. For instance an account of relationships between entire ER-diagrams may take advantage of notions such as sub-algebras and homomorphisms between algebras.

5 Summary

We have compared ER-diagrams with formal grammar formalisms and with relational algebraic logic. The production rules of BNF grammars offer a linear formalism for representing the ER-diagrams, and in addition the generative capabilities facilitate use of ER-diagrams for ontologies. Relational algebraic logic in the form of distributive lattices provides reasoning capabilities for ER-diagrams through axioms and in addition offers an elegant set-theoretic underpinning of ER-models.

Acknowledgement: The research conducted in the OntoQuery project [Andreasen *et al.*, OntoQuery] concerning ontological representation of the conceptual content of nominal phrases for the purpose of text retrieval has been the prime source of inspiration for the present paper. OntoQuery is supported 1999-2004 with a grant from the Danish National Science Foundation under the interdisciplinary informatics programme.

References

Andreasen,T., Nilsson, J. Fischer, & Thomsen, H. Erdman: Ontology-based Querying, in H.L. Larsen *et al.* (eds.), *Flexible Query Answering Systems, Recent Advances*, Proceedings of the FQAS'2000 conference, Physica-Verlag, Springer, 2000.

Andreasen,T. and Nilsson, J. Fischer: Grammatical Specification of Domain Ontologies, forthcoming in *Data & Knowledge Engineering*.

Baader, F., McGuiness, D.L., Nardi, D., Patel-Schneider, P.F.: *Description Logic Handbook*, Cambridge University Press, 2002.

Braüner, T., Nilsson, J. Fischer & Rasmussen, A.: Conceptual Graphs as Algebras -- with an Application to Analogical Reasoning, in Proceedings of the 7th Int. Conf. on Conceptual Structures, Blacksburg, Virginia, July 1999, W. Tepfenhart & W. Cyre (eds.*), Lecture Notes in Artificial Intelligence LNAI 1640*, Springer, 1999.

Brink, C., Britz, K., and Schmidt, R.A.: Peirce Algebras, *Formal Aspects of Computing*, Vol. 6, 1994, pp. 339-358.

Chen, P.P-S.: The entity-relationship model: toward a unified view of data, *ACM Trans. on Database Systems*, 1:1, pp.9-36, 1976.

Chen, P.P-S..: English, Chinese and ER diagrams, *Data & Knowledge Engineering*, 23, 1997. p. 5-16.

Davey, B.A. & Priestley, H.A.: *Introduction to Lattices and order,* Cambridge University Press, 1990.

Garcia-Molina, H., Ullman, J.D., & Widom, J.: *Database Systems: The Complete Book*, Prentice Hall, 2002.

Nilsson, J. Fischer: Are there Conceptual Grammars ?, Panel abstract in Proceedings from the 11th European-Japanese Conference on Information Modelling and Knowledge Bases, Maribor, Slovenia, 2001, reprinted in *Information Modelling and Knowledge Bases*, XIII IOS press, 2002.

OntoQuery: Project net site, http://www.ontoquery.dk

Schmidt, G. & Ströhlein, T: *Relations and Graphs, Discrete Mathematics for Computer Scientists*, Springer-verlag, 1993.

Sowa, J.F.: *Knowledge Representation, Logical, Philosophical, and Computational Foundations*, Brooks/Cole Thomson Learning, 2000.

Spyns, P, Meersman, R. and Jarrar, M.: Data modelling versus ontology engineering, SIGMOD Record, Vol. 31, No. 4, Dec. 2002.

Tarski, A.: On the Calculus of Relations, *Journal of Symbolic Logic*, Vol. 6. No. 3., 1941. pp. 73-89.

Thalheim, B: Foundations of Entity-Relationship Modelling, *Ann. Math. Artif. Intell.* 6, pp. 197-256, 1992.

Information Modelling and Knowledge Bases XVI
Y. Kiyoki et al. (Eds.)
IOS Press, 2005

Acquisition of Requirement Model in Lyee

Tae YONEDA, Takayuki YAMANE, Hiroya SUZUKI, Jun SASAKI, Yutaka FUNYU
Faculty of Software and Information Science, Iwate Prefectural University
152-52, Takizawa-aza-sugo, Takizawa, Iwate 020-0193 Japan

Abstract. With our information society rapidly progressing, information systems have increased in scale and become more complex. For developing information systems, how to acquire user requirements and how to model those are important issues. In this paper, an acquiring requirements model that consists of connections of words is proposed in Lyee (Governmental Methodology for Software Providence), which is a software development methodology based on human thinking processes. In Lyee, since software is defined based on "words", it might be possible to be designed and developed by the end-user. Additionally, automatic coding has been realized. Our proposed requirement model is made from a "word model" that similar to the "word definition" in Lyee. Thus, it can be expected to realize automatic coding from the requirement model. By applying a demand made to the proposed acquisition method by a medical doctor without any experience of system developing, we confirmed that acquisition of the requirement can be realized with the model.

1. Introduction

Within our rapidly progressing information society, the related information systems have increased in scale and become more complex; it seems difficult to succeed in projects that seek to develop information systems. According to an investigation by an Information Technology research organization, the Standish Group, 70 percent of software development projects are reported to be operationally challenged or impaired [1]. Cost and time overruns, content deficiencies, and similar problems result from these failures. As a major factor contributing to the failure of such projects, the lack of user input, incomplete requirements and specifications, and changing requirements and specifications are raised in the Standish Group report. This means that the difficulty in acquiring a user's requirements and defining those requirements correctly may have led to the failure of many projects.

Recently, various design methods, modeling techniques and system development processes, including object-oriented software development, have been investigated as environments supporting system development. UML (Unified Modeling Language), ER (Entity Relationship) and ORM

(Object Role Modeling) are design and modeling techniques for system engineers experienced in system development [2]. XP (eXtreme Programming) is an agile system development process that can promote developments utilizing dialogue between the intended user and the developer [3]. Since these environments supporting system development improve the efficiency of the development, they have been used in various aspects of system developments. However, since an environment supporting conventional system development accompanies a complicated programming regime, there exists much complexity in such system development for users. A fundamental cause can be considered as being the large gap between the processes of system development based on a conventional programming regime and the process of a human's thinking.

On the other hand, Lyee (Governmental Methodology for Software Providence) is a software development methodology based on human thinking processes [4] that has been confirmed by philosophy [5] and mathematic [6][7]. The characteristics of Lyee software are "it can be designed using word definitions which an end-user can define easily" and "it has realized automatic coding." Thus, we expect Lyee may realize a software system developed by the end-user. In a paper [8], we experimented with end-user development of Lyee software systems by using a Lyee tool (Lyee All 2), where developing logs were recorded. A problem with the development by end-users is the difficulty of the definition in the boundary software between Lyee software and other software systems. The other issue is how to detect user requirements. In another paper [9], we modeled the boundary software and investigated a technical contrivance through which an end-user can easily deal with the boundary software. In respect of the requirement analysis in Lyee, some research results have been reported. A method of detecting word definitions in Lyee from requirements expressed in UML using First Order Logic was investigated [10]. However, this approach is on the assumption that it is not for an end-user but for a systems engineer (SE) to design and develop software systems. Rolland et al. of the University of Paris 1 developed a requirement definition tool that is focused on the goal and scenario, and the end-user may be able to properly define [11]. The mapping rule from the requirement definition utilizing the tool to the word definition of Lyee has also been investigated. However, since the original definition does not correspond directly to the word definition in Lyee, such mapping has been difficult.

In this paper, by focusing on the words of natural language, a methodology of an acquiring requirements model that consists of connections of the words is proposed. The requirement model is made from a "word model" that is similar to the "word definition" in Lyee. Thus, it can be expected to realize automatic coding from the requirement model. By applying a demand made by a medical doctor, without any experience of system developing, to the proposed acquisition method, we confirmed that acquisition of requirement model can be realized.

In section 2, the peculiarity and the tendency of Lyee and the Lyee tool for developing a Lyee software system are described. In that, we assume the software development environment of the user.

In section 3, we devise a requirement acquisition scenario introduced from the needs of the requirement acquisition. In section 4, the word model is shown as the key to the proposed requirement model; being similar to the word definition of Lyee. In section 5, a method for acquiring the requirement model using the word model is proposed.

2. Possibility of end-user development environment

2.1 Peculiarities and Tendencies of Lyee and the Lyee Tool

Lyee is a software development methodology that is based on software providence discovered by Fumio Negoro, president of the Institute of Computer Based Software Methodology and Technology Inc. Software providence is the basic structure of software, where a word-unit is a basic factor as shown in Figure 1, and is called a Predicate Structure (PS). A word defined by formula and conditions (which is denoted as a word definition) can be converted to software expressed as a PS form. Another word that is related to the word is automatically referenced by the contrivance of the PS. The referenced word can be also converted to software expressed as a PS form. Thus, the software is constructed with the repetition of PS (which is denoted as Lyee software). In Lyee, the word definition can be converted to Lyee software automatically and this software operates a software system. So, if an end-user can define a requirement of a software system as a word definition, the software system may be created automatically. This means that, Lyee has that possibility by using it an end-user without any programming knowledge could develop a software system.

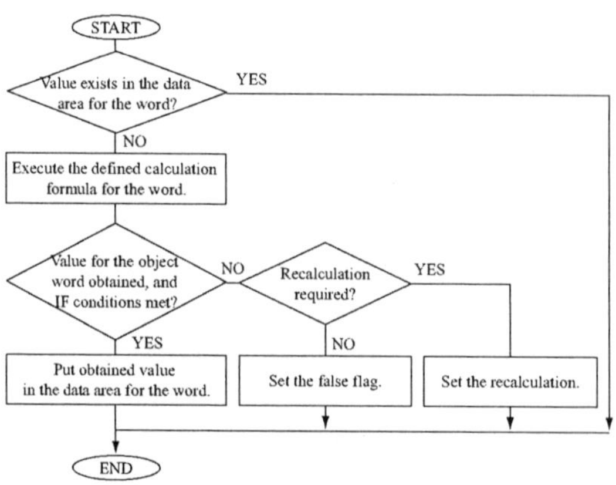

Figure 1 Predicate Structure (PS).

Presently, Lyee All2 is offered as a support tool to develop Lyee software. In the theory of Lyee, it is thought that any software can be described in PS form, that is, it is estimated that any system function can be expressed as a word definition. However, in Lyee All 2, the function described in PS form is restricted to the following functions: business requirements described in formula and conditions, items of screens or lists, and process routes expressed by conditions. Other functions are offered as some lumps of functions which are not described in PS form, by using template. Since using templates loses the flexibility of system development, how to express those functions to PS form has been investigated. Presently, the next version of Lyee All 2, Lyee All 3 has been developed. Further, it is desired to develop Lyee All 4, Lyee All 5, etc ..., all of which aim for an end-user being able to define functions as word definitions for system development.

2.2 User-development environment

The purpose of our research is to establish a software-development environment (a user-development environment), in which users can develop their own software. As we describe in 2.1, software can be elicited automatically from word definitions expressed with the formula and conditions. Presently, though functions expressed by word definitions are restricted, the process has been well researched and developed. In the near future, more functions for system development will be expressed by word definition. Then, software embodying more functions will be created form those word definitions by automatic coding with Lyee tools. At that time, since the word definition corresponds with the word form of near natural language used in usual expressions, end-users may be able to develop software systems by themselves using Lyee tools. This is the reason why we consider that building user-development environments using Lyee is possible.

Though, system functions can be expressed as word definitions and automatic coding from the word definition realized, how to acquire the user requirement still remains an issue. Our research target, shown in Figure 2, is to detect the user requirement from the user demand, which can be ambiguous and incomplete, and to create a model of the required system (the requirement model).

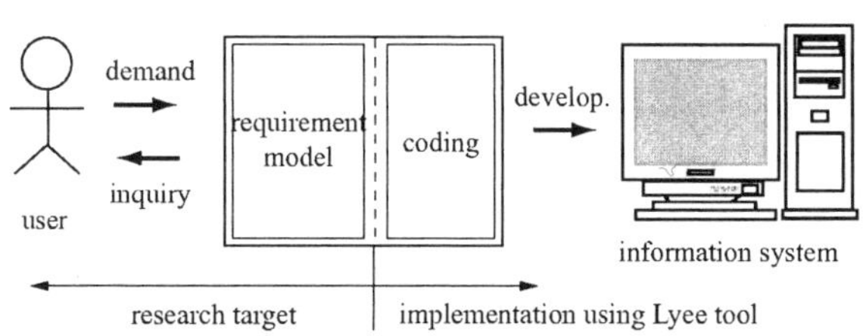

Figure 2 User-development environment.

The preconditions for our research are:

- The function for the developed system can be expressed by word definition,
- Using a Lyee tool, all word definitions can automatically be coded to a software program.

3. Requirement acquisition scenario

The following areas need to be clarified and overcome to detect the user requirements:

(1) For the user, it is difficult to explicitly define the requirement,

(2) It is difficult to express all the requirements in the user's mind,

(3) At the stage of requirement definition, it is necessary to check whether the definition is, or can be, given correctly.

We consider the next issues are important to concretely solve these dilemmas:

- obtaining the emerged requirement that the user recognizes and has expressed,
- eliciting a potential demand that the user has not recognized and has not expressed,
- checking the acquired requirement with the user.

From the above three issues, the following requirement acquisition scenario is considered, as shown in Figure 3.

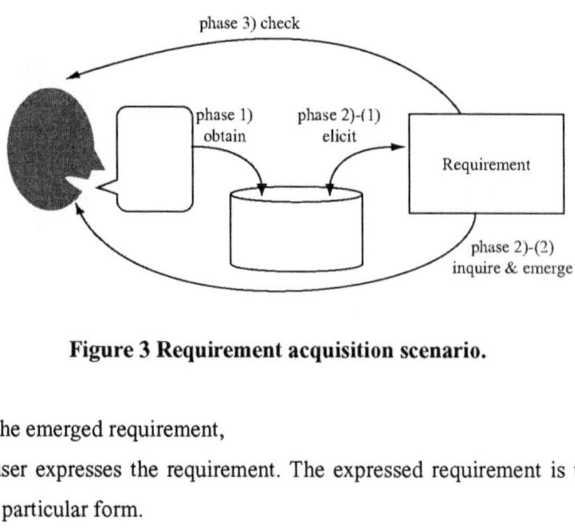

Figure 3 Requirement acquisition scenario.

phase 1) obtaining the emerged requirement,

For a start, the user expresses the requirement. The expressed requirement is thus obtained and accumulated in a particular form.

Phase 2) eliciting the requirement,

The requirement is complemented (added to) by automatic acquisition based on the original requirement (Figure 3 phase2)-(1)). Additionally, when the requirement cannot be complemented, a potential requirement is elicited by further inquiring of the user and making the user recognize the requirement; by this an additional requirement is acquired (Figure 3 phase2)-(2)).

Phase 3) checking the acquired requirement with the user,

The acquired requirement is presented for the user and checked by ensuring that it is easily recognized. After the user has checked it, if it is necessary, additions and corrections to the requirement can be further carried out.

4. Word model

To concrete the scenario for acquiring the user requirement mentioned in section 3, we focus on a word definition of Lyee, which is constituted by formula and conditions. We treat a "word model" that is similar to the "word definition." The word model is constituted by words and relationships between those words. That is called a word model (figure 4). The words in the word model are suitable for terms of the formula in the word definition. The relationships in the word model are suitable for conditions and operators of the formula in the word definition.

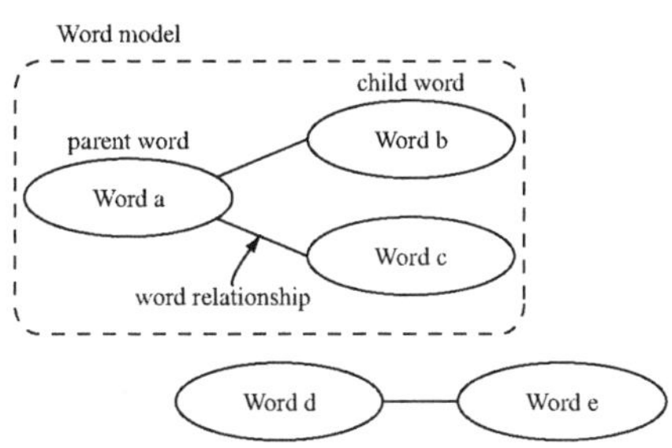

Figure 4 Word model.

In addition, the following conditions of the word model are settled. The 1st condition; a word model consists of double layers of words and the relationship between them. The word that corresponds to the left side member of a conditional expression is called a parent word; the word which corresponds to the right side member of a conditional expression is called a child word. The 2nd condition; a child word has only one parent word. The requirement is regarded as a set of this word model.

We estimate that the acquired requirement using the proposed word model will be linked to the word definition in Lyee.

5. Acquisition of Requirement Model

We propose a requirement acquisition method using the word model (figure 5).

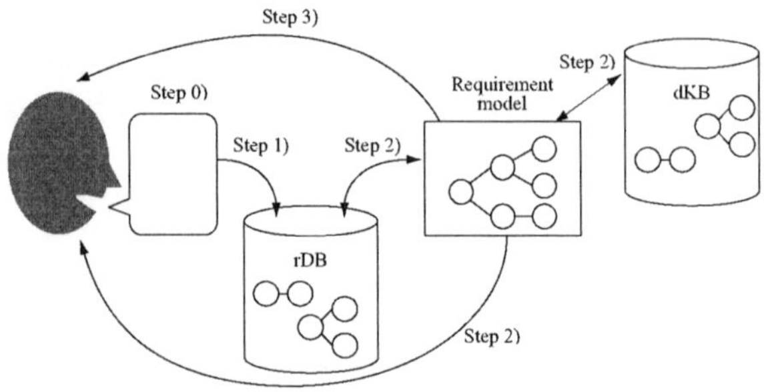

Figure 5 Acquisition of requirement Model.

The proposed process of requirement acquisition is as follows. The domain knowledge base (dKB) is a knowledge base that accumulates sets of word models used in the past by the information system development of the domain.

Step 0) a user expresses a demand using natural language.

Step 1) the expressed requirement is modeled with word models and these are accumulated to the requirement data base (rDB), which stores the word models temporarily. This is equivalent to the 1st phase of the scenario for acquiring the user requirement mentioned in section 3; obtaining the emerged requirement.

Step 2) a word model accumulated at the rDB is arbitrarily abstracted, and words of the abstracted word model are verified with the words of word models in the rDB and the dKB whether the words are matched or not. When a word is matched, the word model is extracted from the rDB or the dKB and connected. The requirement model, which is a set of word models, is created by extracting word models one after another from the rDB and the dKB. When a word does not match, the user is asked again, and the word model is added to the rDB, and the requirement model is made. This is equivalent to the 2nd phase of the scenario for acquiring the user requirements; the elicitation of a potential requirement.

Step 3) the acquired requirement model is presented to the user in a representation that is easy for the user to recognize, and the user in this way checks the requirement. This is equivalent to the 3rd phase; checking the acquired requirement.

The requirement model was acquired by the word model which modeled the user demand made in words, and the relation between those words. The requirement model of a medical examination system was acquired by the proposed process; the requirement which the medical doctor considered in his mind could thus be expressed. Further, the requirement that the medical doctor couldn't express explicitly, could soon also be expressed.

6. Conclusion

The Lyee software development methodology has the possibility that it can realize the end-user development environment. We comment on the characteristics of Lyee and describe the following issues in this paper:

- The proposed method of our acquiring requirement model based on the word model is similar to word definition of Lyee.
- We apply the proposed method to the demand of an end-user (a medical doctor) to a sample system (a medical examination system) as an example. As a result, it is confirmed that requirement model can be detected and the proposed method can express the user's emerged requirement and potential demand.

Some issues remain to be addressed in the future:

- To design, construct and evaluate a system for the acquiring requirement model based on the proposed method.
- To investigate how to support users in facilitating an easy understanding of the requirement model. We consider that a support tool will be necessary for checking the requirement model.

References

[1] The Standish Group, "The Chaos Report," http://www.standishgroup.com, 2001.
[2] I. Jacobson, Grady Booch and James Rumbaugh, The Unified Software Development Process, Addison-Wesley Publishing Company, 1999.
[3] K. Beck, Extreme Programming Explained: Embrace Change, Addison-Wesley Pub Co, 1999.
[4] The Institute of Computer Based Software Methodology and Technology 'Inc, Lyee GovernmentaL MethodologY for SoftwarE ProvidencE, http://www.lyee.co.jp, 2003.
[5] F. Negoro, Lyee's Hypothetical World, Proc. of Lyee International Workshop, 2002, pp.3-22.
[6] O. Arai and H. Fujita, A Word-unit-based Program: Its Mathematical Structure Model and Actual Application, Proc. of Lyee International Workshop, 2002, pp.63-74.
[7] S. Gorlatch, The Lyee Programming Model: Analysis and Correctness in a Fixed-Point Setting, Proc. of Lyee International Workshop, 2003, pp.214-224.
[8] Y. Funyu, J. Sasaki, T. Nakano, T. Yamane and H. Suzuki, An Experiment on Software Development of Hospital Information System, Proc. of Lyee International Workshop, 2002, pp.328-340.
[9] H. Suzuki, T. Yamane, T. Yoneda, J. Sasaki and Y. Funyu, A framework for user accessible boundary software, Proc. of Lyee International Workshop, 2003, pp. 157-166.
[10] B. Amon, L. Ekenberg, P. Johannesson, M. Munguanaze, U. Njabili and R. Tesha, Automated Word Generation for the Lyee Methodology, Proc. of Lyee International Workshop, 2002, pp. 357-374.
[11] C. Souveyet, F. Missonnier and C. Rolland, L'E-Lyee: Coupling L'Ecritoire and Lyee, Proc. of Lyee International Workshop, 2003, pp. 303-319.

Information Modelling and Knowledge Bases XVI
Y. Kiyoki et al. (Eds.)
IOS Press, 2005

An Analysis of ISD as Knowledge Work – How is the Body of Systems Development Process Knowledge Used?

Per BACKLUND
School of Humanities and Informatics, University of Skövde
P.O. Box 408, SE- 541 28, Skövde, Sweden

Abstract This position paper characterises information systems development (ISD) in terms of knowledge work. We extend and apply a framework which models ISD work in a hierarchy of paradigms, approaches, methods, and techniques. The extension of the framework comprises six different perspectives of knowledge. The extended framework is applied on an empirical case in which a new systems development method is utilised in a project. The conclusion of the paper is that the perception of the extent to which the new method has actually been implemented in the organisation is highly dependent on the perspective of knowledge that we adopt in our analysis.

1 Introduction

Information systems development (ISD) has recently been characterised as knowledge work which is based on a distinct body of knowledge [1, 2]. Iivari [2] proposes a framework for ISD as knowledge work which consists of four levels: paradigms, approaches, methods, and techniques. These levels represent an increasingly concrete representation of the body of ISD knowledge; paradigms being the most general and abstract parts, and techniques being the most concrete and specific part of the body of knowledge.

According to Iivari [2] the detailed use of ISD techniques, methods, approaches, and paradigms is relatively unexplored. In this paper we propose that an ethnographically inspired approach [3] is suitable to investigate the phenomenon in an empirical setting. We view commercial development methods as cognitive artefacts, i.e. explicit development process knowledge, which can be introduced and used in development organisations. There are reports on method adaptation [4-6]. However, these reports tend to focus on method adaptation from an organisational point of view and of method use in organisations. There is also the field of method engineering [7, 8] with a focus on the method as such. Typically, method engineering is concerned with how new methods are constructed by combining method fragments, thus creating a new method or method configuration.

The aim of this paper is to highlight and discuss the interplay between explicit and tacit development process knowledge. We do this by combining the framework of Iivari [2] and the views of knowledge presented by Alavi and Leidner [9]. We analyse method use from a knowledge point of view, i.e. how is the knowledge embedded in a method used. Thus we aim at highlighting the interplay between explicit development process knowledge (i.e. the knowledge represented in an ISD method) and the tacit dimension of development process knowledge (the craft- like, professional, and creative aspects of ISD).

In the work presented here we focus on the phase subsequent to the introduction and organizational implementation of the development method, i.e. in its situation of use. In order to better understand the adoption of methods and ISD as knowledge work, we propose that the perspective of knowledge has an impact on how we model knowledge, i.e. the level of adoption of the knowledge embedded in the new development method depends on the perspective of knowledge that we take. In order to investigate this phenomenon we combine the framework of Iivari [2] and the knowledge perspectives of Alavi and Leidner [9].

2 Knowledge Work and Perspectives of Knowledge

We view knowledge work in a similar fashion as do for example Nonaka and Takeuchi [10], Leonard [11], and Davenport and Prusak [12]. This view can be described in terms of the two sub-areas knowledge-building and knowledge-use. Modern organisations spend a large effort in organising their knowledge and its use so that knowledge work can be facilitated. An IS development method is an example of such an effort.

Alavi and Leidner [9] describe six different perspectives of knowledge: *Knowledge vis a vis data and information,* data is facts. Information is processed/interpreted data. Knowledge is personalised information. *Access to information,* knowledge is a condition of access to information. *Capability,* Knowledge is the potential to influence action. *Process* knowledge is the process of applying expertise. *Object,* knowledge refers to objects, which can be stored and manipulated. *State of mind,* knowledge is the state of knowing and understanding.

The perspectives are motivated by the fact that ISD comprises all these views of knowledge. According to Alavi and Leidner [9] the knowledge vis a vis data and information view is common in IS and IT literature and constitutes one of the fundaments for how information systems are perceived. This view also has relevance for the perception of knowledge as objects in that it implies some sort of representation of knowledge. The process view of knowledge is central since knowledge work implies the skill of utilising the theoretical subjects of the area in a practical manner, i.e. applying expertise knowledge in order to build information systems. The process view of knowledge is essential in the perception of knowledge as an ability to apply expertise. Hutchins [13] presents a model to describe how knowledge is mediated via language to become 'automated' without any reference to the explicit representation of either the steps or their sequential relations. This model can be used to describe the process of becoming skilled in a process in the sense of doing a task without reflecting on every step. In the access to information view of knowledge the main issue is that information must be organised to facilitate access to and retrieval of content. The view of knowledge as potential to influence action may be viewed as an extension to the access view in that it recognises that learning and experience result in an ability to interpret information and judge what information is necessary for decision making.

The relation between the knowledge perspective and the approach, method, and technique characterises what it means to use an approach, a method, or a technique respectively.

3 The Case

The study was made at the IT department of a car manufacturer. In order to study how development process knowledge was utilised in an actual ISD project we followed an internal development project during the autumn of 2003. The project is staffed by the following roles: one project manager, one architect, four analysts, four developers five implementers, and one test designer; with a total number of eight people involved.

Data was collected by observing project meetings and work sessions taking detailed field notes. Furthermore, the observation data was complemented by informal discussions with project members. We also had access to internal project documentation in different versions which has given us an opportunity to review how the different RUP artefacts have evolved over time. The different objects of observation provide different views of how the development method was used in the project. Moreover, the different sources cater for source triangulation [3].

4 An Empirical Investigation of ISD as Knowledge Work

We identified a set of indicator classes (see Table 1) which comprise a number of different instances of knowledge work indicators. We use the term artefacts to comprise the notion of artefacts in the Rational Unified Process [14], documents, graphical models, code, and other deliverables. We use the term valuation to denote situations where we observed team members expressing opinions on routines as well as on their use of routine knowledge work. An example of an indication of using templates is when the analyst fills in a document template with project specific information, whereas updating an artefact refers to a situation where a model or a document in use is changed. We note that the concepts of craft- like and professional knowledge work are similar. The main distinction is in the more reflective character of professional knowledge work. An example of an indicator of "Balance total situation" is that a team member tries to reach synergisms by coordinating efforts from other projects. An illustration of a strategy indicator is the effort to build autonomous systems.

Table 1 Classes of knowledge work indicators identified in the observation data

Routine KW	Craft-like KW	Professional KW	Creative KW
Use templates	Pattern similarities	Professional integrity	Informal artefact use
Use artefacts	Work habit	Strategy	Use of similes
Update artefacts	Estimation	Valuation of benefit	Brain storming
Create artefacts	Fluent work	Valuation of work habits	Excited discussion
Organise artefacts	Explicit use of routine	Valuation of approaches	
	Balance total situation	Valuation of routine	

The indications of routine knowledge work are closely connected to an explicit use of the ISD method whereas craft- like, professional, and creative knowledge work rather characterises implicit method use since they presuppose that the method has been learned and internalised. Hence we can describe the interplay between implicit and explicit method use.

5 Discussion and Conclusion

We claim the different indicators of knowledge work can be mapped to the framework and connected to the different perspectives of knowledge. An overview of the findings is presented in Table 2. The extent to which the method can be said to be implemented in the organisation decreases when the capability and process perspectives of knowledge are used. Our interpretation of this is that it is harder to change a process dimension of knowledge since it involves the internalisation of new knowledge as opposed to an objectified view which does not acknowledge such problems.

Table 2 The level of method incorporation depends in the perspective of knowledge.

Perspective of knowledge / level	Object	Data/ information/ knowledge	Access to information	Capability	Process	State of mind
Technique	Yes. Templates are incorporated and used.	Not analysed	Partly. Everything is not used.	Partly. Increased adaptation of templates.	Partly. Discussions on what to do and in what order.	Not analysed
Method	Yes. Process guidelines are available.	Not analysed	No. Perceived as large and complex.	No. Perceived as large and complex. Not actually supporting work.	Partly. Work often proceeds according to old habits	Not analysed
Approach	Yes. Organisationally adapted guidelines available	Not analysed	Partly. Organisationally adapted guidelines available	No. OO is hard to apply when it is known that a relational db is to be used.	No. Frequent discussion about the use case driven approach.	Not analysed

We can also interpret Table 2 horizontally in order to highlight the explicit/tacit dimensions of knowledge. The method and approach leve ls comprise the tacit dimension of ISD. The further down we move (more towards method and approach) the harder it is to change a work habit, since it involves changing already internalised knowledge and essentially skilled based abilities. We propose that the perspective of knowledge has an impact on how we model knowledge, i.e. the level of adoption of the knowledge embedded in the new development method depends on the perspective of knowledge that we take.

Acknowledgements: This research is sponsored by the Swedish Knowledge Foundation (KK-stiftelsen).

References

[1] R. Hirschheim and H. Klein, "Crisis in the IS Field? A Critical Reflection on the State of the Discipline," *Journal of the Association for Information Systems*, vol. 4, pp. 237-293, 2003.

[2] J. Iivari, "Information Systems Development as Knowledge Work: The body of systems development process knowledge," presented at Information Modelling and Knowledge Bases XI, 2000.

[3] K. Williamson, *Research methods for students, academics and professionals Information management and systems*, Second ed. Wagga Wagga: Centre for Information Studies, 2002.

[4] N. L. Russo, R. Hightower, and M. Pearson, "The Failure of Methodologies to Meet the Needs of Current Development Environments," presented at The Fourth Conference of the British Computer Society Information Systems Methodologies Group, Cork, Ireland, 1996.

[5] B. Fitzgerald, "The Use of Systems Development Methodologies in Practice: A Field Study," *The Information Systems Journal*, vol. 7, pp. 201-212, 1997.

[6] B. Fitzgerald, N. L. Russo, and T. O'Kane, "Software Development Method Tailoring in Motorola," *Communications of the ACM*, vol. 46, pp. 64-70, 2002.

[7] S. Brinkkemper, M. Saeki, and F. Harmsen, "Assembly Techniques for Method Engineering," presented at CAiSE'98, Pisa, 1998.

[8] J. Ralyté and C. Rolland, "An Assembly Process for Method Engineering," presented at CAiSE, Interlaken, Switzerland, 2001.

[9] M. Alavi and D. E. Leidner, "Review: Knowledge Management and Knowledge Management Systems: Conceptual Foundations and Research Issues.," *MIS Quarterly*, vol. 25, pp. 107-133, 2001.

[10] I. Nonaka and H. Takeuchi, *The knowledge-creating company: How Japanese companies create the dynamics of innovation*. New York: Oxford University Press, 1995.

[11] D. Leonard, *Wellsprings of Knowledge Building and Sustaining the Source of Innovation*. Boston: Harvard Business School Press, 1995.

[12] T. H. Davenport and L. Prusak, *Working Knowledge: How Organisations Manage What they Know*. Boston, Mass.: Harvard Business school, 1998.

[13] E. Hutchins, *Cognition in the Wild*. Camebridge, Massachusetts: MIT Press, 1995.

[14] P. Kruchten, *The Rational Unified Process An Introduction Second Edition*. Reading, Massachusetts: Adison-Wesley, 2000.

Information Modelling and Knowledge Bases XVI
Y. Kiyoki et al. (Eds.)
IOS Press, 2005

Explicit Considerations of the Process of Designing Information Systems

Eva NERO

School of Humanities and Informatics
University of Skövde, Box 408, 541 28 Skövde Sweden
eva.nero@ida.his.se

Abstract This is report on research in progress. In the process of designing information systems, there are problems of different types that occur at different times during a design process. These problems arise due to complexity that mainly emanate from the fact that people are involved. This kind of complexity may cause system failures which will have organisational, social and economic consequences. There exists no panacea to handle this sort of problem. One approach to handle the problem is to focus on the complexity and reflect on the alternative ways to specify design structures and processes. John P. van Gigch has introduced the Metamodeling (MM) level and distinguishes it from the Model level and Reality level. At the MM level the MM process takes place. At the MM level reflections on the design situation are performed. The purpose is to understand the design situation and conditions related to it in order to be able to specify how the design process should be performed. Since the design process is iterative and complex, MM processes should be performed in a continuous manner. Van Gigch means that MM processes seem to be neglected in design processes, and that this is one reason why system failures occur. This project will investigate the existence of MM processes, what trigger and hamper MM processes, and designer attitudes to MM processes. The application area for this project is Information System Design. Some results exist based on a preliminary study.

Key words: Information System Design, Metamodeling, Reflections

1. Introduction

There arise problems of different types in the process of designing information systems. Some of the problems have their cause in methods, people and in changed conditions. There are many methods available, and they are usually not transparent in their epistemological foundation [1], [9], [15], [2], [8], which makes it hard for the designers to select an appropriate method for the current problem at hand. Complexity exists if human beings are considered in information systems, since people have different pre-knowledge and perceptions of the world and on the design situation [10]. The conditions for the design will change during a project. Functionality requirements may be altered, economical and time constraints or project members will be exchanged. This dynamics gives rise to problems that have implications for the design of the information system. If we want to avoid system failures it is reasonable to propose that more reflections ought to be made on the design situation and its conditions. Van Gigch [7] has introduced a normative model that guides a designer how a reflection process should be performed. A short description of his idea follows below. Van Gigch [7] has in mind a general system design process and the ideas could therefore as well be applied in a specific situation as in a design of information systems. He has introduced the concept of metamodeling (MM). At the MM-level the MM-process takes place, it is here the reflections on the design situation are performed. The purpose is to consider the design situation, the conditions related to it, and its consequences in order to be able to specify how the design process should be designed. It is a three level architecture for system design, consisting of the reality, modeling, and metamodeling levels. Van Gigch [7] states that when a system is designed the work takes place on the level of reality and on the level of modeling, see Figure 1. Modeling is one step or one level above reality. The MM-level is thus two steps or two levels from the reality level. Different approaches will have impacts on the design process of information systems; the design processes for the information systems will be different. It is, according to van Gigch [7], at the MM-level the epistemological foundations is considered and dealt with.

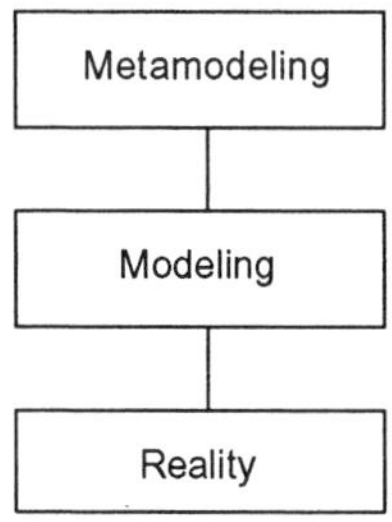

Fig. 1. The three levels of system design [7].

The designer can be guided as to approaches and methods that are adequate to apply considering the system which will be designed. When van Gigch [7] proposes the use of the metamodelling level he is to a great extent applying systems approach (systems design). According to Malmsjö [11], the effect of applying a systems approach does not mean that a holistic perspective only should pervade the process of understanding a situation. Instead there should be a shift in perspectives in a process where the designer gets a better understanding of the situation. When metamodelling is applied in a design process, there is also a shift in perspectives. The relations between the metamodeling, modeling and reality levels and the MM-process are shown in Figure 2. The MM-process in Figure 2 can be described as, at first (1) it is important to describe what kind of system that will be designed. What are the important aspects to consider? (2) When the designer has a good understanding of the situation, the next question is: Does any of the methods known by the designer apply in this situation or should other approaches or methods be used instead? (3) The designer makes a choice of an appropriate method for the current design situation. (4) The designer starts collecting material from reality as a basis for the design.

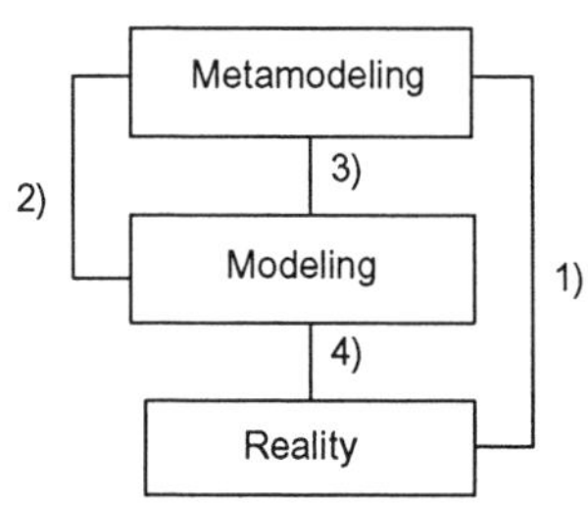

Fig. 2. The MM-process [7], [11].

There exist other theories which also stress the need for reflections in problem solving, where design of information system can be considered as one kind of problem solving, see for instance Inquiring Systems [4], Double Loop Learning (DLL) [3], Total System Intervention (TSI) [6]), and Reflective Systems Development (RSD) [12].

Inquiring Systems concerns decision making in the design process and is based on how knowledge is obtained according to different philosophers and how their different epistemologies have consequences for the reflections at the MM-level. The epistemology and ontology affect the design situation, and therefore must the designer, according to Churchman [4], be aware and make active choices. The similarity with van Gigch' MM-processes are the reflections at the meta-level. Argyris and Schön [3] stress in Double Loop Learning the need to questioning the governing variables in order to solve the underlying problem and not just cure the symptoms. To question is to critically exam, to ask questions why something is the way it is and to study the underlying epistemology. The best way for change and learning is, according to Argyris and Schön [3], detection and correction of errors. Both DLL and van Gigch' MM-processes point at the need to question and to make reflections on the current problem situation. The TSI is a problem solving meta-methodology that is performed at the MM-level. The evaluation of the problem situation and the evaluation and selection of an appropriate methodology is important [6]. The underlying epistemologies of the possible problem solving methodologies are examined. The Reflective Systems Development is based on reflections on the design situation, questions asked are: What has to be done, how should it be done, and why should it be done. Reflections and dialogues are used to generate the necessary insights to carefully prepare the selection of methodology or work practice [12]. RSD and van Gigch' MM-processes both stress the need to carefully prepare the selection.

2. Research aim

There is a need for further research in this area. Based on observations by van Gigch [7], Flood and Carson [5], Nero [13] and Nero and Malmsjö [14] it should be of interest to study more closely what MM-processes are performed in the process of designing information systems, what trigger and hamper MM-processes, and what are the designer attitudes towards MM-processes are.

The overall aim of the research is to get a better understanding of the MM-processes in the ISD process. The study will also show how different actors in the design process will have different attitudes to MM-processes. The improved understanding of MM-processes will probably give arguments for the importance of using MM-processes both in education and profession.

3. Method

The planned research methods are interviews and group discussions. The interviews are performed in order to get data/information about designers' attitudes and what triggers and hampers MM-processes in a design situation. A triangulation is planned in the data collection, in order to collect data from different perspectives a richer and more nuanced picture will hopefully be obtained, and it is also possible to get a validation of the findings from the interviews. The seminars are going to be performed with some of the designers from the interviews. The interviewees are information system designers (ISD) that work in different settings in different organizations. The settings are in-house ISD and consultant ISD at small as well as large Swedish organizations. The selection of the interviewees are made with the purpose to get an understanding of the differences between the different settings, and is to identify what they are and why.

4. Expected results

The expected results of this work are first of all an improved understanding of how MM-processes are performed in the area of ISD. Also the designers' attitudes to MM-processes can be captured. The result can hopefully show what factors hampers, triggers and encourage the use of MM-processes. As a final result of the research, it could be possible to formulate some normative statements to stimulate the use of MM-processes in the area of information system design.

This work have pointed out that there is much more to do in this area in order to grasp the whole picture and to get the aimed understanding. There is a need for further research; the next planned action is to collect more empirical material by extending the interview study with information system designers. Discussions with other researchers in this area point at that the next step in improving the information system design work is to adopt a more reflective approach. With these ideas in mind there ought to be an interest in what the further research in this area brings to mind. This project is a start, and it shows that some designers are reflecting on the design situation not just for philosophical matters but for functional work approaches and economical reasons. This can be summarized in a well known quote: 'Time is money', and it holds in this area as well, because by reflecting on the design situation and its conditions the designers save time and money when they prevent system failures and misdesigns.

5. Acknowledgement

This research is sponsored by the Swedish Knowledge Foundation (KK-stiftelsen). I would like to thank my supervisor Anders Malmsjö and my colleagues for their support, comments and encouragement.

References

[1] Andersen, E. S.: Systemutveckling-principer, metoder och tekniker. Studentlitteratur, Lund (1994)
[2] Avison, D. E. and Fitzgerald, G.: Information System Development: Methodologies, Techniques and Tools. McGraw-Hill, London (1995)
[3] Argyis, M. and Schön, D.: Organizational Learning: A Theory of Action Perspective. Addison Wesley (1978)
[4] Churchman, C. W.: The design of Inquiring Systems. Basic Books, Inc. (1971)
[5] Flood, R.L. and Carson, E.R.: Dealing With Complexity: An introduction to the Theory and Application of Systems Science. 2nd edition. Plenum Press, New York (1993)
[6] Flood, R.L. and Jackson, M.C.: Creative Problem Solving – Total Systems Intervention. Wiley, New York (1991)
[7] van Gigch, J. P.: System Design, Modeling and Metamodeling. Plenum Press, New York (1991)
[8] Iivari, J. and Maansari, J.: The Usage of System Development Methods: Are We Stuck to Old Practices?. Information and Software Technology 40 (1998) 501-510
[9] Jayaratna, N.: Understanding and Evaluating Methodologies: NIMSAD – A Systemic Framework. McGraw-Hill (1994)
[10] Langefors, B.; Essays on Infology. Studentlitteratur, Lund (1995)
[11] Malmsjö, A.: Evaluations- Some Methodological Considerations. Systems Research and Information Systems 8 (1998) 61-80
[12] Mathiassen, L.: (2003) Reflective Systems Development - Position Summary [online]. Aalborg, Denmark: AAlborg University, Department of Computer Science, 1997 [cited 3 December 2003]. Aviable from World Wide Web: (http://www.cs.auc.dk/~larsm/rsd.html)
[13] Nero, E.: The Process of Metathinking in the Area of Information System Design. Master Thesis, University of Skövde, Sweden, HS-IDA-MD-00-009 (2000)
[14] Nero, E. and Malmsjö, A.: A Study of Metamodeling in the Area of Information System Design. In: Wilby, J. & Allen, J. K. (eds.) Proceedings of the 45th Annual Conference of the International Society for the Systems Sciences, July 8-13, 2001. Asilomar, California CD-ROM (2001)
[15] Nilsson, A. G. (1995) Evolution of Methodologies for Information Systems Work - A Historical Perspective, in Dahlbom, B. (ed.) The Infological Equation: Essays in Honour of Börje Langefors. Göteborgs Universitet.

Information Modelling and Knowledge Bases XVI
Y. Kiyoki et al. (Eds.)
IOS Press, 2005

An Access Control Mechanism for SetNS, a Name Service based on Set of Symbols

Yasushi SHINJO
Department of Computer Science, University of Tsukuba
1-1-1 Tennoudai, Tsukuba, Ibaraki 305-8573, Japan
`yas@cs.tsukuba.ac.jp, http://www.cs.tsukuba.ac.jp/~yas/`

Abstract.

SetNS is a name service in which a name is specified as a sequence of element strings or a set of symbols. Unlike conventional tree-based name services, SetNS allows changing the order of element strings and abbreviating a long name.

This paper proposes an access control mechanism for SetNS. The main feature of the proposed mechanism is the introduction of pseudo symbols that describe access control. A user can reorder and omit pseudo symbols in a name as well as normal symbols. Therefore, pseudo symbols are suitable for the syntax and semantics of SetNS. The proposed mechanism has been implemented in a file system of Unix by using a process trace facility.

1 Introduction

Naming is one of the most primitive and essential activities for expressing semantics and knowledge. One of the most important challenges in naming is to realize both flexibility in expressing names and simplicity in describing access control policies, simultaneously. An access control mechanism archives a selective access to contents for providing confidentiality. Simplicity is essential in access control when ordinary users have to use it. For example, access control of a file system must be simple. Otherwise, a complex mechanism causes human errors, and access control becomes pointless.

Currently, most name services provide *flat name spaces* or *tree-based name spaces*. A flat name space is simple and easy to use, and its implementation is trivial. However, a flat same space is not scalable, and a user cannot manipulate a very large flat name space. A tree-based name space has more power of expression than a flat name space. A tree-based name space is most suitable when the target has a tree structure, such as addresses of surface mail and organizations of companies. Efficient implementation methods of tree-based name spaces are also known. However, tree-based name spaces have a problem of *a single point of view*. For example, although the term `"Computer"` and the term `"Education"` do not have "is-a" relations, we have to choose either the view `"Computer/Education"` or the view `"Education/Computer"` in a tree-based name space. Furthermore, names become long when we have to classify a target precisely.

To supplement the limitations of the tree-based name spaces, we are developing a name service called *Set Name Service* or *SetNS*[9]. In SetNS, a name is expressed as an unordered sequence of element strings or a set of *symbols*. SetNS has the following features:

- Within a name, element strings can be reordered.
- A long name can be abbreviated when it is looked up.
- It is easy to aggregate names with common elements.

In tree-based name services, we can describe access control for a group of objects by using a directory in a centralized way. On the other hand, since SetNS allows reordering of element strings, SetNS cannot support such a centralized description of access control. The previous implementation of SetNS relied on conventional access control for individual objects.

In this paper, we propose a new access control mechanism for SetNS. The main feature of the mechanism is the introduction of *pseudo symbols for access control (ACPS)*. A pseudo symbol can be added in an arbitrary place in a name as normal symbols. Furthermore, pseudo symbols can be omitted as normal symbols. An object with a name including two or more pseudo symbols for access control can be accessed if all the pseudo symbols allow it. Those features are appropriate for SetNS well, and we can keep the syntax and semantics of names in SetNS. In this paper, we show the implementation of the access control mechanism in a file system.

The rest of the paper is organized as follows. Section 2 shows the overview of the name service SetNS and its problem of access control. Section 3 describes an access control mechanism for a file system based on SetNS. Section 4 shows the implementation of the access control mechanism for SetNS. Section 5 evaluates the access control mechanism by comparing with tree-based name services and related work. Section 6 concludes the paper.

2 The name service SetNS and its problem of access control

In this section, we show the definitions of the name service SetNS. After that, we describe the problem of access control in SetNS.

2.1 The definitions of SetNS

A name is a sequence of one or more element strings with separators. *An element string* is a sequence of characters except characters for separation. *A name service* is a service that manages the bindings of names and object identifiers, and provides the following operations:

1. *Register*. Create a new binding of a name and an object identifier.
2. *Remove*. Delete a registered binding.
3. *Resolve*(*lookup*). Translate a given name into the corresponding object identifier.
4. *List*. Return a list of names or element strings.

SetNS is a name service, and it uses the character `','` as a separator. In SetNS, an element string is also called *a symbol*. A name in SetNS consists of symbols separated with `','`. Within a name, the order of symbols has no meaning, and no duplication of symbols is allowed.

- `Computers,Hardware,Peripherals,Keyboards,Dvorak,index.html`
- `Computers,Hardware,Peripherals,Keyboards,Dvorak,Products,index.html`
- `Computers,Hardware,Peripherals,Keyboards,Qwerty,index.html`

Figure 1: The names that include both the symbols `Computers` and `Keyboards` in the Web directory of the Open Directory Project.

Figure 1 shows three names that include the symbols `Computers` and `Keyboards`. Those names are taken from the Web directory of the Open Directory Project [7]. This project provides an open-source directory of Web in Resource Description Framework (RDF) [8].

A user who is familiar with tree-based name services can regard SetNS as a tree-based name service with minor exceptions. In a file system based on SetNS, a user can browse symbols of SetNS with the commands `ls` and `cd`. For example, a user can show the content of the last file in Figure 1 as follows:

```
% cd Computers
% cat index.html,Qwerty
```

In SetNS, the command `cd` changes the context of the shell. *A context* is a set of symbols, and it is automatically used when a name is manipulated (register, remove, list, and resolve). Contexts are used for abbreviating long names and grouping names with a common part.

A directory in SetNS is very different from a directory in a tree-based name service. A directory in a tree-based name service corresponds to an intermediate node of a tree structure. Making, removing, and renaming directories build and change the tree-based name space. On the other hand, a directory in SetNS is a transient object, and we cannot make a directory explicitly. A directory in SetNS has the following roles:

1. Hold a context. This helps a user to group names with a common part.
2. Show a list of names or a list of element strings. This helps a user to browse names or symbols.

2.2 *An access control problem in SetNS*

A typical file system with a tree-based name service provides two description methods of access control.

1. For individual objects (files).
2. For subtrees rooted by directories.

The latter facility is useful because we can describe access control for a group of files and directories at a single point. For example, Alice can deny retrievals of the following directory.

- `/home/alice/Mail`

In this case, the other people cannot access individual files under the directory even though they have a permission to access the individual files. For example, the other people cannot access the following file.

- `/home/alice/Mail/from-to-bob`

On the other hand, since SetNS does not have the idea of subtrees, it is impossible to use such simple description. For example, consider the following symbols.

- `home,alice,Mail`

Since those symbols specify a virtual directory, and a directory is a transient object in SetNS, Alice cannot save a permission mode to the directory. Moreover, individual files can be specified with an arbitrary order of symbols, and some of them may be omitted, as follows.

- `Mail,from-to-bob,alice,home`
- `alice,from-to-bob`

Therefore, the idea of subtrees does not work for access control in SetNS. In the previous implementation of SetNS [9], we could not describe access control for a group of files, so we relied on access control for individual files. In this paper, we address this problem of SetNS with the idea of pseudo symbols for access control.

3 An access control mechanism for SetNS

In this section, we show the design of an access control mechanism for a file system based on SetNS.

3.1 The access control model

In access control, we have to define mapping of users outside the computer and objects within the computer. In this paper, we deal with individual users and groups of users. A user and a group outside the computer correspond to an integer *user identifier* (UID) and an integer *group identifier* (GID), respectively.

An access control model is defined by the following three terms: subjects (principals), objects, and operations. In this paper, the subjects are processes, the objects are files, and the operations are manipulations of files, such as read and write.

A process must be associated with a user outside the computer. A process sometimes acts as a member of one or more groups. To represent this feature, a process has two attributes: *a UID and a list of GIDs.*

A file is owned by a single user or shared among several users and groups. To represent this feature, a file includes the following attribute: a list of UIDs and GIDs. We call this list *a belonging list* or *blist* for short.

Table 1 summaries the operations for a file system. The operations list and remove are name-level operations, so they also appear when the target is not a file system. The other operations are specific to files, and these correspond to the name-level operation resolve. There is another operation, register, which is performed on creation of a file.

Access control is performed with the name-level operations (register, remove, list, and resolve). In case of resolve, the name service takes a specific file-level operation (read, write, exec, get-attr, or set-attr). For example, when a process tries to open a file for reading, the name service takes the file name, the process attributes, and the operation resolve for read. Based on these pieces of information, the name service looks up a file and determines whether or not the operation is allowed.

3.2 Access control pseudo symbols

In SetNS, a name is represented as a sequence of symbols. To describe access control, we introduce a new concept *called access control pseudo symbols (ACPSes).* Table 2 shows selected ACPSes for files. The system administrators can add new ACPSes. An ACPS starts with the character `':'` to distinguish from normal symbols.

Each ACPS associates with a list of pairs: *a predicate and a permission list.* The predicate `UID-Match` means that the UID of the process is included in the blist of the file. The predicate `GID-Match-Any` means that one of GIDs of the process is included in the blist of the file. In other words, the process and the file share at least one GID. The predicate `GID-Match-All` means that the process has all the GIDs of the file. In other words, the GIDs of the file are included in the GID of the process. The predicate `True` means always true. A permission list is a list of allowed operations and a subset of the operations in Table 1.

The ACPS `:private` means that the owner (an individual user) can perform all operations, and the other people cannot perform any operation. The ACPS `:group-exclusive` is similar to `:private`, but it differs in that `:group-exclusive` works for GIDs.

The ACPS `:group-export` allows group members to perform nondestructive operations. *nondestructive operations* mean read, list, exec and get-attr. The ACPS `:group-export` has two entries. The first entry is for the individual owner. The second entry is for group members.

The ACPS `:public` allows the owner of the file to perform all operations and the other people to perform nondestructive operations. The ACPS `:group-public` allows the owner and group members to perform all operations and the other people to perform nondestructive operations. The ACPS `:hidden` disallows the operation list, so a user cannot browse the file name with the command `ls`.

We believe that ACPSes in Table 1 are enough for everyday use. The fourth column of the table includes the corresponding umasks and default file modes of Unix. In Unix, the process attribute *umask* affects the modes of newly created files. The umask 0, 2, and 7 mean that the permissions of all operations, non-destructive operations, and no operation, respectively. Each digit of a umask corresponds to the owner, the group, and the other people. Frequently used umasks are 077, 022, 007, 027, 002. They corresponds to ACPSes `:private`, `:public`, `:group-exclusive`, `:group-export`, and `:group-public`, respectively.

3.3 Usage of access control pseudo symbols

In this subsection, we show the usage of ACPSes. Note that a user does not have to specify ACPSes on name resolution, that is, on the operations read, write, exec, set-attr, and get-attr.

Table 1: The operations for files

Operations	Description	Operations	Description
list	Lists names or symbols of files	exec	Execute the content of a file
remove	Remove a name of a file	get-attr	Get the attribute of a file
read	Read the content of a file	set-attr	Set the attribute of a file
write	Write the content of a file		

Table 2: Selected access control pseudo symbols (ACPSes) for files

Pseudo Symbols	Predicates	the permission list	Unix umasks and modes
:private	UID-Match	ALL	077 rwx------
:group-exclusive	GID-Match-Any	ALL	007 rwxrwx---
:group-export	UID-Match GID-Match-Any	ALL NONDESTRUCTIVE	027 rwxr-x---
:group-public	UID-Match GID-Match-Any True	ALL ALL NONDESTRUCTIVE	002 rwxrwxr-x
:public	UID-Match True	ALL NONDESTRUCTIVE	022 rwxr-xr-x
:locked	True	NONDESTRUCTIVE	
:hidden	True	ALL-list	

ALL==list|remove|read|write|exec|get-attr|set-attr
NONDESTRUCTIVE==list|read|exec|get-attr

Therefore, a user does not have to worry about ACPSes in these cases.

3.3.1 Files belonging to an individual user

In a conventional file system based on a tree-based name service, files belonging to an individual user are often named with a prefix, such as `/home/alice`. In SetNS, such files typically have the names that include the symbols `alice,:private` or `alice,:public`. The symbol `alice` is a normal symbol, and means that the file is owned by Alice. To express the owner, the file also has to include the UID `alice` in the blist. The following file name is for Alice's mailbox to save e-mail from or to Bob:

- `alice,Mail,from-to-bob,:private`

The following file name is for the web home page of Alice:

- `alice,public_html,index.html,:public`

3.3.2 Files belonging to a single group

If a file belongs to a single group, its file name should include a normal symbol that is associated with the group. Furthermore, the blist of the file must include the GID of the group. The following name is for a file that holds a private key of the host:

- `wheel,secure-shell,deamon,host-key,rsa,rsa-private,:group-exclusive`

In this example, a normal symbol `wheel` means the system administrators group.

3.3.3 Cooperation among groups

If a file is shared by multiple groups, its name should include multiple normal symbols that are associated with these groups. Furthermore, the blist of the file must include all the GIDs of the groups. For example, a file is shared between the group `wheel` and the group `students`, the file should include the symbols `wheel,students,:group-exclusive`. Furthermore, the blist of the file must include the GID `wheel` and the GID `students`. The ACPS `:group-exclusive` can be replaced with `:group-export` or `:group-public`. The following name is for a file that holds a message of a bulletin board of student administrators.

- `wheel,students,:group-exclusive,bulletin-board,article,100`

3.4 Setting access control pseudo symbols

ACPSes are usually attached to files when these files are created. Users can also change ACPSes of existing files. There are two typical methods to attach ACPSes to new files. First, each application specifies ACPSes when the application creates a new file. For example, an e-mail program includes ACPS `:private` when it creates a mail box. Second, we use the process attribute context. Suppose that a user is going to create web pages with a text editor. Before the user executes the text editor, the user changes the context of the shell as follows:

```
% cd alice,public_html,:public
```

After that, newly created files include not only the normal symbols `alice,public_html` but also the ACPS `:public`.

To change ACPSes in existing files, we provide the command `set-acps`. The next script sets the ACPS `:private` to mail box files.

```
% set-acps :private alice,Mail
```

First, this command gets file names that includes symbols `alice,Mail`. Next, for each file name, this command sets the ACPS `:private`.

4 The implementation of the access control mechanism for SetNS

We have implemented the access control mechanism for SetNS in the file system of Solaris [10], a variant of Unix.

4.1 Implementation of SetNS with file name mapping

In Solaris, we do not modify the kernel to implement SetNS, but we use the process trace facility though the `/proc` file system [10]. Using this facility, we can capture the invocations of the system call `open()` and other ones that take file names. On entry of a system call, we replace the file name in the argument of the system call if the name is of SetNS. We map a name of SetNS with a system-generated one that holds the contents of the file. After that, we continue the execution of the system call, so the Solaris kernel reads the mapped file name. Consequently, processes can manipulate file names of SetNS while keeping the kernel unmodified.

In this scheme, access control is performed before mapping a name of SetNS to a name of Solaris. If the access control mechanism of SetNS returns `DENY`, we replace the file name of SetNS with an illegal one and continue the execution of the system call. Since the file name is illegal, the kernel returns an error soon. On exit of the system call, we modify the error code with an appropriate one.

4.2 *Performance*

We have shown that the performance of SetNS is bounded by the number of I/O [9]. Table 3 summarizes the result when Btree is used for indexing. The number of I/O is determined by the following parameters.

l**:** The number of symbols in the given argument of the operation.

n**:** The number of names in the system.

h**:** The number of names that are matched with the given argument on the list operation.

We chose Btree rather than hashing because Btree is useful for browsing. If we use hashing, $O(\log n)$ will disappear.

The number of access control pseudo symbols is small and we can keep all the pseudo symbols in main memory. Therefore, the number of access control pseudo symbols does not affect the number of I/O. In other words, adding access control does not increase the number of I/O compared with the previous implementation without access control.

5 Evaluation and related work

In this section, we evaluate the access control mechanism for SetNS by comparing with tree-based name services and related work.

5.1 *Comparisons with access control in tree-based name services*

We have discussed about the description of access control in tree-base name services in Section 2.2. Since SetNS does not have the idea of subtrees, it is impossible to use such simple description. To supplement this feature of SetNS, we have introduced the idea of access control pseudo symbols. We can add a pseudo symbol in an arbitrary place in a name as normal symbols. Furthermore, we can omit pseudo symbols as normal symbols. These features are appropriate for SetNS, so we can keep the syntax and semantics of the names in SetNS.

As discussed in Section 3.4, we can include access control pseudo symbols in contexts as well as normal symbols. Furthermore, we can set access control pseudo symbols to a group of files with common symbols. Therefore, we can manipulate the names that contain common normal symbols and access control pseudo symbols, easily. This corresponds to the description of access control with subtrees in tree-based name services.

SetNS has been designed for dealing with names that are not suitable for tree-based name services. SetNS is not almighty as a tree-based name service is not almighty. If a target has a tree structure, a tree-based name service is more suitable than SetNS, obviously. There exist some targets that do not have tree structures, and SetNS is more suitable for the targets than

Table 3: The order of the number of I/O when Btree is used for indexing.

Operations	SetNS (Btree)	Tree-based name service
list	$O(h \log n)$	$O(l)$
register, remove, and resolve	$O(l \log n)$	$O(l)$

tree-based name services. If SetNS is suitable for a target, access control with pseudo symbols is also suitable for the target.

SetNS is suitable for organizing HTML files of the Web directory and SGML files of computer manuals[9]. Similarly, SetNS is supposed to be suitable for organizing XML files that do not match with a tree structure. On the other hand, when a single XML document is treated as a tree structure, SetNS is not suitable.

5.2 Comparisons with attribute-based name services

Some attribute-based name services allow reordering attribute-value pairs (AV-pairs) in a name.

Semantic File Systems extract attributes from tree-based file systems, and provide supplementary name resolutions with the attributes [4]. In Semantic File Systems, a user can give a file name with AV-pairs. The order of AV-pairs is meaningless.

Exacting keywords from contents sometimes breaks an access control mechanism and makes a covert channel. For example, in a Web search engine, a user can guess the contents of inaccessible pages from typed keywords. Similarly, content-based indexing for local files must solve this problem. To avoid this problem, the Semantic File System in the paper [4] only indexes public readable files.

Directory services based on X.500 also take AV-pairs to search nodes [2]. LDAP[5] is a lightweight implementation of X.500 on TCP/IP. LDAP is often used for access control based on a tree structure. A directory service based on X.500 allows reordering of AV-pairs [6]. Allowing reordering of AV-pairs disables the normal access control facility based on a tree structure. The article [6] does not discuss about this problem.

In this paper, we have discussed a similar reordering problem in SetNS and shown a solution for the problem in SetNS.

5.3 Simplicity and flexibility

In access control, simplicity is often given priority over flexibility.

NTFS is a file system that is used in the Windows NT operating system and its descendants [3]. NTFS allows a user to attach an access control list (ACL) for each file. ACLs of NTFS are similar to ACLs in packet filters of firewalls, and they have strong power of expression. However, it is hard to write ACLs of NTFS for ordinary users as to write ACLs of packet filters. In NTFS and packet filters, an ACL can include both permissive (positive) and restrictive (negative) entries. The order of entries in an ACL changes the meaning of the ACL, completely. To reduce complexity, Window 2000 limits the order of entries to the restrictive entries following the permissive entries in the graphical user interface. Linux Security Module (LSM) [11] is a framework for providing additional access control in Linux. LSM fundamentally provides only hooks to restrict access to objects.

In SetNS, an ACPS expresses only restrictive access control, and the order of ACPSes does not affect the meaning. It is much easier for ordinary users to use ACPSes of SetNS than to use ACLs in NTFS. SetNS does not have to perform resolving conflicts among positive and negative authorizations in [1].

6 Conclusion

In this paper, we have described an access control mechanism for SetNS, a name services based on set of symbols. SetNS allows changing the order of element strings or symbols, abbreviating a long name, and aggregating names with common element strings.

In this paper, we have introduced a new idea called access control pseudo symbols. A user can reorder and omit access control pseudo symbols in a name as well as normal symbols. Therefore, they fit the syntax and semantics of SetNS. The proposed access control mechanism also supports coordinating multiple groups of users.

The proposed access control mechanism has been implemented in a file system of Unix with a process trace facility. The performance is bounded by the number of I/O. For the operations register, remove, and resolve, the order of the number of I/O is $O(l \log n)$ if Btree is used for indexing. In this case, n is the number of names in the system, and l is the number of given symbols in the arguments. Compared with the previous implementation of SetNS, no extra I/O is needed for adding the access control facility.

Currently, users have to choose access control pseudo symbols from the existing global table. Power users desire to use their own access control pseudo symbols. We would like to extend our access control mechanism to support this facility. We are also interested in implementing SetNS in Web servers with access control.

Acknowledgments

This research was supported by Japanese Government / Ministry of Education, Culture, Sports, Science and Technology, Grant-in-Aid for Young Scientists (A) No.14702062.

References

[1] Elisa Bertino, Sushil Jajodia, and Pierangela Samarati. A flexible authorization mechanism for relational data management systems. *ACM Trans. Inf. Syst.*, 17(2):101–140, 1999.

[2] CCITT. The directory, part 1: Overview of concepts, models and services. CCITT Draft Recommendation X.500, 1988.

[3] Helen Custer. *Inside the Windows NT File System*. Microsoft Press, 1994.

[4] David K. Gifford, Pierre Jouvelot, Mark A. Sheldon, and James W. O'Toole, Jr. Semantic file systems. In *Proceedings of the 13th ACM Symposium on Operating Systems Principles*, pages 16–25, 1991.

[5] J. Hodges and R. Morgan. *Lightweight Directory Access Protocol (v3): Technical Specification*. RFC3388, September 2002.

[6] G. W. Neufeld. Descriptive names in X.500. In *Proceedings on Communications Architectures & Protocols*, pages 64–71, 1989.

[7] The Open Directory Project. `structure.rdf.u8.gz`. `http://dmoz.org/about.html`, 2001.

[8] W3C Proposed Recommendation. *RDF/XML Syntax Specification*. `http://www.w3.org/RDF/`, 2003.

[9] Yasushi Shinjo, Katsumi Nishio, and Kozo Itano. SetNS : A name service based on set of symbols. *IPSJ Transactions on Advanced Computing Systems (ACS)*, 44(SIG 11 (ACS 3)):201–214, 2003.

[10] Sun Microsystems, Inc. *Solaris 8 Reference Manual*, 2000.

[11] C. Wright, C. Cowan, James Morris, S. Smalley, and G. Kroah-Hartman. Linux security modules: General security support for the linux kernel. In *USENIX Security Symposium*, 2002.

Information Modelling and Knowledge Bases XVI
Y. Kiyoki et al. (Eds.)
IOS Press, 2005

eTrium: Concepts and Knowledge in Practice

Zdenko STANÍČEK, Filip PROCHÁZKA
Faculty of Informatics, Masaryk University,
Botanická 68a, 602 00 Brno, Czech Republic

Abstract: The development of rule and model based systems has currently become a hot topic of Information Technology. In this paper we present the Diamond Modeling Tool (DMT) for knowledge management and its extension – the eTrium technology, which can be seen as an architecture of model-based IS development. Applying this technology we demonstrate the way of building up a business-rule based information system. This system does not fix a current state of a company, but supports a permanent business tuning according to the needs of the company.

1. Motivation

The development of rule and model based systems has currently become a hot topic of Information Technology [10]. Such a framework makes it possible to create extremely flexible information systems. A model of a business domain plays a crucial role of a *theory* of the IS; having an adequate theory, behavior of the system does not have to be 'hardwired' in a program code, but it is flexibly given by an interpretation of the theory. Hence we need an expressive logical apparatus for defining conceptual models, including ontologies of business domains controlled by well-specified logical rules. Transparent Intensional Logic (TIL) and its applications, namely non-traditional theory of structured concepts and HIT conceptual scheme, is such a fundamental tool, [5], [4] and [7]. In this paper we present the Diamond Modeling Tool (DMT) for knowledge management, the underlying logical apparatus of which is based on TIL and HIT. An extension of DMT – the eTrium technology, which can be seen as a kind of model-based architecture of IS development, is introduced. Applying this technology we demonstrate the way of building up a business-rule based information system.

2. The Diamond Tool Definition

The crucial point of an IS supporting business activities is its ability to *adjust and improve business functioning*. Typically we need to represent and utilize knowledge of *business rules* (i.e. rules that govern the business – cf. [1], [2], [3]) and knowledge of *rules of ICT support*. These rules concern connections between objects, operations on them and ways of organizing these into certain categories. This can be seen on Fig. 1—the Diamond Modeling Tool—a universal tool for modeling and simulation of any conceptual system.

Each concrete/abstract object of our interest is represented in the tool by a *dmt-object*. The role of entity classes of classical conceptual models is played by Diamond *containers*. Elements of any kind can be inserted into the container as well as removed from it.

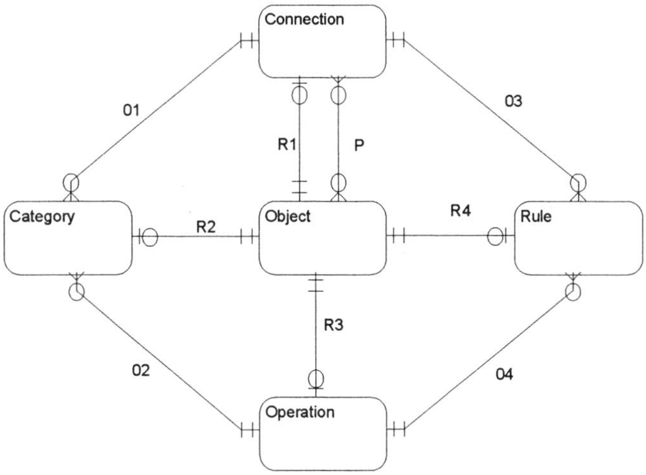

Figure 1. The Diamond Tool conceptual data scheme

Definition 1: *Container* (#Object) contains all the dmt-objects that can be mentioned in DMT, i.e. which can be assigned certain properties.

If the object we focus on turns to be a connection or an operation or a category or a rule, and we want to use it, then we link it to (#Connection) or (#Operation) or (#Category) or (#Rule), respectively, by an edge R1 or R3 or R2 or R4. Then we say that this object from the class (#Object) represents this connection or operation or category or rule—cf. Fig. 1. Note: an object may represent only one of the four mentioned possibilities.

Definition 2: *Container* (#Connection) contains *n*-tuples of dmt-objects; every element of the container (#Connection) is called a *connection*. Each element of the *container* (#Operation) is an algorithmically computable transformation from Diamond Tool states to a new DMT state. By the state of Diamond Tool we mean one particular filling of the Diamond Tool (as a container) by elements-instances that may dwell in the Diamond Tool. The elements of the container (#Operation) are called *operations*.

Definition 3: Each element C of the *container* (#Category) is a container for dmt-objects such that a pair <Cn, Op> is uniquely assigned to C, Cn $\in$ (#Connection), Op $\in$ (#Operation); the operation Op serves as a characteristic function that recognizes objects belonging to the container C. The elements of the container (#Category) are called *categories*. The connection Cn is called a *defining connection* of this category. The operation Op is called a *defining operation* or an *evaluator* of this category.

Definition 4: Each element R of the *container* (#Rule) is a dmt-object such that a pair <Cn, Op> is uniquely assigned to R, Cn $\in$ (#Connection), Op $\in$ (#Operation); the operation Op performs a test on the validity of the rule R using the connection Cn. The elements of the container (#Rule) are called *rules*. The connection Cn is called a *specifying connection* of this rule. The operation Op is called a *specifying* (or *testing*) *operation* of this rule.

The edges 01 and 02 in Figure 1 link a category with, respectively, a *defining connection* and an *evaluating operation* of this category. The edges 03 and 04 link a rule with a *specifying connection* and the *testing operation* of this rule. The Diamond Diagram contains, moreover, the so-called P-edges (projection edges) that link individual objects with a connection that describes the given relationship between these objects.

The Diamond Tool contains a family of operations, which enable us to create, delete, and update dmt-objects (CREATE, DELETE, WRITE, READ, OBTAIN, LABEL), to "travel" around the circumference of the Diamond (GET_CONNECTION, GET_OPERATION, GET_CATEGORY, GET_RULE), and to work with P-edges (projection of a static connection onto its component, adding and removing an element to/from a dynamic connection, testing if an element belongs to a connection, etc.).

The most important are those operations that make the Diamond Tool a *universal modeling tool.* Edges R1, R2, R3 and R4 are jointly called R-edges. By means of R-edges switching between mentioning (operation MENTION) and using objects (operation USE) is realized. If we stand on an object of the class (#Object), which has been mentioned, then in order to use the object (operation USE) we cross along the respective R-edge to the represented connection (#Connection) or operation (#Operation) or category (#Category) or rule (#Rule), which can be directly used. Operation USE may fail on objects belonging to (#Object) that are "mere" objects, i.e., do not represent any category, operation, rule or connection. An example of such an object can be a concrete document.

If we are in a vertice of the Diamond Diagram, then the operation MENTION takes us, along the respective R-edge, to the representing object which can be directly mentioned. Operation MENTION applied to an object belonging to (#Object) returns the same object without any change. Thus, this operation cannot fail.

Operation USE applied to a connection belonging to (#Connection) makes it possible to use the connection: it returns a list of objects—components—of the given connection. Similarly, operation USE applied to a category belonging to (#Category) makes it possible to use the category (to inform on its current content, or possibly, to add further elements to the category). Operation USE applied to an operation belonging to (#Operation) simply executes this operation. Finally, operation USE applied to a rule belonging to (#Rule) performs a test that checks the validity of the given rule.

Categories are constructed in an extremely universal and flexible way. We can work with categories specified simply by the list of their elements, or with categories that use their relationships to other categories for the evaluation of their own content, with fuzzy categories, etc. Similar flexibility and universality has been applied in case of rules.

Due to representative edges R1, R2, R3 and R4, we can mention any DMT object—relate it with other entities, categorize it and fill its attributes. We can create connections, e.g. between categories (more accurately, between representative objects of these categories), or between categories and rules, we can categorize the categories, etc. This flexibility is of a key importance. Unlike traditional modeling tools, which strictly separate the modeled world from the "tool world", the Diamond Tool works within one *universal world.* Using the Diamond it is possible to integrate models on various levels of abstraction. For instance, we can create a model of reality first, then develop another model used for tuning the former, and still another model above the two, which can learn by "observing" their interactions. Switching between USE / MENTION operations makes it possible to use the knowledge represented in the center of the Diamond as well as to develop it (by mentioning in the diamond vertices).

The Diamond Tool is implemented in Prolog, and it is used to develop other systems in the form of plug-in units of the Diamond.

3. The eTrium Architecture

To obtain a practical efficient tool it is necessary to embed the Diamond into such architecture that provides a communication interface, data store facilities, and a kind of inference. This architecture is called the eTrium architecture. In what follows we use as an example the Universal Information Recorder (UIR), implemented in 2003 [11].

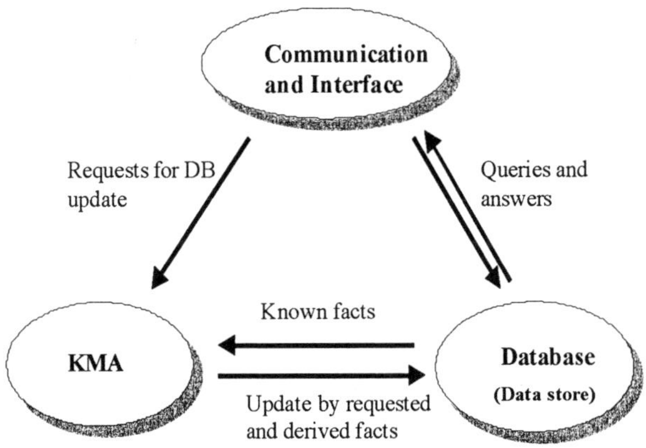

Figure 2. eTrium Architecture

Here are the key components:
Communication and Interface is realized as a web solution.
Database is a simple store of facts that can be easily reported on user requests.
KMA (knowledge management agent) is a program that realizes the whole business logic, using a centralized knowledge base. It has been built-up using the Diamond Tool.

The system is *universal*. It can be used both as an IS to support business operative processes, and as a modeling tool in which it is possible to develop and modify on the fly (without re-programming) its own functioning. There are three cases of communication (co-operation) between components [11]:

Case A corresponds to routine queries on an operative execution of business processes.
Case B corresponds to database updates.
Both in case A and in case B the system is used in order to *use* the information process.
Case C describes a business and model tuning. The system is used here in order to *mention* (i.e. to analyze and develop) changes in operative processes.

Case A: Answering a query
Query processing is realized by the Communication and Database components. The Communication interface component provides query forms. The Database component performs the query transaction.

Case B: Updating the database
The KM agent is fully responsible for the 'logical update' of the database. Obtaining a request from the Communication component, it performs all the logical inferences following from the required update operation. As a result the KM agent executes both the required and the inferred updates of the database.

Case C: Modifying business rules and/or their IT support

This case involves an update of knowledge base facts and rules controlled by the KM agent. The change of the KM agent is carried out as follows: first, the knowledge base is updated, and then facts stored in the database are made consistent with the new knowledge base state. The simplest way to perform such an update transaction is: first, drop out all the facts that have been inferred by the previous version of the KM agent, and second, use the new version of the KM agent to infer the new database state. It takes few minutes contrary to days or weeks of re-programming in a classical approach.

4. The Knowledge management agent (KMA)

A fundamental knowledge on a specific business area is determined by the HIT conceptual model, see [5], [6], [4], where HIT attributes over the base of sorts and integrity constraints connected with these attributes are specified. On the basis of the HIT schema we define object classes as containers for objects (instances of entity sorts), and relationship classes as containers for instances of particular associations between objects. Properties of objects are modeled by grouping objects into categories. Categories are grouped by categorizations.

Example: Consider a class Document, to which a categorization Document_State has been defined. According to business demands we insert categories Developed_Document, Authorized_Document, and Archived_Document as instances of the Document_State. If some other document states are needed, we simply add (on the fly) new categories into the categorization.

The distinction between classes and categories can be characterized as the distinction between supported and unsupported containers, respectively:

Definition 5: An (object or relationship) *class* is a special kind of category such that:

- The ISA relationship between classes can be defined; the KMA makes use of the inheritance between subtype/supertype classes.
  Class A is *a subclass* of class B iff the defining property of the class B is a pre-requisite of the defining property of the class A (i.e., necessarily, each individual that has the property A has the property B as well). Classes form a tree structure with respect to the ISA relationship.
- Each object has to be a direct element of just one class (the most inner subclass in the tree hierarchy containing the object); membership of the object in all the super-classes is a derived one.
- Each class introduced by user has to be a *specialization* of an existing system super-class.

The general categories are characterized as follows:

- An object can be a member of n categories, $n \geq 0$.
- The system does not implicitly support relationships between categories.
- A user can specify relationship between categories. In this case the user defines rules which provide a desirable behavior.

The KMA communication language is based on the HIT functional approach [4]. Example:

```
for class of relation partners_related_to_given_product create new
subclass with features:
```

```
* name = product_guarantor
* elements:
   ** guarantor from class partner
   ** product from class product
* rotations:
   ** guarantor <- product = guarantor_of_given_product
   ** product <- guarantor = products_which_are_guaranteed
                            _by_given_partner
```

The example illustrates a specification of a new relationship class using specialization of the existing class partners_related_to_given_product. First, roles of relationship participants are defined (guarantor, etc.) by specifying the direct super-class. Second, rotations of the relationship are functionally defined together with their semantics. For instance, the first rotation expresses a function that supplies as an output the partner who is a guarantor of the product – input argument.

A new categorization is defined as follows:

```
for class document create new categorization with features:
   * name=document_state
```

The categorization is filled by categories using the following instruction:

```
for categorization document_state create new categories:
  * category with features:
      ** name=developed_documents
  * category with features:
       ** name=authorized_documents
```

Definition 6: The **KMA-*rule*** (abbreviated simply rule) is a tuple (list_of_conditions, list_of_actions), where conditions from list_of_conditions are rules according to Definition 4 and actions from list_of_actions are operations according to Definition 2. All the actions from the list_of_actions are executed provided all the conditions from the list_of_conditions are valid.

Note that the notion of a rule is used within KMA in a wider sense than it has been defined in the Diamond Tool (where a rule is a simple True/False condition); this is a common meaning of the term "rule" in the area of Business-rule based information systems.

Three types of (KMA-)rules are distinguished in our system:

- categorization rules
- constraints
- activity rules

Categorization rules express a membership of objects in categories. An example is as follows: "Each project in which one to three delayed (customer) payments occur is rather problematic." KMA specification is:

```
for each P:project which satisfy
  number of payments with features:
      * is delayed
      * is in payments_in_given_project(P)
 is between 1 and 3
arrange:
  P has features:
      * is rather_problematic
```

Constraints are restrictions on the acceptable states of the system. An example is: "For each project that is a loan, and for which an activity of business plan approval is planned, there must exist a document Business Plan and it must be completed." KMA specification:

```
for each P:project which satisfy
  P has features:
      * is in loan
  and
  there exists task with features:
      * is business_case_approval
      * is in tasks_in_given_project(P)
must hold:
  there exists document with features:
        * is business_case
        * is complete
        * is in documents_in_given_project(P)
if not report error: "If you want to approve business case, you
                      must have complete one"
```

Activity rules are typically triggered by an event and they execute some actions. They are often used as macro-instructions making the work easier. Example: "When the project is marked as a loan, then it must be connected to an elaborated (incomplete) document which is a rudiment of the Business Plan." KMA specification:

```
on event P:project was assigned to category loan
do actions:
- create document with features:
   * name=business_case_in_a_loan_project
   * is business_case
   * is incomplete
   * is in documents_of_given_project(P)
```

Note that by actions of the activity rules we can specify instructions defining the knowledge base of the given business area (as explained in the beginning of this paragraph). The whole domain is specified by activity rules. Moreover, it is possible to do it in a *dynamic way* in the running system.

The dynamic component of the KMA is its ***inference machine.*** It makes it possible to derive *implicit* knowledge using the explicit facts and rules stored in the knowledge base. From the logical point of view, KMA realizes the general resolution and unification method of automatic deduction.

5. Outcomes and conclusions

We described a new progressive architecture of an IS design – the eTrium. It makes it possible to create highly flexible, dynamic systems of IT support [8], [9]. As soon as the top-management of a client agrees on the business logic, the IT system begins to support an efficient and flexible business tuning. All that has to be done is the specification of basic categories, facts and rules in a user-friendly natural-like language. The whole process of launching a new version of an IS component can be managed within a two-day user/expert seminar.
This process minimizes the need for mutual communication between programmers (the world of IT) and business people (owners of the problem). On the other hand, the role of a "business-IT consultant" is highly stressed. He provides a necessary support, facilitates client's agreements on business logic, decisions on categories, facts and rules.

Concluding we would like to stress the invaluable role of the underlying logical theory –Transparent Intensional Logic. Using the powerful logical tool enabled us to exploit the idea of using and mentioning any entity of the infinite type hierarchy within the theory, avoiding thus many discrepancies and pseudo-problems common to traditional approaches. Last but not least, surprisingly progressive results have been obtained by exploiting the theory of *concepts* viewed as TIL logical constructions assigned to expressions as their *structured meanings*.

References

[1] Business Rules Group (Eds. Ronald G. Ross, Keri Anderson Healy): Organizing Business Plans - The Standard Model for Business Rule Motivation. www.BusinessRulesGroup.org

[2] Business Rules Group (Eds. David Hay, Keri Anderson Healy): Defining Business Rules – What Are They Really? www.BusinessRulesGroup.org

[3] Business Rules Group (Ed. Ronald G. Ross): Business Rules Manifesto. www.BusinessRulesGroup.org

[4] Duží, M., Materna, P.: Intensional Logic as a Medium of Knowledge Representation and Acquisition in the HIT Conceptual Model. In: Information Modelling and Knowledge Bases XIV. Ed. Hannu Kangassalo, Eiji Kawaguchi, Amsterodam, Netherlands. IOS Press , 2003, vol. Vol. 94, p.51-65, ISBN 1 58603 318 2

[5] Duží, M., Pokorny, J.: Semantics of General Data Structures. Information Modelling and Knowledge Bases XI, IOS Press 1998, pp.107-124

[6] Duží, M.: Logical Foundations of Conceptual Modelling Using HIT Data Model. Information Modelling and Knowledge Bases XII, IOS Press, pp.65-80

[7] Duží, M.: Concepts, Language and Ontologies (from the logical point of view). In: Proc. of the 12th European-Japanese Conference on Information Modelling and Knowledge Bases, ed. by Y. Kiyoki, H. Kangassalo, E Kawaguchi. Kitakyushu, Japan, 2003, pp. 197-212.

[8] Hřebíček, J., Staníček, Z., Soukopová, J.: Support of Processing Public Projects for Cohesion Fund Using ICT Tools. Proc. of IADIS International Conference Applied Computing 2004, March 2004, Lisbon, Portugal

[9] Hřebíček, J., Staníček, Z.: New ICT Tools for Regional Crisis Management in the Czech Republic. Proc. of ISESS 2004: A Workshop on Environmental Risk and Emergency, James Madison University, Harrisonburg, Virginia, USA, May 2004

[10] Sinur, J.: The Business Rule Engine 2003 Magic Quadrant. Gartner, Research note Markets, 07 April 2003. www.gartner.com/reprints/fairisaac/114166.html

[11] Staníček, Z., Procházka, F.: The Diamond Tool: A Way of Effective Development and Utilization of Knowledge. KYBERNETIKA, special issue, June 2004. (to appear)

Information Modelling and Knowledge Bases XVI
Y. Kiyoki et al. (Eds.)
IOS Press, 2005

Knowledge Transfer between Networking Organizations – a Case Evaluated and Lessons Learned

Mattias STRAND and Per BACKLUND
Information Systems Engineering Research Group, University of Skövde
Box 408, 542 28 Skövde, Sweden

Abstract. In order for organizations to survive in a competitive and ever-changing marketplace, it is of vital interest to acquire knowledge about the actors in the organizational environment e.g. customers, suppliers, and business partners. One way of acquiring such knowledge is to cooperate with other organizations in knowledge sharing networks. This paper describes the results of a project, targeted towards the transfer of manufacturability knowledge from manufacturing organizations back to designing organizations, within the electronic circuit industry. However, the project was only partly successful, since the needs of the receiving organizations were not in alignment with the source organizations′ willingness to supply relevant knowledge. In order to clarify the deficiencies of the project, ideas from the knowledge transfer framework of Gupta & Govindarajan [7] was applied. Their framework helps to conceptualize and analyze the knowledge transfer between organizations. The application of the framework to the project showed that the motivational disposition of the source units were crucial.

1. Introduction

Organizations have spent large resources on acquiring, storing, managing and distributing their own, internal knowledge, i.e. knowledge that is produced and used within the organizations′ own boundaries. However, besides the need for internally produced knowledge, organizations are nowadays facing an increased need for knowledge about and from their business environment and actors in that environment, e.g. customers, suppliers and business partners [3].

The primary reason for the increased interest in such knowledge comes from the inability of organizations to stay flexible and competitive when only using knowledge produced and stored within the organization itself [1]. The results of their study showed that most of the knowledge domains, which managers in different organizations considered as most important, were external. In order to acquire knowledge about external domains, such as business partners or suppliers, organizations can buy data or information from marketing institutes or from organizations specialized in selling data [2]. Another way for organizations to acquire external data and information is to cooperate with other organizations in different types of network constellations [3].

This paper describes the results of a knowledge management (KM) research project aimed at establishing routines for allowing transfer of manufacturability knowledge, from manufacturing organizations back to designing organizations, within the electronic circuit industry. To facilitate the knowledge transfer, a Web-based knowledge management system (KMS) was also designed and implemented. The project group was consisted of personnel from a major Swedish industrial research institute, a number of multinational companies within the electronic circuit industry, and the Department of Computer Science at University of Skövde. However, the project was only partly successful, and this paper analyses and discusses why. The analysis was strongly influenced by the conceptualised framework for knowledge transfer within multinational corporation (MNC), given by [7]. Since their framework is focused on MNCs and the project consists of participants from separate organizations, the framework is described and adjusted to fit this particular purpose.

The paper is outlines as follows. Section 2 briefly describes the KM spectrum. Section 3 describes the case. Section 4 describes the conceptualized knowledge transfer framework used to analyze the deficiencies of the project and why the project was only partly successful. Section 5 includes the analysis of the knowledge transfer, based upon the framework given in Section 4. Finally, Section 6 includes a discussion on the reasons for the deficiencies of the project.

2. Knowledge Management

Knowledge and knowledge management is difficult to describe, since there exist many descriptions and definitions thereof (e.g. [9], [5], [12]). The fluid mix of concepts, technologies and approaches in knowledge management also contributes to make the whole area almost indefinable. Therefore, instead of defining KM, the different elements constituting KM will be described. The descriptions are based on the work conducted by Binney [3] and his separation of the knowledge management spectrum into different elements is a necessity for comprehending the area and its activities and applications. Binney [3] claims that the concept of KM includes six distinct elements; Transactional KM, Analytical KM, Asset Management KM, Process KM, Developmental KM, and Innovation and Creation KM. In addition, each element has a particular aim to fulfil in order to allow organisations to be able to cover the whole KM-spectrum. Below, the essence of these elements is briefly described.

Transactional KM is focused on supporting the user in day-to-day tasks such as completing a transaction or handling a customer query, by reusing already existing knowledge. When a user is confronted with a problem, applications often support the user by providing solutions to similar problems.

Analytical KM is focused on creating new knowledge. The core of analytical KM is the integration of large amounts of data and information, from both internal and external sources, which is then used to derive previously unknown trends and patterns.

Asset management KM concerns processes associated with the management of knowledge assets. Asset management involves one of the following:

1. Management of codified explicit knowledge.
2. Management of intellectual property.

When these assets have been captured, they are made available to the users in the organization. Binney [3] uses the analogy of a library, since knowledge assets, just as the books in a library, are catalogued and made available to the user. These knowledge assets are often by-products to the ordinary business.

Process-based KM covers, as the name implies, processes. More specific, this element is focused on the codification and improvement of processes and procedures and methodology.

The process-based KM activities often have their origin in total quality management (TQM) and process reengineering activities.

Developmental KM focuses on increasing the competencies and capabilities of organizations knowledge workers. The KM element concerns both the transfer of explicit knowledge and the development of tacit knowledge. The explicit knowledge is transferred via training interventions whereas the tacit knowledge is developed through developmental interventions such as experimental assignments or membership in a certain community of interest.

Innovation/Creation KM applications focus on the creation of a "learning" environment, in which the personnel of an organization or from different organizations can come together and exchange knowledge or create new knowledge. This KM element is given a lot of attention in literature and much research is currently devoted on how to create an organizational learning environment.

The above given elements will be used to describe which KM activities and elements that were focused on in the case. It is essential to be aware of the focus of the projects, from a knowledge management perspective, to be able to comprehend the issues which influenced the outcome of the research project.

3. The Case

The case study included the establishment of routines for knowledge sharing as well as the creation of a Web-based knowledge management system (KMS), aimed at integrating knowledge of design and manufacturability within the electronic circuit industry. The main motivation for the cooperation was the designers' need for knowledge on manufacturing limitations in order to increase the quality of the design of the electronic circuits and thereby reduce the number of product design cycles. Another motivation for the cooperation was the need to implement new manufacturing techniques in short time spans. The implementation of a new manufacturing technique complicates the product development process. The product development process covers stages from initial product development to large volume manufacturing, were the designers were responsible for the initial phases, whereas the manufacturers were responsible for the latter phases. The general idea was to form a feedback loop from the manufacturers, including knowledge and experiences from mounting components, back to the designers (Figure 1).

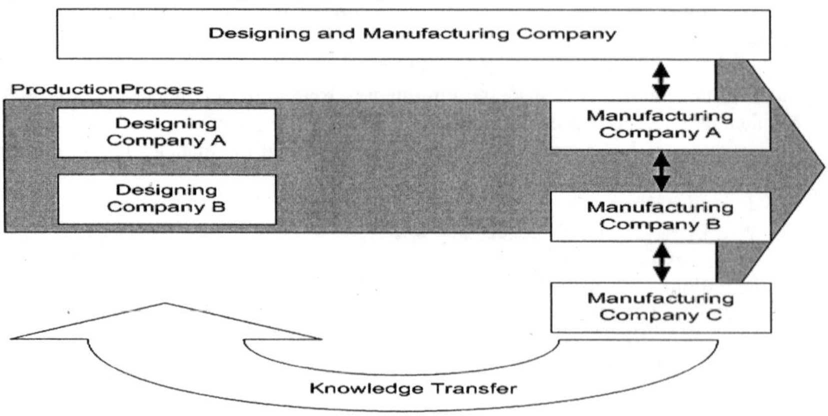

Figure 1. The knowledge transfer from manufacturing back to designing companies. Involving both vertical competition and horizontal cooperation.

Before describing the knowledge management elements applicable for this case, a more thorough description on the denotation of horizontal cooperation and vertical competition in Figure 1 will be given. This description is included, since it strongly influences the analysis on why the knowledge transfer was only partly successful. The horizontal cooperation is depicted by the grey arrow labeled *product process*. This is a rather straightforward process in which the work performed follows the ordinary patterns of production. That is, first design and then manufacturing. The vertical competition is depicted by the double-headed black arrows. The manufacturing companies are contractual manufacturers, meaning that they have to deal for every electronic circuit board to assembly and complete. Since many of the manufacturers have the same production capabilities, they are acting in a strongly competitive market and being able to match the other manufacturers offers are vital for survival. The designing companies participating in the network are having a slightly different situation. In the network, their situation was not exposed to such competition as the manufacturers, since the designing companies were targeted towards different types of products. This is also the reason why only the manufacturing companies are considered when claiming the vertical competition.

If relating the KM elements covered by the project to Binney's work, the following three KM elements were covered: 1) *Asset management KM,* 2) *Innovation and creation KM and 3) Analytical KM.* Of course, one may argue that most of the KM elements are covered by the case and that may truly be so. However, the elements pinpointed above were the most obvious ones.

Finally, the knowledge transfer was formalized via a Web-based KMS. The motivation for using a Web-based system to support the knowledge transfer was four-folded. Firstly, since the designers and manufacturers reside in different, distributed companies, there was a need to transfer knowledge between organizations and this transfer is strongly facilitated by utilizing Web-technology. Secondly, by utilizing the Web as the distribution channel, the system was based on well-established and rather inexpensive technology [4]. Thirdly, by utilizing a Web-based system, the knowledge transfer was formalized. This was important since there was a need to unease the pressure on key personnel in each participating organization. Previously, all knowledge transfer had been performed informally, based upon social networks. Finally, many researchers interested in KM claim that Web-technology is a prerequisite for efficient KM in organizations (e.g. [10], [11], [3]).

4. The Conceptualized Knowledge Transfer Framework

As a baseline for analyzing the knowledge transfer between the designing and manufacturing companies, the knowledge transfer framework for MNCs, given by Gupta and Govindarajan [7], was applied. However, their framework is intended for examining knowledge transfer between subsidiaries in a MNC, while the organizations involved in this particular case are individual organizations cooperating in a network. Still, we find the framework applicable, since it is based on general communication theory ideas and therefore easily applied to other contexts of knowledge transfer. In addition, to further motivate the applicability of the framework, the following motivations are given. Firstly, the studies conducted by [7] are focused on the same type of knowledge relevant as in this particular case. That is, procedural knowledge such as knowledge on product designs. Secondly, their framework is based upon the idea that knowledge transfer takes place in the context of networks. The importance of networking was also pinpointed in the description of the case.

The framework given by [7] is based upon the following 5 factors:

- Value of the source unit's knowledge stock
- Motivational disposition of the source
- Existence and richness of transmission channels
- Motivational disposition of the target unit
- Absorptive capacity of the target unit

Value of the source unit's knowledge stock. This factor concerns the value of the sources knowledge to the whole network. The greater the value of the source's knowledge is, the greater would its attractiveness for the whole network be. In addition, [7] pinpoint the importance of making a distinction between duplicative and non-duplicative knowledge. The non-duplicative knowledge is the most important knowledge in networks and the amount of outflows from the source unit is strongly related to its amount of non-duplicative knowledge. If the source only is able to distribute duplicative knowledge, its existence in the framework is relevant to question. Finally, [7] give the following two important characteristics of the non-duplicative knowledge; 1) the source unit must create the non-duplicative knowledge on its own, and 2) the non-duplicative knowledge must be of relevance for the whole network.

Motivational disposition of the source. As the name implies, this factor is focused on the willingness of the source unit to share it knowledge. This is important to pay attention to, since many organizations are unwilling to share their non-duplicative knowledge, in fear of loosing competitive advantage. Therefore, it is of utmost importance to establish reward systems, which may be used to increase the sources' motivation for sharing their knowledge.

Existence and richness of transmission channels. This factor concerns the abilities to share the knowledge within the network. In order to share the knowledge, communication channels must be established, allowing the units to distribute the knowledge between them. These channels may be formal or informal. With formal is referred to e.g. task forces or permanent committees, linking the units together. The informal channels concerns those communication channels base upon familiarity with personnel in other units and social networks.

Motivational disposition of the target unit. This factor is focused on the willingness of the target unit to adopt and integrated the acquired knowledge in the work. According to [7] one has to be aware of the possibility that the target unit is unwillingly to incorporate the work routines, due to its "external" origin. They refer to this as the "Not-Invented-Here (NIH)" syndrome.

Absorptive capacity of the target unit. As the name implies, this factor is focused on the ability of the target unit to recognize the value of the acquired knowledge, assimilate it, and apply it in the daily work. Gupta and Govindarajan [7] have identified the following two reasons why the absorptive capacity may differ between organizations: 1) the extent of prior knowledge and 2) to what extent the units are alike with respect to culture, social structures etc.

The description of the factors is directed towards the case context, since the framework originally is directed towards another type of organizational constellation, i.e. multinational corporations. However, the only adjustment that was made relates to the naming of the participants. Originally, the term subsidiary was suggested by [7], but in this paper the term unit was applied, since it is more general and may be used to describe individual organizations participating in the network. Nevertheless, the core of the factors is kept consistent with the original ideas of [7].

5. Analysis and Results

In this section we will analyze why the knowledge transfer in the case described was only partly successful. The analysis of the knowledge transfer in the dynamic network will be based on the factors given by [7]. The analysis of the knowledge transfer is divided into 5 parts, were each part is devoted to one factor each.

However, before analyzing the factors, one general remark will be given. That is, some of the problems associated with MNC units sharing knowledge have not been relevant in this case. Primarily because of the fact that the organizations participating in the cooperation describe in the case are volunteering in the cooperation. This is not the case for units in the MNC. They are in the MNC and will be compared to other equal units in the MNC. For them it is more important to be competitive and show the managing board of the MNC that they are skillfully and create a lot of value, than it is for the participants in the case described to show the broker. Being aware of this is important, since it gives the frame for the analysis with respect to the organizations willingness to share data in and their willingness to participate in the network. Since they were not forced into the network, by internal organizational legislations, they had not much to loose in an initial participation. While keeping this in mind, the factors will be analyzed in the following sections.

Value of the source unit's knowledge stock. The analysis showed that the value of the knowledge of the manufacturers was vital for the ability of the designers to be able to compete in a strongly competitive industry. Therefore, the value of the source units′ knowledge stock was very high. Especially since the knowledge were of a non-duplicative kind, fulfilling the requirements of being both created by the source unit and of high relevance for the whole network. For example, when mounting components to the printed circuit board, one must have the knowledge on how reliable such mounting is, since the product may be used in many different types of applications, ranging from simple toy given for free when purchasing hamburgers to advanced systems in cars and trains. In order to be able to classify the product to the different application areas, testing must be performed. The result of the testing process, conducted by the manufacturer, is mounting knowledge created within the unit and the knowledge is very valuable to the whole network, particularly the designing companies, since they than have the knowledge on manufacturing limitation and might thereby decrease the number of design loops.

Motivational disposition of the source. The analysis gave at hand that the manufacturing companies were generally rather unwilling to share their knowledge within the network. For many of the participating manufacturers, it was more important to keep their non-duplicative knowledge secret and thereby sustain some competitive edge in relation to their competitors, than to share their knowledge within the network. Another matter that also was identified during the analysis is the relevance of what Gupta and Govindarajan (2000) refers to as *information monopoly*. They claim that in MNCs there is always a risk that information monopoly appears, since the source uses its non-duplicative knowledge as a mean to sustain its position in the MNC and therefore are highly unwilling to transfer that knowledge to the receiving units. In the network, information monopoly appeared, but not between the sources and the receiving units. Instead the monopoly was appearing between the manufacturing companies. Due to the organization of the network, with multiple competing manufacturers, they were very restricted with the knowledge being transferred. This is well aligned with the ideas of horizontal cooperation and vertical competition (Figure 1). Another reason for the unwillingness of the manufacturers to share their knowledge, besides the value of the knowledge in it self, is the resources spent on creating such knowledge. The testing processes of, e.g. mounting reliability is rather expensive and if sharing the such knowledge, the supplying organization is not only contributing with valuable knowledge, but also the resources spent on creating such knowledge. Finally, the analysis of this factor also showed that there were no reward systems established between the organizations or between the organizations and the broker. The underlying idea was that the value of the network should attract the participants to such an extent that different reward systems were considered as unnecessary.

Existence and richness of transmission channels. If concerning the existence and richness of transmission channels the analysis gave at hand that the channels were well established due to the broker. Since the broker was assigned the responsibility to cater for cooperation in the network and for the maintenance of the underlying system, the communication channels were formal. As a matter of fact, one underlying intention of the whole network was to formalize the communication channels. Previously, all sharing of knowledge was conducted informally, between key personnel in each organization. In order to decrease the workload on these personnel, the knowledge sharing was formalized with help from the web-based KMS.

Motivational disposition of the target unit. As described previously, the designers' motivational disposition of the manufacturing knowledge was very high, due to its contribution to their ability to compete in a competitive marketplace. Therefore, the designing organizations' willingness to adopt and integrate the acquired knowledge in their work-routines was very high. The main idea behind the project also made the designing organizations the first group of organizations that could clearly apprehend the value of the cooperation. Since they acquired very valuable knowledge, which they without any major adjustments could integrate in the design process, they were more the willing to cooperate. Furthermore, [7] claim that the NIT syndrome may occur, when units are to integrate external knowledge within their work routines. However, the analysis showed that no such syndrome occurred. On the contrary, as described above, the designers were more than willing to acquire the manufacturability knowledge from the manufacturers. This is also well aligned with the ideas presented in section 1, in which it was clearly shown on the need for organizations to acquire knowledge from e.g. their business partners.

Absorptive capacity of the target unit. The analysis gave at hand that the absorptive capacity of the designing organizations was high. This high absorptive capacity is mostly dependent on the outline of the cooperation, in which the knowledge transfer is a feedback loop from the manufacturers back to the designers. Since the knowledge transferred is closely related to the work conducted by the designers, in which the designers is responsible for the "theory" and the manufactures are responsible for the "practice", the ability of the designing companies to integrate the knowledge in their work-routines, is high. The designers is actually suggesting a design and the manufacturers is testing if it is the best design or even a possible design, the designers are thereby able to learn from its mistakes, with help from the manufacturers. It is not unrealistic that the absorptive capacity had been much lower, if the knowledge transfer was the other way around, without the influence of feedback. If the designers had transferred designing knowledge to the manufacturers, the knowledge would not have been grounded in the work conducted by these and therefore its value and relevance would have been lower. This ides is also well aligned with the ideas of [7] since they claim that the extent of prior knowledge is one major reason why the absorptive capacity may differ between organizations.

Finally, [7] state that besides the extent of prior knowledge, the degree of equality of e.g. social structures and culture between the distributing and receiving unit may influence the absorptive capacity. If organizations are alike in culture and social structure, the absorptive capacity is considered to benefit from this. This factor is also difficult to analyze, for the same reason as the influence of prior knowledge. However, since the participants in the network are separate, multinational organizations, it is likely that the social structures and the culture are differing. That in turn, may negatively affect the absorptive capacity.

6. Discussion

In this section, we conclude the paper by discussing the results. It became obvious that the factor that revealed most deficiencies was the *motivational disposition of the source*. Afterwards, this may seem rather natural due to the outline and the set-up of the network, which by no means showed the manufacturers the benefits of their participation in the network. Therefore, on may argue that the project management should have been more detailed upon the expectations of each participant and also worked stronger for establishing power and mandate, allowing the project manager (i.e. the broker) to steer the network into predefined paths. In addition, the project management should also been more careful with the establishment of a knowledge sharing culture within the network. From our perspective, such culture must be the result of a win-win situation, which may be measured and made concrete to each participant. For being able to show on the benefits in concrete terms, metrics should have been established, allowing each organization to measure and report e.g. the number of design cycles reduced in average, a reduced level of cassation of components on the manufacturer side, the number of inputs into the KMS by each participant. Such metrics would have contributed by showing on the benefits of the network and the KMS. Thee last example should also allow the broker to identify the contribution of each manufacturer, with respect to knowledge shared and thereby supply the broker with metrics that would have been applicable when exercising control and influence of the participants.

References

[1] Alavi, M. and Leidner, D. (1999). Knowledge Management Systems: Emerging Views and Practices from the Field. In *Proceedings of the 32nd Hawaii International Conference on System Sciences* (HICSS), 5-8 January, Maui, Hawaii.

[2] Asbrand, D. (1998) *Making money from data*. Datamation article, wysiwyg://56/http://datamation.earthweb.com/dataw/11tran.html. Printed 000621.

[3] Binney, D. (2001). The KM spectrum – Understanding the KM Landscape. *Journal of Knowledge Management*, **5(1)**, 33-42.

[4] Connolly, T., Begg, C. and Strachan, A. (1999) *Database Systems, A practical approach to design, implementation and management*. 2nd Edition, Addison-Wesley.

[5] Davenport, T.H. and Prusak, L. (1998). Working Knowledge. Harvard Business School Press, Boston.

[6] Franke U. J. (1999). The virtual web as a new entrepreneurial approach to network organizations. *Entrepreneurship & Regional Development*, 203-229.

[7] Gupta, A.K., and Govindarajan, V. (2000). Organizing for knowledge flows within MNCs, *International Business Review*, **3**, 443-458.

[8] Meso, P. and Smith, R. (2000). A Resource-Based View of Organizational Knowledge Management Systems. *Journal of Knowledge Management*, **4(3)**, 224-234.

[9] Nonaka, I. and Takeushi, F. (1995). *The Knowledge Creating Company*. Oxford University Press, Oxford.

[10] Tiwana, A. (2000). *The Knowledge Management Toolkit*. Prentice Hall PTR, Upper Saddle River.

[11] Tiwana, A. and Balasubramaniam, R. (2001). A design knowledge management system to support collaborative information product evolution. *Decision Support System*, **31**, 241-262.

[12] Wiig, K.M. (1993). *Knowledge Management Foundations, vol. 1*. Schema Press, Arlington.

Information Modelling and Knowledge Bases XVI
Y. Kiyoki et al. (Eds.)
IOS Press, 2005

To search and summarize on Internet with Human Language Technology

Hercules DALIANIS
Department of Computer and System Sciences
KTH and Stockholm University, Forum 100, 164 40 Kista, Sweden
Email: hercules@kth.se

Abstract. More and more text are available on the Internet and we need tools to tame this flow. Automatic text summarization is one solution, a text is given to the computer and it returns a non-redundant shorter text. Automatic text summarization can also be used in search engines to decrease time finding documents. To further improve search engines one can use human language technology in form of word analysis as stemming and spell checking. Other methods that can be used are multilingual or cross language information retrieval in searching and finding documents written in other languages than the languages one has knowledge in. In understanding foreign languages one can use machine translation techniques that today had become good enough for practical use. Machine translation (MT) is the technique where the computer translates automatically between natural languages. The MT-techniques have been developed since the early 50'ies.

1. Introduction

The rapid change of our environment in form of more and more information available on the Internet increased the speed of development of highly advanced tools to extract, filter, retrieve and translate documents. Three research areas are automatic text summarization, information retrieval tools and machine translation.

In automatic text summarization, the most relevant parts of a document are extracted and put together into a non-redundant summary that is shorter than the original document. A good overview of the area can be found in [1]. A more advanced form of summarization is multi-text summarization where several documents are condensed into one summary.

2. Application areas of automatic text summarization

The application areas for automatic text summarization are extensive. As the amount of information on the Internet grows abundantly, it is difficult to select relevant information. In for example Business Intelligence one can by using automatic text summarization easily access the most relevant part of the found news article in the abundant news flow.

Automatic text summarization is also extremely useful in combination with a search engine when managing large document collections, as for example, the Web. By presenting summaries of retrieved documents to the user, it is easier to assess the relevance of the search results without having to access, read and skim the full documents. Here the summaries are user adapted depending on the search keywords provided by the user, resulting in a more advanced version of Google's hitlist.

Furthermore, information is published simultaneously on many media channels in different versions, for instance, a paper news paper, web news paper, WAP news paper, SMS message, radio transmission, or a spoken news paper for the visually impaired.

Customization of information for different channels and formats is an immense editing job that notably involves shortening of original texts. Automatic text summarization can automate this work completely or at least assist in the process by producing a raw summary for the editor to work with.

Also, documents can be made accessible in other languages by first summarizing the document and then translate the summary, which in many cases would be sufficient to establish the relevance of a foreign language document. The translation can be made manually or in some cases by using machine translation tools.

Automatic text summarization can also be used to summarize a text before it is read using an automatic speech synthesizer, thus reducing the time needed to absorb the essential parts of a document. It can also aid the listener in the navigation of the document being read aloud by lessening the amount of time being spent on listening to a part of a document before deciding if it is relevant or not, much as in the search engine scenario.

In particular, automatic text summarization can be used to prepare information for use in small mobile devices, which may need considerable reduction of content size.

The techniques used in automatic summarization have interesting spin-off effects in the area of advanced search engine technologies in form of document extraction, stemming, query expansion, the use of synonym dictionaries, as well as spell checking of the query. Other techniques are indexing, clustering and categorization of texts.

3. SweSum

Here follows a description of our summarization tool SweSum and the evaluation process. SweSum is in its current form is built on both statistical and linguistic methods as well as heuristic methods. SweSum is available for eight languages Swedish, Danish, Norwegian, English, French, Spanish, German and Farsi (Persian), [2,3].

3.1 The architecture of SweSum

SweSum works basically by performing three passes, (see figure 1). In the first pass tokenization is performed and sentence boundaries of a text are found. Simultaneously the keywords are extracted from the text. In the second pass is each sentence is ranked according to the keywords and scoring values and finally in the third pass the summary is created by extracting the highest scoring sentences above a certain threshold or up to a certain cut-off value. A cut-off value can for example be to keep a certain given percentage of the original text or a specified number of characters, words or sentences.

Scoring/ranking parameters in SweSum

- Title: Words in titles and in the immediately following sentences are given a high score.
- Position score: The assumption is that certain genres put important sentences in fixed positions. For example, newspaper articles usually have most important terms in the beginning of the article. Reports on the other hand have important sentences evenly spread out and maybe in the beginning and at the end of the document. This means that SweSum in news mode gives a higher score to sentences in the beginning than in the end of the newspaper article. For reports there are no position scores at all applied by SweSum.

- Average lexical connectivity: Number terms shared with other sentences. The assumption is that a sentence that share more terms with other sentences is more important.
- Numerical data and formatting tags: Sentences containing numerical data and bold tagging are scored higher than the sentences without numerical values or emphasis.
- Sentence length: Long sentences tend to obtain higher scoring because they contain more keywords, therefore is sentence length normalized in such a way that weights for keywords are inverse proportional to sentence length.
- The only language dependent parameter is keyword detection and query signature carried out by finding and counting the keywords or open class terms.
- Term frequency *tf*: Key words (or open class terms) that are high frequent in the text are more important than the less frequent
- Query signature: The query of the user can be used to affect the summary in the way that the extract will contain these words if present. This will result in a slanted summary that also can be called a user adapted summary.

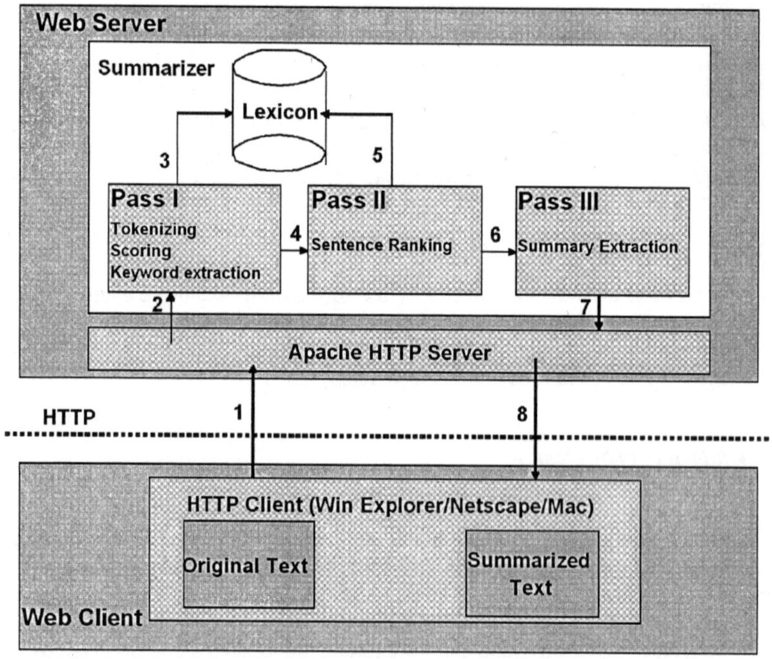

Figure 1 Architecture of SweSum (figure by Nima Mazdak, 2004)

SweSum for Swedish uses a 700 000 key word entries dictionary that tells if the word belongs to the open word class group and specifies the base form (lemma). The FarsiSum (SweSum for Farsi) uses a Persian stop list (containing high frequent insignificant words, as *and, or, under, with, etc*) and verb removal and GerSum (SweSum for German) uses only detection and stemming of nouns.

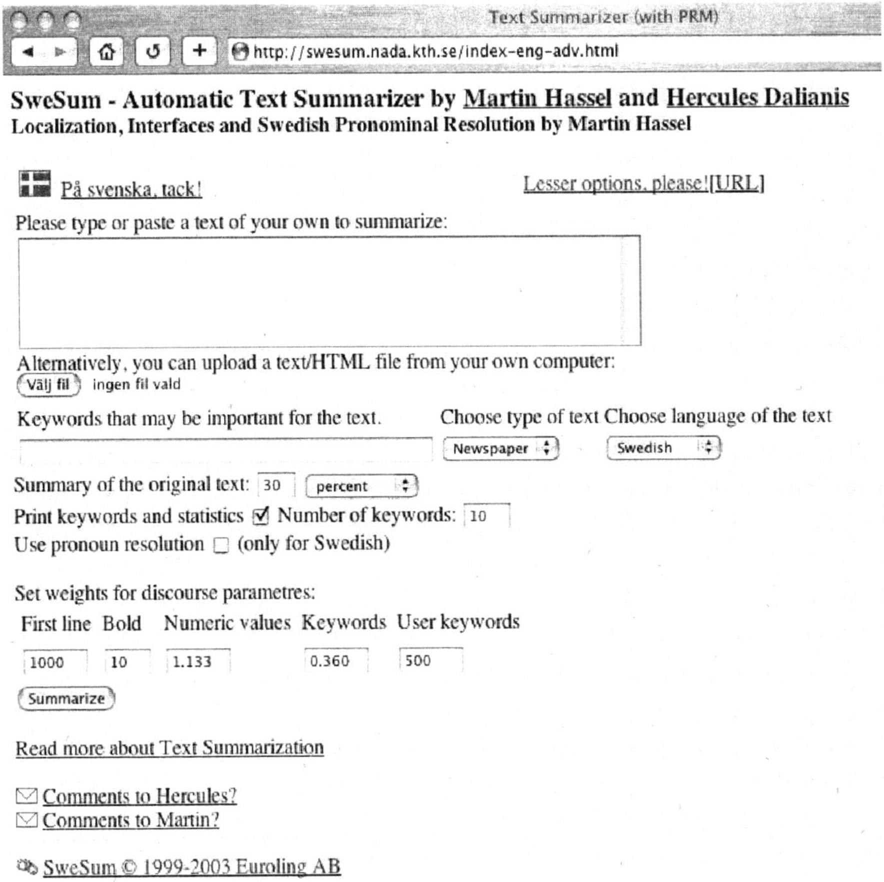

Figure 2. SweSum's English interface, but for Swedish texts

All the above parameters are normalized and put into a naïve combination function with modifiable weights for each parameter, (see figure 2). The idea is that high scoring sentences in the original text are kept in the summary, the scores are calculated according to the criteria above.

The domain of SweSum is Swedish HTML tagged newspaper text. SweSum ignores HTML tags that control the layout of the page but processes the HTML tags that control the formatting of text. The summarizer is currently written in Perl. A nice overview of the architecture of SweSum can be found in [3].

On-line demos in all above mentioned languages are available on the Internet [4]. The site has around 2 200 visitors per month, where around 100 are unique.

3.2 Evaluation of text summarizers

One of the most difficult tasks in the research of automatic text summarization is to evaluate the summarization systems. There have been various attempts to evaluate text summarizers. A thorough overview of the area can be found in [5]. [5] describes also various attempts to evaluate SweSum. Generally speaking SweSum produces both coherent and readable summarized news text when compression rate is up to 70 percent. This means preserving 30 percent of the original text. But one can observe that evaluation of text summarizers is a very difficult task since one does not really now what encompasses a "good" summary. Two persons can create two different summaries of a text and consider their own summary as the best one.

4. SiteSeeker search engine

SiteSeeker is a powerful search engine for web sites and intranets. Siteeeker has built-in human language technology, such as stemming for Swedish, English and Danish as well as compound joining. Stemming makes it possible to search using a given word and find both the word and all inflections of it.

Stemming improves precision and recall with around 15 and 18 percent for Swedish and should be about the same for Danish but less for English as English has a less complex morphology [6]. High precision and recall means that the user obtains more and better hits when searching.

SiteSeeker has also built-in dynamic spelling support where the index is the lexicon. It is well known that around 10 percent of all search queries are misspelled in various ways. SiteSeeker corrects around 90 percent of these misspellings, [7]. Evaluation results indicate also that the spelling support improves both precision and recall with 4 and 11.5 percent respectably [8].

SiteSeeker uses extraction of the most relevant context around the search words from each found document. The extracts are presented together with the high lighted search words and presented in the hit list. The extracts are also called snippets or KWIC (Key Word In Context). This extraction feature makes search fast and efficient while the user does not need to click on every hit to see if the found document were relevant. Except for the traditional term weighting model SiteSeeker also uses search word proximity ranking.

Word proximity ranking is that a document, or passage of a text, that contains the query words close to one another scores higher than a document or passage where the words are far apart. SiteSeeker uses also web page structure as well link validation to obtain the best relevance ranking. SiteSeeker also has a language recognizer for 40 European languages. SiteSeeker can index text-, html-, PDF-files and MS Office files. SiteSeeker is currently used at over 70 public and company websites as well as intranets in Sweden.

SiteSeeker, for example, is used at Nordoknet [9] that is a portal for Language Technology Information in the Nordic countries. Nordoknet encompasses five different countries: Sweden, Denmark, Norway, Finland and Iceland with information in six different languages: Swedish, Danish, Norwegian, Finnish, Icelandic and English on ten different web servers.

We are in the aim to soon start research around constructing cross-language information retrieval engine. Where one can search in one language and obtain answers in documents written in other languages on Nordoknet. Many of the Scandinavian language are closely related and the speakers have passive knowledge in the other languages that means they are able to read and understand but not to write and speak the other language. So by developing tools to find information in the other language we could cross the language barriers and give Scandinavians access to neighbouring languages.

5. Machine Translation

Automatic translation between languages using computers is also called Machine Translation (MT) and started already in the early 50'ies. MT is considered as one of the most difficult research areas with in human language technology. Machine translation uses either linguistic and statistical methods or both. The source text is given to the MT-system that makes syntactical, semantic and pragmatic analysis on the text, then is various transfer or generation rules used to make the target text readable. For the interested reader see [10]. One can say that now 50 years later the systems somehow have become usable for bulk translation of websites such that one can have a clue what a page can mean. for a sample of the machine translation system Systran see figure 3, and for a demo of Systran go to [11].

Figure 3. Daily japanese news paper Yomiuri On line translated to English online in "real time" by Systran. One can surf around the website obtaining all information in English.

6. Conclusions and future directions

We have just seen the beginning in using human language technologies in searching and summarizing text on the Internet. Text summarization will for sure become more sharp, flexible and commonly used than it is today. Text summarization has interesting spin off techniques that will play key-role in the future.

Regarding search we also believe that multilinguality will find its place in that one will search in one language and obtain hits on relevant document in other languages and then translate these documents to your mother tongue using MT. This will be extremely valuable for us belonging to small language groups that are dependent on information in other languages and influences beyond our language domains.

References

[1] Mani, I. and M. T. Maybury (eds) 1999. *Advances in Automatic Text Summarization,* Cambridge, MA: The MIT Press.

[2] Dalianis, H. 2000. *SweSum - A Text Summarizer for Swedish*. Technical report TRITA-NA-P0015, IPLab-174, NADA, KTH, October 2000 http://www.nada.kth.se/~hercules/Textsumsummary.html

[3] Mazdak, N., 2004. *FarsiSum - A Persian text summarizer.* Master Thesis. Department of Linguistics, Stockholm University. http://www.dsv.su.se/~hercules/papers/FarsiSum.pdf

[4] SweSum 2003. *SweSum demo at Internet.* http://swesum.nada.kth.se/index-eng.html

[5] Hassel, M. 2004. Evaluation of Automatic Text Summarization: A Practical Implementation. Licentitate thesis. KTH, May 2004.

[6]Carlberger, J., H. Dalianis, M. Hassel, O. Knutsson 2001. *Improving Precision in Information Retrieval for Swedish using Stemming*. In the Proceedings of NODALIDA '01 - 13th Nordic Conference on Computational Linguistics, May 21-22, 2001, Uppsala, Sweden.

[7] Dalianis, H. 2002. *Evaluating a Spelling Support in a Search Engine,* in Natural Language Processing and Information Systems, 6th International Conference on Applications of Natural Language to Information Systems, NLDB 2002 (Eds.) B. Andersson, M. Bergholtz, P. Johannesson, Stockholm, Sweden, June 27-28, 2002. Lecture Notes in Computer Science. Vol. 2553. pp. 183-190. Springer Verlag, 2002.

[8] Mansour Sarr: 2003. *Improving precision and recall using a spell checker in a search engine*. In the proceeding of NODALIDA 2003, the 14th Nordic Conference of Computational Linguistics, Reykjavik, May 30-31, 2003.

[9] Nordoknet 2004, http://www.nordoknet.org/

[10] Hutchins, J. & Somers, H. 1992. An Introduction to Machine Translation. Academic Press Limited

[11] Systran 2004, http://world.altavista.com

Author Index